Self-tuning Systems

Self-tuning Systems

Control and Signal Processing

P. E. Wellstead and M. B. Zarrop

Control Systems Centre, University of Manchester,
Institute of Science and Technology, UK

JOHN WILEY & SONS

Chichester • New York • Brisbane • Toronto • Singapore

Other Wiley Editorial Offices

John Wiley & Sons, Inc., 605 Third Avenue,
New York, NY 10158-0012, USA

Jacaranda Wiley Ltd, G.P.O. Box 859, Brisbane,
Queensland 4001, Australia

John Wiley & Sons (Canada) Ltd, 22 Worcester Road,
Rexdale, Ontario M9W 1L1, Canada

John Wiley & Sons (SEA) Pte Ltd, 37 Jalan Pemimpin 05-04,
Block B, Union Industrial Building, Singapore 2057

Library of Congress Cataloging-in-Publications Data:
Wellstead, P. E.
 Self-tuning systems : control and signal processing / P.E. Wellstead and M.B. Zarrop.
  p. cm.
  Includes bibliographical references and index.
 ISBN 0 471 92883 6—0 471 93054 7 (pbk)
  1. Adaptive control systems. 2. Self-tuning controllers.
 I. Zarrop, Martin B. II. Title.
 TJ217.W38 1991
 629.8'36—dc20   90-13074

British Library Cataloguing in Publication Data:
Wellstead, P. E.
 Self-tuning systems : control and signal processing.
 1. Control systems. Design
 I. Title II. Zarrop, M. B.
 629.8312

 ISBN 0 471 92883 6
 ISBN 0 471 93054 7 pbk

Typeset in 10/13 pt Palatino by Photo·graphics, Honiton, Devon, England
Printed and bound in Great Britain by Bookcraft (Bath) Ltd.

For:
Jane, Stephen, Sarah
and the memory of
Sarah Ainsley Wellstead

Claudette, with love

Contents

PART 1: SYSTEM IDENTIFICATION FOR SELF-TUNING

PART 2: SELF-TUNING CONTROLLERS

PART 3: SELF-TUNING SIGNAL PROCESSING

PART 4: SPECIAL TOPICS

Preface

PURPOSE OF THE BOOK

The aim of this book is twofold. First, we want to present the subject of self-tuning control and signal processing from a user's viewpoint. In particular, the intention is to provide the reader with an access point to the main methods and techniques used in constructing a self-tuning system and, to this end, the treatment of the subject is not as deep as in other texts. Rather we focus upon the algorithms involved in self-tuning and describe their functional properties. The second major aim is to bring together in one volume the main application areas influenced by self-tuning systems. For the most part, texts on self-tuning or adaptive systems will treat either the control or the signal processing application. We feel that it is important that all aspects of adaptive information engineering should be dealt with in a unified way. Such an approach has significant advantages for the student engineer in that it emphasizes the common framework within which control and signal processing can be considered.

The motivation for collecting together material for a book on self-tuning came from a wish to document the research on self-tuning and self-adaptive processes which has been carried out at the Control Systems Centre, UMIST over a fifteen year period. Many researchers have participated in this activity and their individual contributions are referenced within the text. The key point, however, is that this research effort has been, and continues to be, application driven. Specifically, our philosophy is that technical advance should be motivated by a study of application needs. As a result of this we have covered rather a wide range of different types of self-tuning algorithm. For example, the development of self-tuning extremum control (Chapter 13) was motivated by practical problems of automotive engine control. The frequency domain self-tuner (Chapter 14) was developed specifically to meet the needs of environmental testing. Other elements of the book originated as reactions to the perceived deficiencies of other design techniques. For example, pole assignment was originally conceived as a means of avoiding problems with minimum variance methods. Likewise, reduced variance pole assignment was developed in order to improve the disturbance rejection properties of the basic pole assignment algorithm. The advantage which this approach brings to the book is that each self-tuning method or technique presented here was developed in direct response to a perceived practical need.

The specific application aspects of the book are counterbalanced by treatments of standard self-tuning techniques drawn from the general literature on self-tuning control. Again, however, the treatment is slanted toward the algorithmic issues rather than theory. Theory, however, should not be completely avoided, and in this connection we include a chapter on the convergence of adaptive algorithms, together with specific subsections introduced as required throughout the book to explain the convergence properties of particular algorithms. These parts may be omitted on a first reading but they should be viewed nonetheless as important in building support for the text as a whole.

STRUCTURE OF THE BOOK

As noted above, the aim of this book is to give a self-contained presentation of the main algorithmic components of a self-tuning system. The theoretical questions of stability and convergence of self-adaptive systems are dealt with (Chapter 6), but only to the extent that is necessary to provide a basic understanding of algorithmic behaviour.

The text is divided into four distinct parts in order to reflect the different aspects of a self-tuning system. Part 1 describes the identification algorithms used in self-tuning. Part 2 covers controller algorithms and in Part 3 the area of signal processing is discussed. The book concludes with Part 4 consisting of a set of special topics which cover some important self-tuning algorithms which do not fit directly into Parts 2 and 3.

The material presented in this book has been used for several years at the Control Systems Centre, UMIST as the basis of an introductory course on self-tuning systems. It has also been used in special courses for industrial engineers who required a review course on self-tuning systems. However, some of the chapters are more advanced and can be omitted on a first reading. In particular, Chapters 5 and 6 in Part 1 refer to special computational methods and convergence analysis for recursive estimation and these are not important for an introduction to the area. Likewise, Chapter 9 in Part 2 deals with a relatively complex self-tuning control algorithm and should be skipped on a first reading.

The special topics of Part 4 contain a mixture of advanced and straightforward material. The self-tuning extremum controller (Chapter 13), for example, is a very simple algorithm and can be used to introduce the concept of self-tuning in an easy way.

We suggest the following alternative paths for either self-study or organizing courses based on the book.

Introduction to Recursive Estimation

Chapters 2, 3, 4 from Part 1.

Introduction to Self-tuning Control

Chapter 1.
Chapters 2, 3, 4 from Part 1.
Chapters 7, 8 from Part 2.

For greater depth the following is suggested.

Self-tuning Control

Chapter 1.
All of Part 1.
All of Part 2 and Chapter 13 from Part 4.

Introduction to Self-tuning Signal Processing

Chapter 1.
Chapters 2, 3, 4 from Part 1.
All of Part 3.

Self-tuning Signal Processing

Chapter 1.
All of Part 1.
All of Part 3 and Part 4.

The text should also be useful for presenting self-tuning application lectures. The three special topics in Part 4 can each be used in this way. The extremum controller (Chapter 13), for example, relates to the self-tuning engine controller mentioned in Chapter 1. The frequency domain self-tuner (Chapter 14) is widely used for environmental vibration testing including the very appealing and worthy application of earthquake testing of structures. The two-dimensional signal processing algorithms of Chapter 12 are typical examples of the cross-fertilization between self-tuning control and image restoration. The authors use this as an example to students to show that there are no frontiers in engineering research and that many of the most useful 'new' algorithms arise as a result of cross-fertilization of ideas from more than one discipline.

ACKNOWLEDGMENTS

As noted above, this book is the result of many years of research and development at the Control Systems Centre. Our research on self-tuning systems has been a collaboration with numerous research students and associates who have passed through our M.Sc. programme in the 'Theory and Practice of Automatic Control' and its successor the current M.Sc. programme 'Control and Information Technology'. It is not possible to mention all the researchers who were involved in this effort. However, certain postgraduate students and technical staff made significant contributions and these we now list in alphabetical order:

Alnoor Allidina, Alex Bozin, Rogiero Caldas-Pinto, Francisco Carvalhal, Ray Davies, Barry Dwolatzky, John Edmunds, Ricardo Fernandez-Del-Busto, Michael Fischer, Will Heath, Dino Lelic, Roy Moody, Dennis Prager, Steve Sanoff, Peter Scotson, Nigel Sym, Tony Truscott, George Wagner, Din Wahab, Phil Zanker.

It is gratifying to note that our ex-colleagues now occupy influential positions in engineering research and development.

This book has passed through many forms for which we owe a debt of gratitude to the typing and wordprocessing skills of Vera Butterworth, Tita Fernandez-Del-Busto and Janice Prunty. Artwork for the book was draughted with skill and professionalism by Mrs M. Greenhalgh. Photographic work was produced with corresponding professionalism and good humour by John Howe.

Finally, we acknowledge the organizations that have allowed us to reproduce commercial material supplied by them. These are Schlumberger Technologies, Eurotherm Ltd., A.B.B., TRT, Lucas Automotive and the National Laboratory for Civil Engineering of Portugal.

Glossary of Symbols

The following glossary covers all the main symbols and nomenclature used in this book, together with the section in which each symbol first appears. Usage is standard except in the self-contained Chapters 12, 13 and 14 which deal with three important 'special topics' not usually treated in mainstream self-tuning research. Hopefully, this will create no significant problem for the reader.

The notation for polynomial degrees and coefficients is standard, e.g.

$$B(z^{-1}) = b_0 + b_1 z^{-1} + \ldots + b_{n_b} z^{-n_b}$$

and usually we omit the dependence upon z^{-1}, e.g. $B = B(z^{-1})$. The presence of a circumflex (e.g. $\hat{B}(z^{-1})$) indicates an *estimate*. A tilde (e.g. $\tilde{\theta}$) usually indicates an *error* object.

a_0	curvature parameter	13.2
A	system denominator polynomial (1D/2D)	2.2, 12.3.2
B	system numerator polynomial	2.2
B^{+}	minimum phase factor of B	7.5.2
B^{-}	nonminimum phase factor of B	7.5.2
C	noise numerator polynomial (1D/2D)	2.3.2, 12.3.2
d	offset	2.3.1
D	known noise numerator polynomial	2.3.1
	innovations model numerator polynomial (1D/2D)	11.3.2, 12.5
$\mathbf{D}(t)$	diagonal matrix (U–D factorization)	5.2.1
$\mathscr{D}(t)$	drift signal	2.3.1
$\mathscr{D}(m,n)$	2D interference process	12.7.2
$e(t)$	white noise sequence	2.3.2,
$e(m,n)$	white noise sequence (two-dimensional)	12.3.2
$\hat{e}(t)$	modelling error	3.2
$\mathbf{e}$	stacked white noise vector	3.2
$\hat{\mathbf{e}}$	stacked modelling error vector	3.2
E	expectation operator	3.2.1

$\bar{E}$	generalized expectation operator	6.3.3
E	GMV polynomial	8.3
E_i	GPP polynomial	9.2.2
E_t	expectation conditioned on data up to time t	9.2.1
f	frequency (Hertz)	14.2
f	GPP stacked vector	9.2.2
$f(\cdot)$	ODE function	6.3.2
$f, \mathscr{F}$	controller polynomial	7.2.2, 9.2.2
F_i	GPP polynomial	9.2.2
$F_1(w)$	$= F(w,1)$	12.4
$G, \mathscr{G}$	controller polynomial	7.2.2, 9.2.2
G	2D prediction polynomial	12.4
$\mathbf{G}(\cdot,\cdot)$	ODE matrix	6.5.4
$\mathbf{G}(\cdot)$	ODE matrix in RLS	6.6.1
G	GPP stacked vectors	9.2.2
G_i	GPP polynomial	9.2.2
h	optimization step length	13.3
H	precompensator polynomial	7.4.1
$\mathbf{H}(\cdot)$	ODE system matrix	6.5.4
H_c	channel transfer function	11.4
H_n	filter transfer function for noise reference	11.5.1
$\mathbf{I}_m$	$m \times m$ unit matrix	3.2.1
j	$\sqrt{-1}$	2.3.2
J	cost function for estimation	3.2
k	time delay (discrete)	2.2.1
k_p	proportional gain (PID control)	7.3
k_d	derivative gain (PID control)	7.3
k_i	integral gain (PID control)	7.3
L	Laplace transform operator	2.2.1
$\mathbf{L}(t)$	RLS gain vector	3.4
m	number of estimation parameters	3.2
m_X	mean of X	3.2.1
$n(t)$	noise source	11.5.1
$n_r(t)$	noise reference	11.5.1
N	number of data points	3.2
	output horizon	9.2.1
$N1$	(lower) output horizon	9.2.1
NU	control horizon	9.2.3
$\mathcal{N}$	GPP controller polynomial	9.2.2
P	overparametrization polynomial (reduced variance)	7.5.3
	GMV pseudo-output polynomial	8.3

	predictor polynomial	10.4.2
$\mathbf{P}(t)$	covariance matrix in RLS	3.3
$P_{yy}(f)$	output power spectrum/periodogram	14.3.2
Q	GMV pseudo-output polynomial	8.3
	predictor polynomial (1D/2D)	10.4.1, 12.4
$r(t)$	reference or dither signal	3.3
$\mathbf{r}$	stacked reference vector	9.2.2
R	GMV pseudo-output polynomial	8.3
$\mathbf{R}$	random walk matrix	4.5.5
$\mathbf{R}(t)$	ODE Hessian matrix	6.5.1
$R(M,N)$	2D support region	12.3.2
$R_{xy}(\tau)$	cross covariance function	7.6.3
s	Laplace operator; differential d/dt	2.2
$s(t)$	signal	2.3.1,
$s(m,n)$	signal (two-dimensional)	12.3
$s_f(t)$	cumulative sum in variable forgetting factor	4.5.9
S	GDP pseudo-output polynomial	9.2.1
$S(m,n)$	natural past of $y(m,n)$	12.4
$S_{ss}(w)$	power spectrum (spectral density fn) of $s(t)$	2.3.2
t	discrete time index	2.2
$\mathbf{T}$	desired characteristic polynomial	7.2.1
$u_c(\tau)$	input signal (continuous time)	2.2
$u(t)$	input signal (discrete time)	2.2
u_{fact}	factory setting	13.8
u_{adj}	adjustment to u_{fact}	13.8
u_{pert}	perturbation signal	13.8
$u_f(t)$	filtered input	3.6
$U(\tau)$	unit step function	2.2.1
U_s	input saturation level	4.4.1
U_d	dead-zone magnitude	4.4.1
$\mathbf{U}(t)$	upper triangular matrix (U–D factorization)	5.2.1
$v_c(\tau)$	known noise source (continuous time)	2.3.1
$v(t)$	known noise source (discrete time)	2.3.1
$V(t)$	accumulated loss (theoretical)	10.4.1
$V_a(t)$	accumulated loss (actual)	10.4.1
w	unit forward column shift operator	12.3
$\mathbf{w}(t)$	zero mean vector white noise	4.5.5
$x_c(\tau)$	noise-free output (continuous time)	2.2
$x(t)$	noise-free output (discrete time)	2.2
$x(z)$	Laplace transform of $x(t)$ sequence	2.2
$\mathbf{x}(t)$	regression (or data) vector	3.2
$\hat{\mathbf{x}}(t)$	AML regression vector	6.7.1

$\bar{\mathbf{x}}(t)$	filtered $\hat{\mathbf{x}}(t)$	6.7.1
$\mathbf{X},\mathbf{X}(t)$	stacked data vectors (matrix)	3.2
$y_c(\tau)$	output (continuous time)	2.4
$y(t)$	output (discrete time)	2.4
$y_f(t)$	filtered output	3.6
$\mathbf{y},\mathbf{y}(t)$	stacked output vector	3.2
$\hat{y}(t|t-k)$	k-step-ahead output predictor	10.2
$\tilde{y}(t)$	output prediction error	10.2
z	unit forward shift operator; z-transform argument	2.2
	unit forward row shift operator	12.3
α	support region for 2D C polynomial	12.3.2
β	support region for 2D A polynomial	12.3.2
γ,γ_P	2D support region	12.3.3
$\gamma(t)$	scalar/matrix gain sequence	3.5
Γ_i	GPP polynomial	9.2.2
$\mathbf{\Gamma}$	stacked GPP vector	9.2.2
δ	difference operator (delta model)	2.5
Δ	difference operator $(1-z^{-1})$	2.3.2
$\Delta\mathbf{u}$	vector of control increments	9.2.2
$\epsilon(t)$	(*a priori*) output prediction error	3.3
$\eta(t)$	residual (*a posteriori* prediction error)	3.2
$\mathbf{\eta}$	stacked residual vector	3.2
θ	2D support region for $G(w,z)$	12.4
$\mathbf{\theta}$	true parameter vector	3.2
$\mathbf{\theta}(t)$	true time varying parameter vector	4.5.5
$\hat{\mathbf{\theta}}(t)$	estimated parameter vector based on data available at time t	3.3
λ	constant forgetting factor	4.5.6
	lambda controller weighting	8.8
λ'	directional forgetting parameter	4.5.8
λ_i	eigenvalue	7.7
$\lambda_1(t)$, $\lambda_2(t)$, $\lambda_v(t)$	variable forgetting factors	4.5.9
$\nu(t)$	sensor noise	11.3.1
$\nu(m,n)$	sensor noise (two-dimensional)	12.3.3
ξ	damping factor	7.2.1
$\xi(t)$	correlated disturbance	3.2
$\pi(t)$	backward time prediction error	5.3
ρ	smoothing factor	14.3.2
ρ_{ij}	dither statistic	13.9
σ_c^2	variance of $e(t)$	2.3.2
τ	continuous time index	2.2
	ODE time variable	6.3.1

τ_s	sample interval	2.2
τ_d	time delay	2.2.1
τ_f	forgetting factor time constant	4.5.9
$\phi(t)$	GMV pseudo-output	8.3
$\Phi_i(t)$	GPP pseudo-output	9.2.1
Φ	GPP stacked vector	9.2.2
$\mathbf{\Phi}(t)$	regression vector in RELS	3.6
$\hat{\mathbf{\Phi}}(t)$	backward time parameter estimates	5.3
$\phi(t)$	regression vector in AML	3.6
ω,Ω	angular frequency (rad/sec)	2.3.1
ω_b	system bandwidth	4.3.1
ω_n	natural frequency	7.2.1

ABBREVIATIONS

A/D	analogue to digital
AML (RML$_2$)	approximate maximum likelihood
ARIMA	autoregressive integrated moving average
ARIMAX (CARIMA)	ARIMA with exogenous input (control)
ARMAX (CARMA)	autoregressive moving average with control
ARX	autoregressive with control
cov	covariance matrix
D/A	digital to analogue
det	determinant
Im	imaginary part of
int	integer part of
LHS	left-hand side
lim	limit
ln	natural logarithm
log	logarithm to base 10
ODE	ordinary differential equation
plim	probability limit
pr	positive real
PRBS	pseudorandom binary sequence
prob	probability
Re	real part of
res	sum of residues at stable poles
RHS	righthand side
rv	random variable
s.p.r.	strictly positive real
tr	matrix trace

var	variance
w.l.o.g.	with loss of generality
w.r.t.	with respect to
ZOH	zero order hold

1 Introduction

The aim of this chapter is to provide some background information on control and signal processing and to motivate the use of self-tuning systems. The sequence of ideas is arranged as follows. Section 1.1 reviews conventional control and signal processing systems. Section 1.2 builds on this review by introducing the concepts of self-tuning and its relationships with other adaptive methodologies. The structure of a self-tuning system is detailed in Section 1.3 and Section 1.4 discusses various applications and embodiments of self-tuning systems. The chapter concludes with a review of the technical aspects of self-tuning and uses this to justify the layout of the book.

1.1 CONVENTIONAL CONTROLLERS AND SIGNAL PROCESSORS

The conventional procedures for control and signal processing involve systems with fixed coefficients. In this section we review the forms which conventional fixed coefficient controllers and signal processors take and discuss their various roles.

1.1.1 Control systems

A standard form for a feedback control system is shown in Figure 1.1 in which the *system* is a dynamical process which produces an *output*. The aim of the control system is to ensure that the output follows in some way the *setpoint*

Figure 1.1 A feedback control system.

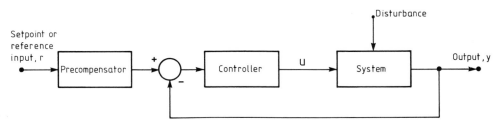

(or *reference*) signal and rejects the influence of the unwanted *disturbance* signal which corrupts the system output.

Generally, the control system will consist of two parts:

(i) a precompensator which is used to shape the transfer function between the setpoint and the output;

(ii) a controller which is used to determine the closed loop stability, disturbance rejection and sensitivity characteristics of the control system.

When the setpoint changes infrequently and the disturbance signal is large, then achieving the controller design objective is said to be a *regulator* problem. Conversely, if the disturbance signal is negligible and the setpoint changes frequently, the design objective is said to be a *servo* problem.

A classical regulator problem is Watt's flyball governor which was used to govern (or regulate) the speed of steam engines at a constant setpoint against disturbances caused by engine load variations and fluctuations in steam supply and quality.

A classical servo mechanism problem is associated with automatic gun-aiming systems. Here the system disturbances are relatively small and the key objective is to maintain a large gun barrel aiming at a target which is taking rapid evasive action. The reference signal in this situation is the measured target position.

1.1.2   Signal processing systems

The signal processing problems considered in this book take a variety of related forms each depending upon the application.

The basic signal processing problem is that of noise (or disturbance) removal. This is illustrated in Figure 1.2(a) in which the objective is to remove the noise (or disturbance) from a measured signal. The signal processing mechanism for this is a *filter* which is designed to reject signal components associated with the noise signal and to pass undistorted the components associated with the original signal.

A further important signal processor is the *predictor* illustrated in Figure 1.2(b). Here the task is to use measurements of a signal and construct an estimate (or prediction) of its future value.

A famous example of the predictor type of signal processor is again in the gun aiming problem. Since it takes a finite time for a gun shell to reach its target, it is important that the gun should be pointed, not at the current target position, but at where the target will be when the shell converges with it. In this case a predictor is applied to a signal derived from the measured target position and used to predict its *future* position.

Figure 1.2
Signal
processing
systems. (a)
Signal filtering;
(b) Signal
predictor.

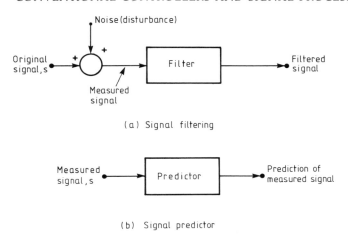

(a) Signal filtering

(b) Signal predictor

There is a linkage here with the control problem, since the output of the target predictor would be used as the setpoint on the gun aiming control system. Thus the overall gun-aiming system is a combination of a signal processing algorithm (predictor) and a control system (the feedback controller). This is not a fortuitous example. In fact almost all control system applications will require elements of control and signal processing in order to form the overall system.

1.1.3 Some characteristics of conventional controller and signal processing systems

The conventional control signal system forms described above are based upon certain assumptions and embody special technical characteristics. It is important to mention these since they are relevant to self-tuning algorithms developed in this book.

(a) *Time-invariance* The greater part of conventional system theory is based upon the assumption that the systems and signal generating mechanisms have constant coefficients. Moreover, the design procedures for control and signal processing lead to algorithms which have constant parameters. This assumption of *time-invariance* is fundamental to conventional design procedures. In practical situations, however, system parameters often vary with time. Self-tuning is one design philosophy for monitoring time variations and incorporating them into the design process.

(b) *Linearity* Alongside time-invariance the major assumption in conventional design is that the underlying systems are *linear dynamical processes*. In fact, this is always an approximation which is tolerated because a complete nonlinear theory does not exist and the fact that linear design methods still work rather well even when applied to nonlinear processes.

Self-tuning techniques are also based upon linear design methods. They have the potential practical advantage that they can cope with some nonlinear phenomena by changing (retuning) parameters to create a new locally linear approximation.

1.2   WHAT IS A SELF-TUNING SYSTEM?

1.2.1   Basic idea

The previous section stressed that conventional design methods produce constant coefficient algorithms based upon a linear time-invariant assumption for the system and signals under consideration. In self-tuning the main philosophical step from this is to introduce control and signal processing algorithms with coefficients which can vary with time. Specifically, the basic idea of a self-tuning system is to construct an algorithm that will automatically change its parameters to meet a particular requirement or situation. This is done by the addition of an *adjustment mechanism* which monitors the system (in a control setting) or the signal (in a signal processing setting) and adjusts the coefficients of the corresponding controller or signal processor to maintain a required performance. Figure 1.3 illustrates this situation. In the remainder of this section we will discuss the interpretation, forms and uses of such adjustment mechanisms.

1.2.2   Self-tuning and adaption

Opinions are divided as to the correct way of describing the automatic adjustment mechanism illustrated in Figure 1.3. Mainly this is a question of nomenclature. For example, the adjustment mechanism of Figure 1.3 can be said to provide an adaption mechanism or an automatic tuning mechanism. Hence, the terminology: *self-tuning* system and *adaptive* system. In essence, both these terms convey the same idea. However, for historical reasons they convey different meanings to different people. Specifically, the idea of a self-tuning adjustment mechanism was conceived originally as a means of handling the initial tuning up of a controller or filter for systems and signal sources which are time-invariant but unknown. Thus, after the initial (self) tuning, the adjustment mechanism is not required and can be disabled. It is obvious that, if the adjustment mechanism is *not* disabled, it can provide a means of continually *adapting* to changes in the system or signal sources.

Figure 1.3 (a)
Self-tuning
control concept,
based on an
automatic
adjustment or
tuning
mechanism for
the controller
parameters
linked to
monitoring the
system. (b) Self-
tuning signal
processing
concept, based
on an automatic
adjustment
mechanism for
signal processor
parameters
linked to
monitoring the
measured signal.

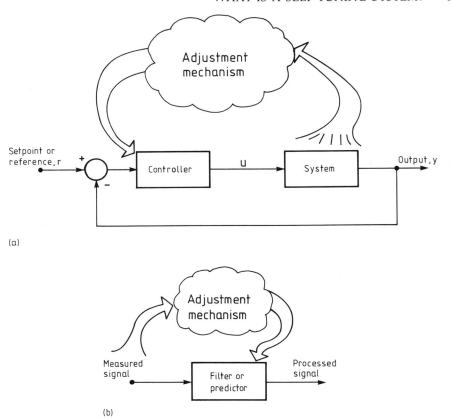

Figure 1.3 (a) Self-tuning control concept, based on an automatic adjustment or tuning mechanism for the controller parameters linked to monitoring the system. (b) Self-tuning signal processing concept, based on an automatic adjustment mechanism for signal processor parameters linked to monitoring the measured signal.

The nuance here is that an adaptive system is often viewed as a method for *continuous* adjustment, whereas self-tuning has been viewed as a mechanism for *initial* adjustment. In practice, the difference between these two ideas is rather small and depends upon some algorithmic details. For this reason we will not distinguish between the two terms. However, because of the way in which our work has developed we will use the term *self-tuning* and apply it to initial and continuous algorithm adjustment mechanisms.

In addition to the semantic distinctions raised in the foregoing paragraphs it is also important to note that the term *adaptive* is sometimes applied (erroneously) to fixed coefficient feedback systems. For example, vehicle suspensions which incorporate feedback are referred to as adaptive. Like beauty, adaption seems to lie in the eye of the beholder!

1.2.3 Approaches to self-tuning and adaption

There are a number of ways of approaching the task of providing the adjustment mechanism illustrated in Figure 1.3. Here we briefly describe some of the important approaches as they emerged from the control discipline. They are further discussed in the context of historical developments in Section 1.2.6.

Self-tuning controller

The self-tuning controller philosophy for obtaining an automatic adjustment mechanism is to identify the system using measured input and output data and then to form an appropriate controller using the identified system. This is illustrated in broad terms in Figure 1.4.

Figure 1.4
Overview of a self-tuning controller.

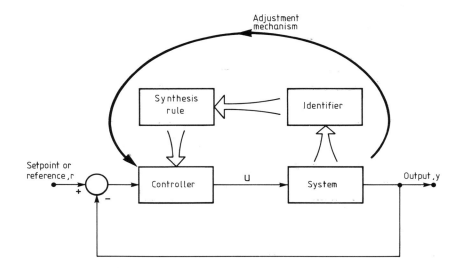

Model reference adaptive controller (MRAC)

The main idea in this approach is that some reference model exists which specifies the desired performance of the control system. Again, measured input/output data are used to monitor the system performance. These are then combined with the reference model output according to an adaption rule and the result is used to adjust the controller. The overall aim is to force the output to correspond to the model output. Figure 1.5 illustrates this approach.

Expert tuning systems

The field of expert systems has had a large impact upon control engineering practice. One area of impact has been the use of heuristic rules to provide information for an expert adjustment mechanism. In this approach measured input/output data are compared in an expert system against qualitative

Figure 1.5
Overview of a
model reference
adaptive
controller.

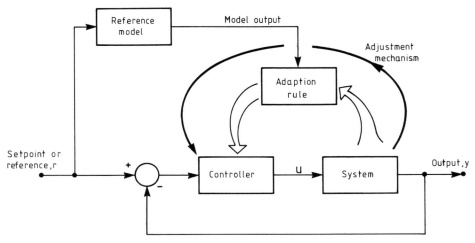

performance criteria. The results of these comparisons are used to adjust the controller settings. Figure 1.6 illustrates this form of adjustment mechanism.

A number of other methods exist for implementing parameter adjustment mechanisms. However, those mentioned above cover the most important approaches. Moreover, the different philosophies can be shown to be formally equivalent. The philosophy adopted in this text is based upon the self-tuning control formulation. In the remainder of this section we discuss alternatives to self-tuning, indicate where self-tuning can be useful and provide some historical perspectives.

Figure 1.6
Overview of an
expert tuning
system.

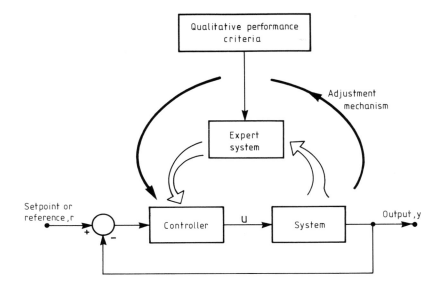

1.2.4 Alternatives to self-tuning and adaption

Alternatives exist to self-tuning and adaption, avoiding the use of an adjustment mechanism.

Gain scheduling

A central feature of self-tuning and adaption is that the adjustment mechanism allows changes in the system to be tracked and compensated for. In many cases, however, the system changes are related in a known way to system operating conditions. In such circumstances the controller coefficients can be adjusted directly as a function of these conditions. This procedure is termed *gain scheduling*. It is widely used in the process industries where the process dynamics are a nonlinear function of one or two operating parameters. Locally linearized controller designs are carried out at different values of these operating parameters. The controller is switched between these controller settings as the operating parameters vary. It is usually assumed that the operating parameters change relatively slowly and that the system nonlinearities are not severe.

High gain feedback

One of the traditional motivations for feedback, particularly in amplifier design, is that it is possible to reduce the sensitivity of the overall system to parameter changes in the feedback loop by using high feedback gains. It is therefore reasonable to ask why we should use a potentially complex adjustment mechanism to compensate for system change when high gain feedback can desensitize the overall performance to such changes. Original work on high gain feedback has led to the area known as *robust control*. The proponents of robust control argue strongly against the need for self-tuning and the debate on the valid areas of application for self-tuning and robust control will undoubtedly run and run. For the purpose of this text we assert that self-tuning is, at least, a valid alternative to robust control and, at best, it can handle applications which are not tractable using robust design.

1.2.5 Uses of self-tuning

As pointed out previously, the basis of self-tuning is to provide an adjustment mechanism for varying controller and signal processor parameters. However, it has also been pointed out that alternatives exist, particularly in the control engineering area. The aim of this subsection is to argue that a rational basis for self-tuning methods exists which is quite independent of the comparative merits of 'rival' techniques.

System commissioning

One of the most time consuming aspects of a control or signal processing engineer's job is the installation and commissioning of algorithms. When a conventional off-line design procedure is applied, the commissioning engineer almost inevitably finds that retuning and further on-line adjustments are required. This is completely natural since no off-line design procedure, based on idealized models, can account for all the features of the real system. Section 1.3 will continue this discussion and argue that self-tuning is a natural process whereby the automation of the design cycle may be achieved. In the jargon of the day, self-tuning is a personal productivity tool for the engineering systems analyst.

It follows that, should the system be changed at some future stage, the controller/signal processor can be easily and automatically retuned. For example, one of the most common control loops in the process industries is flow rate control. It is not uncommon for the valve in such loops to require replacement and the replacement valve to have a different specification. The self-tuning controller on such a loop would retune automatically to the new valve specification.

Compensation for system change

It has been said that the English politician Baldwin spent his declining years studying the 'Irish Question'. Unfortunately, whenever he thought that he had found the answer, the Irish changed the question. The situation is rather similar for the control/signal processing design process. Dynamical systems and signal sources are subject to change. Often this change is unpredictable and of a type which cannot be compensated for by robust design. However, the adjustment mechanism of a self-tuning or adaptive system can provide the means of adapting to system change.

System monitoring

One of the main features of a self-tuning system is that it builds and maintains a model of the system/signal under examination. Thus, in addition to allowing for compensation for system change, the model also allows the nature of the change to be monitored. This monitoring function is extremely useful for plant management purposes and fault diagnosis.

1.2.6   Some background to the development of self-tuning and adaptive systems

The idea of a self-tuning system is not new. The realization of the idea into practical algorithms, however, is a recent development made possible by low cost digital computing equipment. For example, Kalman described a self-optimizing controller in 1958, but the algorithm was impractical at the time due to digital computer limitations of cost, speed and size. The current interest in self-tuning control was first stimulated by (amongst others) the Czech researcher Peterka who showed how system identification and controller synthesis could be combined into one iterative procedure for process control. This idea was timely since the research area of system identification was at that time rather lacking in direction. The idea of combining an identifier with a control law gave identification research both motivation and direction. In the 1970s the results of this new impetus became evident with the publication of several key papers presenting self-tuning controllers for various design criteria (see Section 1.8 Notes and references). Subsequent research has focused upon the theoretical convergence and stability properties of these and other self-tuning algorithms. The theoretical investigations have been complemented by an impressive number of engineering applications.

Running parallel with and predating self-tuning research was the activity on adaptive control. In particular, research into model reference adaptive control (MRAC) has been active for many years. For a time the self-tuning and adaptive research schools ran in parallel with the USA dominating the former and Europe dominating the latter. It is now realized that the two approaches are technically closely related and that the differences are of philosophy and nomenclature. Nonetheless, model reference adaptive control continues to exist as an independent and vigorous research area with numerous successful applications. The MRAC approach will not be treated in this text. In the same spirit, self-tuning and model reference adaptive methods can be approached both from a transfer function and state-space approach with equivalent results. This text is based upon a transfer function approach to system representation. State space methods are only used where they are necessary for analytical purposes.

The area of self-tuning signal processing has a more diverse background than that of control. In particular, the area of self-tuning filters is usually associated with communication research, a popular algorithm being the noise cancelling filter associated with Widrow. The related self-tuning signal processor algorithms for smoothing and prediction arose independently from recognition of the duality with self-tuning regulator algorithms.

In addition to conventional control and signal processing, other independent areas of self-tuning have developed. In order that the reader be aware of such developments, which are usually neglected in mainstream self-tuning research and teaching, we have included material under the heading of Part 4: Special Topics. The distinguishing feature of the self-tuning algorithms included in this part is that the Control Systems Centre at UMIST has played a significant role in their theoretical development and practical use.

1.3   STRUCTURE OF A SELF-TUNING SYSTEM

1.3.1   Sequence of the design process

The aim of self-tuning and adaptive systems is to automate in some way certain activities of the control system and signal processing engineer. As a preparation for explaining how this is done, consider first what these activities might be.

The principal tasks involved in control system and signal processor engineering consist of the following:

  (i) *modelling* of a system or signal generating mechanism;

 (ii) *design* of a controller or signal processor;

(iii) *implementation* of the controller or signal processor.

The engineer will conventionally approach a project in the sequence shown in Figure 1.7. In Stage 1 he or she will first construct a mathematical representation of the system relevant to the problem in hand. In Stage 2 the mathematical representation will be used in a design algorithm to provide a specification for a controller or signal processor which will realize a selected experimental objective. In Stage 3 the specification is then implemented and validated against the design objective. The validation phase of this sequence is vital to success and is one of the most important arguments in favour of self-tuning. To be specific, in conventional off-line design the validation phase often proves unsatisfactory. This leads to a time-consuming repetition of the

Figure 1.7 The three stages of control system and signal processor design: (1) denotes the modelling stages; (2) the design stage; and (3) the implementation stage.

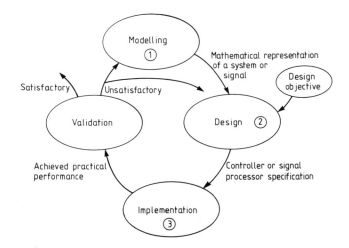

whole sequence of modelling, design and implementation. A central advantage of self-tuning is that the sequence is performed on-line, usually in real-time, in such a manner that the validation process is much faster.

The *modelling* task concerns the construction of a mathematical representation of a system or signal in a form suitable for design. There are two ways of achieving a mathematical representation or model. The first is to apply the laws of physics and chemistry to the system or signal source and hence construct a model based upon first principles. The alternative is to use empirical methods and construct a model from observed behaviour of the system or signal source. The use of observed input and output signals to form a model is termed *system identification* or estimation. The modelling techniques used in self-tuning systems will be drawn from the identification approach to system/signal representation. Part 1 deals with these issues.

The *design* task within Figure 1.7 is intended to exploit the mathematical model of the underlying data generator in order to synthesize a controller or signal processor. The synthesis is performed according to some design objective and may follow any one of a number of design procedures or algorithms. The most important control design procedures are described in Part 2 of this book. Signal processor design in terms of filtering objectives is covered in Part 3. The design procedures used in self-tuning are usually true synthesis methods. That is to say they are techniques whereby a specific model representation gives rise to a specific controller or signal processor specification. This should be compared with the literal interpretation of a design procedure as one in which some degree of qualitative choice rests with the designer. In a synthesis method there is usually only *one* outcome for a given model and design objective.

The *implementation* of a self-tuning controller or signal processor will be assumed to be carried out in terms of a digital algorithm. The parameters produced by the design procedure are then inserted directly into the digital algorithm.

It is the modelling, design and implementation tasks in Figure 1.7 which are automated within a self-tuning or adaptive system. The design objective and the validation issues remain the province of the engineer, although, as previously noted, the process of validation is made much more rapid when self-tuning is used.

1.3.2 Self-tuning controller

The three stages of modelling, design and implementation sketched in Figure 1.7 are shown again in Figure 1.8. This time they are associated with the system identifier, control synthesizer and controller blocks of a self-tuning controller. Note that the validation stage is not shown explicitly since this is

Figure 1.8 Self-
tuning controller
structure.
Showing the
three stages of
engineering
associated with
the boxes 1, 2
and 3.

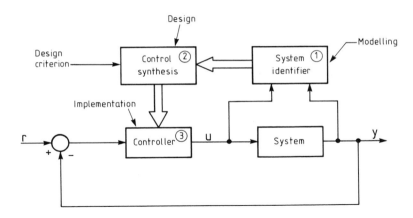

associated with a qualitative assessment of performance and is not an
algorithmic element of the self-tuning system.

1.3.3   Self-tuning signal processor

The sketch in Figure 1.9 shows the basic components of a self-tuning signal
processor. Once again the three stages of engineering are marked for cross-
referencing to Figure 1.7. They are associated with a signal identifier, a signal
processing synthesizer and signal processor blocks in a self-tuning or adaptive
signal processor.

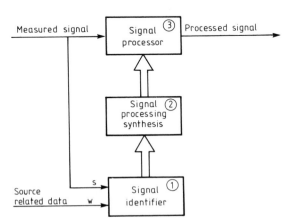

Figure 1.9 Self-tuning signal processor structure. Showing the three stages of
engineering associated with the boxes 1, 2, 3. Note: because of the diversity of signal
processors this structure is only representative.

1.4 SOME SELF-TUNING APPLICATIONS AND INSTRUMENTS

The basic idea of a self-tuning algorithm is now rather widespread in its application. The aim of this section is to describe some applications and commercial self-tuning instrumentation. The intention here is to convey an impression of the diverse uses of self-tuning.

1.4.1 Self-adaptive engine management

The engine of a motor vehicle must be operated in such a manner that it performs safely and efficiently. Most of the responsibility for operating an engine in this way rests with the driver who, by skilled use of throttle and gears, must maintain the engine load and speed within acceptable operational limits. In addition, a significant amount of operational control of automotive engines is performed automatically; for example, the selection of spark ignition timing in the distributor and air/fuel ratio in the carburettor. In previous generations of vehicles such control action was performed by mechanical devices. The level of engine management achieved by mechanical means was necessarily limited.

In modern motor vehicles the management of engine performance is performed electronically using an electronic control unit (ECU) which senses engine data and responds with appropriate control actions (Figure 1.10). At the heart of an ECU is a microcomputer and as microcomputer technology evolves we can expect future generations of motor vehicles to incorporate increasingly sophisticated ECUs. An impression of current technology can be gained from Figure 1.11 which shows a current technology ECU.

Figure 1.10 A spark ignition engine with electronic engine control unit (ECU).

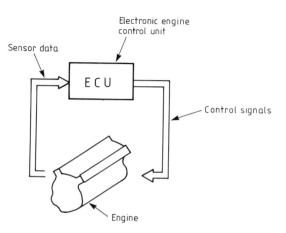

Figure 1.11   A
current
technology
electronic engine
control unit.
(Photograph
courtesy of
Lucas
Automotive Ltd.)

One of the basic ECU functions is to control the timing of the ignition spark for the engine. The timing of the ignition spark is conventionally measured in terms of degrees before top dead centre (°btdc) and this is also referred to as the spark advance. It is most important to select the correct angle of advance at which to ignite the fuel/air mixture in the cylinders. An incorrect choice can cause a significant reduction in the efficiency of the combustion process and reduce the performance of the vehicle. In addition, the production of polluting exhaust gases is a function of the spark ignition angle. Figure 1.12 shows the way in which the engine performance varies with spark angle at a particular load and speed.

By judicious control of the spark angle advance the engine output torque can be maximized by operating it near the peak of the performance curve. The position of this peak varies as a function of the engine load and speed. Thus an ECU will contain a matrix of spark angle values, one for each of a set of engine load and speed combinations. This matrix is called an engine map and is stored in read only memory (ROM) in the ECU. Figure 1.13 shows a typical engine spark angle advance map for a current production engine. At each revolution of the crank the ECU determines the appropriate spark ignition angle by sensing the engine speed and load (by measuring the inlet manifold

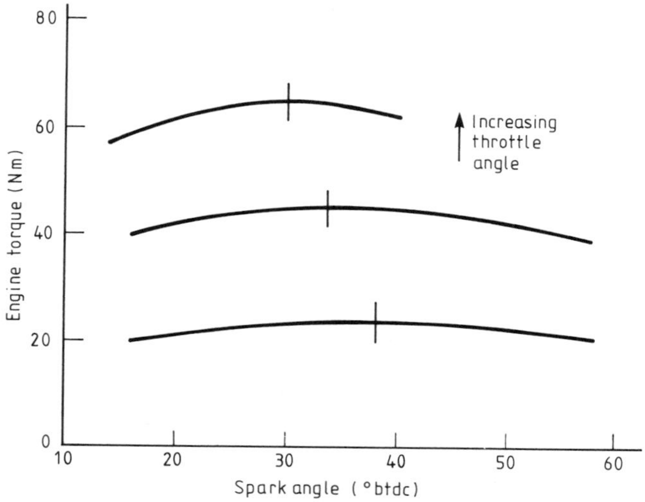

Figure 1.12 Showing the variation in engine torque with spark ignition timing. Note that by convention the spark ignition timing is measured as spark advance. This refers to the crank angle degrees before top dead centre (°btdc).

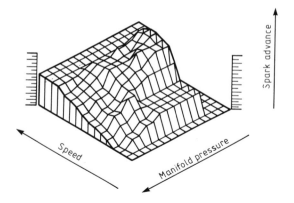

Figure 1.13 A typical engine spark map for a current production engine. Note that the term 'spark advance' is an alternative way of expressing spark angle in degrees before top dead centre (°btdc). Also, the 'manifold pressure' is a measure of engine load.

pressure), selecting the corresponding matrix entry and outputting the appropriate spark angle to the electronic ignition system.

The engine spark ignition map is produced in a factory engine test cell and based upon an average of a number of tests on production engines. Even nominally identical engines, however, differ significantly in their performance. In addition, engine characteristics change as the engine condition and operating environment alter, so that the fixed map based upon the factory averaging procedure is not the best for any one particular engine. A self-tuning system is required which adapts the spark angle map so as to optimize the ignition timing map for each particular engine (Figure 1.14). Using such a system the

Figure 1.14 An Adaptive Engine Control Unit (AECU) with a self-tuning mechanism which adapts the ECU engine maps to reflect changes in the engine.

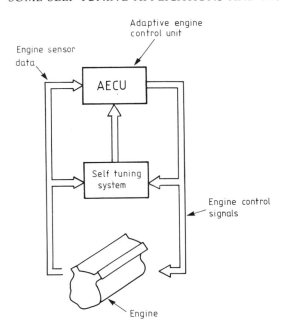

performance of an engine will be continually optimized to meet the needs of the individual vehicle. Even with new vehicles of the same type the improvements obtained by self-tuning each engine can be significant. In a trial on a fleet of four such vehicles, the spark advance angle maps adapted to quite different shapes. Figure 1.15 shows the results of this test. With reference to Figure 1.15(a), the centre map is the original fixed map for the vehicles. The four maps which surround it show the adapted maps for vehicles A, B, C and D. Two points can be made from this illustration:

(i) The general shape of the original fixed map (as determined in factory tests) is clearly different from the general shape of the adapted maps. This would indicate that factory fixed mapping does not reflect the requirements of actual road use.

(ii) The detailed shape of the adapted maps is significantly different for each vehicle, indicating a clear need for individually adapted spark angle mapping for each individual vehicle.

The potential performance benefit of self-tuning engines can be measured in terms of increased fuel efficiency. In the trial just described the vehicles with adapted maps used approximately 9% less fuel than the vehicles with a fixed spark ignition map. The improvement in fuel economy for each vehicle in the trial is shown in Figure 1.15(b).

Figure 1.15
Showing the
results of fleet
trials of self-
tuning engine
management.
(Results courtesy
of Lucas
Automotive
Ltd): (a)
comparing the
adapted engine
spark maps
obtained from
four different
cars (A, B, C, D)
of the model
and
manufacturer;
(b) the
improvement in
fuel
consumption
with self-tuning
of spark angle.

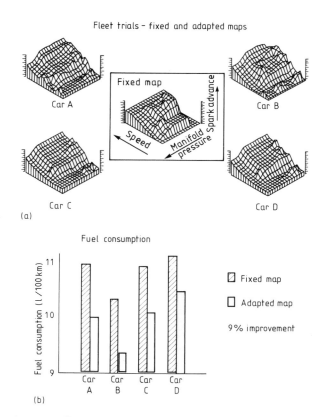

The self-tuning application outlined above is an example of self-tuning extremum control. The theory and algorithms for self-tuning extremum control are explained in detail in Chapter 13.

1.4.2   Self-tuning process control

The automatic tuning of industrial controllers is one of the most widespread and important uses of self-tuning. Numerous companies have produced self-tuning process control instruments for this purpose. In this section we describe two such instruments which together typify the main approaches adopted by commercial process control companies to self-tuning. The instruments considered are the Eurotherm Communicating Controller and the ABB Novatune controller.

Eurotherm self-tuning controller

The Eurotherm instrument described in this section is applied in situations where it is necessary to automatically tune the initial settings of a three-term

(PID) control algorithm. The Eurotherm 815 and 818 Communicating Controllers (Figure 1.16) are designed to self-tune three-term controller settings on the basis of time and amplitude response measurements. The approach adopted is particularly suited to temperature control and is included here as an example of how the area of application affects the design of a self-tuning control algorithm. In particular, there are special considerations which apply in temperature control which must be accounted for in the self-tuning algorithm. An important consideration is that many temperature control systems require that the system output temperature (often referred to as the process variable (PV)) must not overshoot the setpoint (SP). Similarly, temperature control systems often have different dynamical responses when they are heating and cooling.

The natural engineering response to these special features of temperature control is to tailor a self-tuning algorithm to the constraints of the problem in hand. It is important to note, however, that the Eurotherm algorithm also works when the requirement for zero overshoot is removed and directional dynamics are absent.

The Eurotherm self-tuning algorithm analyzes data during the start-up of a process. With reference to Figure 1.17, the process control input is kept at zero for an initial period of one minute. During this period the system output $y(t)$ is monitored to check for plant noise or disturbances from neighbouring control loops. The control input $u(t)$ is then switched fully on and the system dead time t_1 measured and stored for use in controller parameter calculations. As the system output increases, the rate of change is measured and used with the dead time t_1 to set an artificial setpoint level which will ensure that the output $y(t)$ will not overshoot the true setpoint. As the output passes the artificial setpoint, the control input is set to zero and the magnitude and time (t_2) of the peak overshoot noted. As the system output drops below the artificial setpoint, full power is reapplied. The time t_3 and value of the undershoot are then measured and noted. As the system output rises above the artificial setpoint again, the control input is set to zero and the time (t_4) and magnitude of the peak overshoot which follows is noted and recorded. Based upon the transient response measurements made during these tests, heuristic rules can be used to set the coefficients of a three-term controller. Note that the test includes a phase during which the system output is reducing (cooling) and increasing (heating). This allows the algorithm to discriminate between different heating and cooling dynamics. A further interesting feature is that the information required for self-tuning the controller parameters is obtained during the start-up of the system or during the system's transition between setpoints. This makes sound engineering sense in thermal control systems since the start-up phase is often performed 'open-loop' on full power. It is not until the system output is within a reasonable range of the setpoint

Figure 1.16   The Eurotherm self-tuning controller family; (a) the 815 and (b) the 818 Communicating Controllers. (Photographs courtesy of Eurotherm Ltd.)

(a)

(b)

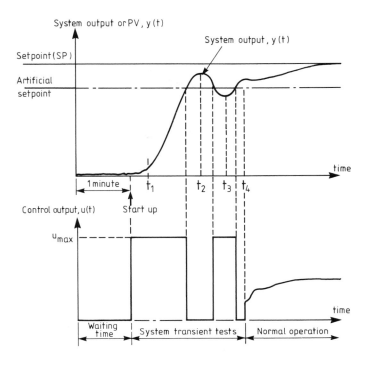

Figure 1.17 The start-up self-tuning cycle of the Eurotherm 815/818 communicating controllers.

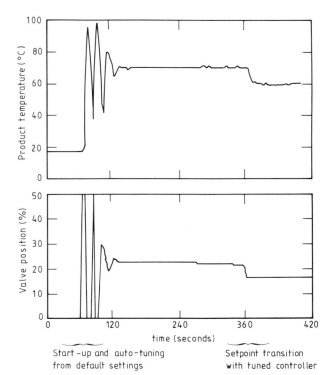

Figure 1.18 Showing the response of a Eurotherm 818 controller applied to the auto-tuning of a PID process temperature control loop. The valve position is the control input $u(t)$, the product temperature is the process output, $y(t)$. (Courtesy of Eurotherm Ltd.)

that the PID control is brought into action. By gathering the information required for self-tuning during this initial phase, the Eurotherm algorithm makes effective use of the open loop start-up phase. Figure 1.18 shows a typical practical performance of the 818 controller applied to the self-tuning of a process product temperature control loop. The product temperature is controlled by the flow rate of coolant through a valve. The control input is therefore the valve position. The figure shows the initial start-up and self-tuning of the 818 controller. For the purposes of comparison the step decrease in product temperature at 360 seconds shows the control loop functioning correctly. This example is an excellent example of how a relatively low cost self-tuning instrument can be used to routinely and automatically tune the coefficients of a process flow loop. This is important because there are many hundreds of thousands of simple control loops used in the world's industrial complexes. However, a significant percentage of the associated controllers are incorrectly tuned leading to reduced process efficiency and product quality. The use of self-tuning PID controllers provides a low cost solution to this problem.

The Eurotherm 818 controller also has an adaptive tuning capacity which acts when plant disturbances occur. In this mode the controller gathers overshoot, undershoot and time data from the disturbance response and retunes the PID controller parameters according to a set of heuristic rules. The use of heuristic rules in the Eurotherm 818 device is an important feature. These rules are based upon expert control engineering experience with temperature control. This expertise is translated into a set of rules for controller design and operation. In effect, the Eurotherm device is a Rule-Based Self-Tuner or Expert System Self-Tuner of the sort mentioned in Section 1.2.3 and illustrated in Figure 1.6. The Eurotherm controller is described here because it is representative of the Rule-Based Self-Tuning algorithms used by a number of process control instrument manufacturers. Other examples include the Foxboro' EXACT controller and the Control and Readout 452 Auto-Tuner.

Rule-Based Self-Tuners are not treated elsewhere in this text since they tend to be geared to expert knowledge of process applications. For commercial reasons instrument companies are reluctant to reveal this expert knowledge and how they use it in self-tuning.

ABB Novatune self-tuning controller

The Novatune device is a general purpose self-tuning process controller which forms part of a process control station sold by ABB as the Masterpiece 200 (Figure 1.19). The self-tuning control principles used in the Novatune are very different from the Eurotherm device. The first and most immediate difference is that the Novatune self-tuning algorithms are used to adjust the parameters

Figure 1.19 The Masterpiece 200 process control station incorporating the Novatune self-tuning controller. (Photograph courtesy of ABB.)

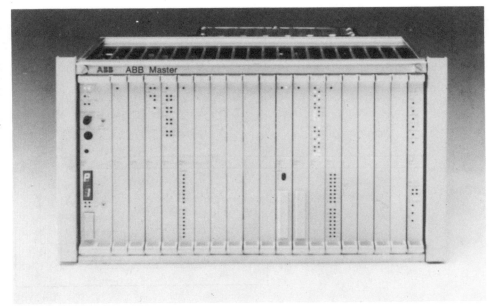

of a general discrete-time controller, rather than a three-term controller as in the Eurotherm 818 communicating controller. At a more fundamental level the Novatune gathers information concerning a system in a different form from the Eurotherm device. Specifically, the 818 controller uses special knowledge of a control problem in order to determine qualitative rules which characterize the system dynamics. By contrast the Novatune uses statistical estimation methods to model the system and disturbance dynamics associated with a process and its environment. The process to be controlled is represented by a linear discrete-time transfer function and the Novatune estimates the parameters of this discrete time model. The parameters of the estimated model are then used to calculate the appropriate setting of the discrete time controller. As will be clear from the foregoing remarks, the Novatune controller is a sophisticated instrument and is typical of those self-tuning controllers which are based upon statistical system estimation and optimization. A typical Novatune application is sketched in Figure 1.20 and shows a simplified representation of a Novatune control scheme used on a lime kiln. The function of the kiln is to convert limestone to quick-lime. The control problem involves regulating the kiln temperature in two zones (the burning zone and exhaust gas zone) in order to reduce the fluctuations in the final temperature. Note that the control problem is a relatively difficult one, with two interacting Novatune controllers, complex time-varying system dynamics and an optimal process regulation objective. The cost benefits to be achieved by a good

Figure 1.20
Simplified
diagram of a
Novatune
installation for
the control of a
lime kiln.
(Courtesy of
ABB.)

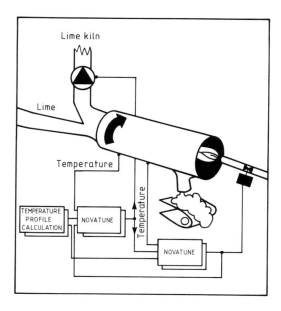

solution to such a problem are high and therefore merit the use of a
sophisticated self-tuning system.

The Novatune is a highly refined instrument which embodies within it in
an explicit form all the self-tuning controller components outlined in Figure
1.8. This can be appreciated by an examination of the Novatune schematic
diagram reproduced in Figure 1.21. In this figure the distinctive features of a
self-tuning control system (system identification, control synthesis and
controller) have been labelled in a way which corresponds with the labelling
of Figure 1.8. The Novatune is based upon a controller synthesis technique
known as minimum variance control. The basic aim of this approach is to
minimize the variance of the output tracking error. Novatune incorporates an
additional feature whereby a pole of the closed loop system can be assigned
to a desired location. The desired location is input to the Novatune by the
parameter PL in Figure 1.21. A penalty on excessive control action can also
be introduced via the parameter PN in Figure 1.21. The control algorithms
associated with minimum variance control are described in Chapter 8 of this
book. The assignment of closed loop poles in a self-tuning framework is
described in Chapter 7.

The system identifier used in the Novatune adaption block is a recursive
least squares algorithm (see Figure 1.21). This is the most widely used
estimation method in self-tuning. A full description of this form of estimator
is given in Chapter 3. Note that the basic self-tuning features are augmented
in the Novatune by a number of additional functions and input parameters.

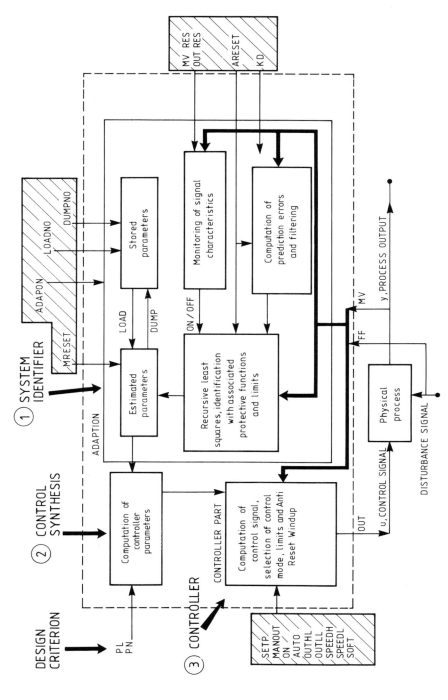

Figure 1.21 The Novatune schematic diagram (courtesy of ABB). Note the shaded boxes contain input variables used to manage self-tuning algorithms. These variables are specific to the Novatune algorithm. General procedures for managing self-tuners are covered in Chapter 4.

These are indicated by the shaded regions of Figure 1.21. Such additional functions are present in one form or another in all general purpose self-tuning controllers. Indeed, they are necessary to make a self-tuner robust in a practical situation.

The additional functions indicated in Figure 1.21 include the following features:

   (i) monitoring of signal characteristics;

  (ii) protective functions and limits during identification; and

 (iii) control mode select, signal limiting and reset in the controller.

These functions, which are vital to the proper running of a self-tuning system, are associated with self-tuning algorithm management. The management of self-tuning systems is covered in Chapter 4.

1.4.3 Adaptive echo cancelling

In this section we discuss an application of self-tuning signal processing in telecommunications. Long haul telephone links suffer from special problems associated with echoes. Typically, and if no special precautions are taken, a telephone conversation will be corrupted by echoes of each speaker's voice being returned to them. When the distance between speakers is short, the echo return is almost instantaneous and the speaker perceives no delay. As the distance between speakers is increased, however, the echo delay increases and can cause alarming results. A student trick is to ask a colleague to speak into a microphone and then to replay his words over earphones after a delay. The delay is achieved using a variable speed tape recorder with offset read and write heads. As the delay between speaking and hearing is increased, the victim usually becomes incoherent to the great amusement of all bystanders.

One approach to echo removal over long haul telephone circuits is to use a technique known as voice-activated echo suppression. In this technique the speaker's voice is used to locally activate a connection circuit, such that each local telephone circuit is only sending information when the corresponding speaker is talking. This has the effect of breaking the echo return circuit but also prevents speakers talking at the same time and in a natural manner.

Self-tuning echo suppression is an alternative way of removing echo interference. The basic idea is shown in Figure 1.22. In this system the sent signal S_{in} and received signal R_{in} at a local circuit are fed to an echo path estimator. The estimator constructs a model of the echo path impulse response which relates the two signals. Then a replica of the echo (e_{cho}) is constructed by passing the received signal R_{in} through the model of the echo path impulse

Figure 1.22
Illustrating the
idea of echo
cancellation in
telephone
circuits.

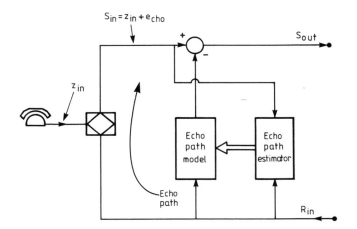

response and subtracting the replica from the signal S_{in} in the send path. The
signal sent (S_{out}) should then have the echo component cancelled.

A commercial instrument used for telephone echo cancellation is shown in
Figure 1.23. This shows a CEN 231 digital adaptive echo canceller, developed
by TRT working in collaboration with the French National Telecommunications
Research Centre (CNET). The CEN 231 echo canceller is for use on pulse code
modulated telephone links transmitting at a rate of 2048 kbit/s and is capable
of cancelling the echoes on all 30 associated telephone channels.

The growing use of self-tuning echo cancellation in telephones is significant
and it means that if you have recently made a trans- or inter-continental
telephone call then you may have unwittingly used a self-tuning device! The
theory behind self-tuning echo cancellation is based upon the interference
cancellation methods presented in Chapter 11.

1.4.4 Self-tuning vibration simulation

Many engineering devices are subject to environmental vibrations during
their use. The vibration may be small and insignificant to the well-being of
the device. In many cases, however, a system's resistance to environmental
disturbances is vital. For example, the flight control computers of an aircraft
are subject to continuous in-flight vibration. If a component of the computer
failed due to vibration induced fatigue, then the safety of the aircraft could
be called into question.

In order to test the vibration resistance of safety critical devices it is accepted
practice that samples should be subject to vibration tests which reproduce the
environmental disturbances which the device in question is likely to encounter.
Often these take the form of random vibrations which can be described by a

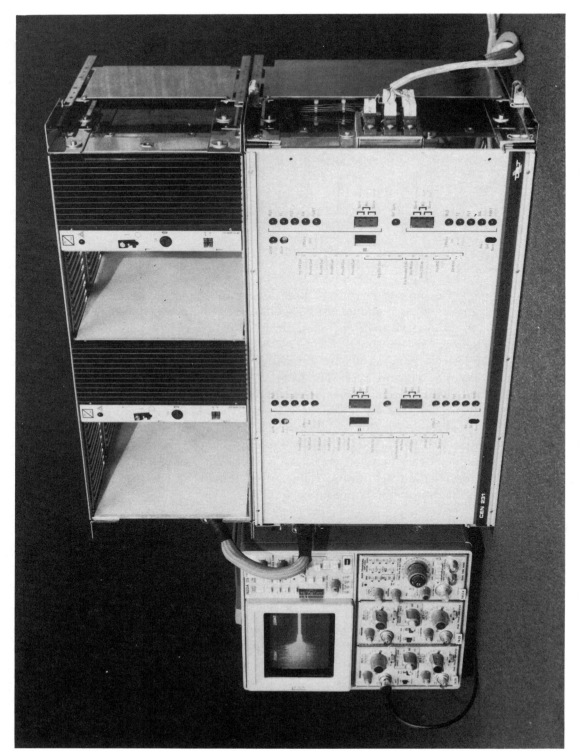

Figure 1.23 A 30 channel self-tuning echo canceller. (Photograph courtesy of TRT.)

Figure 1.24
General view of a model three-axis shaker table. (Courtesy of National Civil Engineering Laboratory of Lisbon, LNEC.)

stochastic process with a specific power spectrum and probability distribution. In such cases, the test device is attached to a shaker table and subjected to vibrations with the power spectrum and amplitude distribution associated with the appropriate environmental disturbances. The shaker table may be a small electromechanically driven system for small test components. At the other end of the scale, however, the shaker table can be a large multi-axis device driven by hydraulic actuators. For example, Figure 1.24 shows a model of a large three-axis shaker table. The full-sized device has a test platform measuring 5.5 m by 4.5 m and is used in the earthquake testing of civil engineering structures. Scale models of civil engineering structures are tested on such tables and the test results 'scaled-up'. It is not unusual to test full size components such as voltage isolator columns or power station control cabinets on such shaker tables.

The vibration testing of systems can be made more efficient by using self-tuning techniques. Specifically, the usual requirement is that the sensed vibration signal, $y(t)$ (see Figure 1.25), has a specified power spectrum. The task of the self-tuning system is to estimate the power spectrum of the sensed vibrations, compare it with the specified power spectrum and to synthesize a test signal $u(t)$ such that the power spectrum of $y(t)$ coincides with the specified power spectrum.

Figure 1.25
Schematic
diagram for a
self-tuning
vibration control
system.

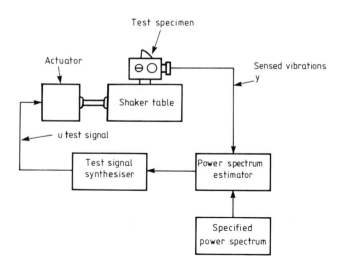

The requirement for vibration testing to a predefined vibration spectrum is now widespread. It is incorporated into many military specifications (for example MIL-STD 810D) and is an increasingly important aspect of consumer product testing. A commercial instrument which automates the self-tuning of spectral properties and test signal generation as illustrated in Figure 1.25 is the Schlumberger SI 1215 Vibration Control System. The SI 1215 (shown in Figure 1.26) is a sophisticated instrument which allows the real-time self-tuning of spectral properties over a bandwidth of 0 to 10 kHz. The computations involved in spectral estimation are extremely demanding and as a result the instrument contains highly developed hardware and software which has been optimized to allow the system to work in real-time. In a typical test a component is attached to a vibration table (for example, Figure 1.27 shows a camera attached to a single-axis vibrator). The SI 1215 begins its self-tuning cycle with a rough estimate of the appropriate test signal. This produces an initial estimated output excitation spectrum which will usually deviate considerably from that required. The initial estimate is then used in real-time to adjust the input excitation signal and the process continues until the input excitation has been tuned in such a way that the output excitation spectrum lies within a specified neighbourhood of the required spectrum. Figure 1.28 shows a typical evolution of the output spectrum from the initial 'untuned' value to the final self-tuned spectrum which lies within specified limits.

The Solartron SI 1215 is an embodiment of a frequency domain self-tuning algorithm. The background theory for this form of self-tuning method is given in Chapter 14.

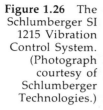
Figure 1.26 The Schlumberger SI 1215 Vibration Control System. (Photograph courtesy of Schlumberger Technologies.)

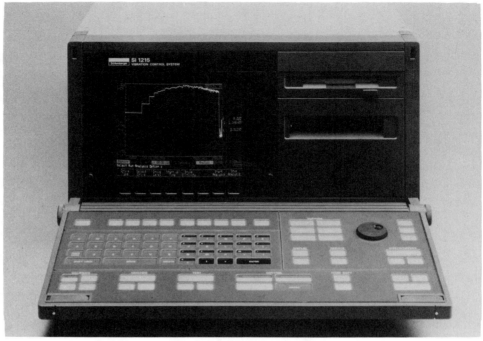

1.5   TECHNICAL FEATURES OF A SELF-TUNING SYSTEM

The study of self-tuning and adaption brings together a number of disciplines in rather a special way. The organization of this book has been motivated by the wish to present the technical components of these disciplines in an ordered manner.

The main technical components for a study of self-tuning are:

Digital representation

Self-tuners are almost all implemented in a digital form. It is therefore important to understand how continuous time dynamical systems and signals are represented in digital (sampled data) form.

Recursive estimation

At the heart of every self-tuning system is a recursive estimator. Accordingly, a study of self-tuning necessarily involves a working knowledge of recursive estimation.

Figure 1.27
Typical small test component (in this case a Hasselblad camera) on a single-axis vibrator. (Photograph courtesy of Schlumberger Technologies.)

Control design

The basis of self-tuning is the on-line combination of an estimator and a corresponding design rule. In self-tuning control the main design rules currently in use are pole assignment and single step and multistep optimization methods. However, other important applications such as extremum control have a simple self-tuning format.

Signal processing

The signal processing algorithms which are used in a self-tuning form are prediction and noise reduction filters.

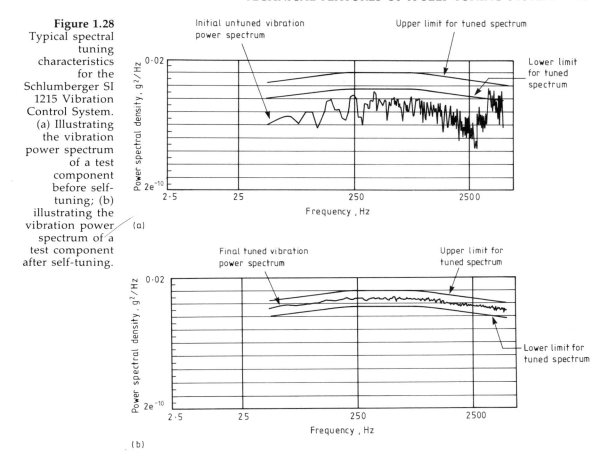

Figure 1.28 Typical spectral tuning characteristics for the Schlumberger SI 1215 Vibration Control System. (a) Illustrating the vibration power spectrum of a test component before self-tuning; (b) illustrating the vibration power spectrum of a test component after self-tuning.

Stability

The self-tuning idea of combining the design process with the implementation process leads to some complex stability problems. These can be approached from a number of theoretical directions. However, within the philosophy of this book, the most logical approach is to consider the stability of a self-tuning system as being associated with the stability of the associated recursive estimator. In general, the global stability of self-tuning algorithms cannot be proven. In certain special cases, however, an almost complete stability analysis can be provided. The extremum controller and frequency domain self-tuning controller are examples of such algorithms.

1.6   SUMMARY

This chapter has introduced the area of self-tuning by a coverage of the following main points:

- A review of conventional control and signal processing.
- A description of the basic idea of self-tuning.
- A discussion of approaches to self-tuning, adaption and rival techniques.
- A description of the components and philosophy of self-tuning systems.
- A presentation of some typical industrial self-tuning applications.

1.7   PROBLEMS

Problem 1.1

It is customary for the introductory chapter of control engineering books to contain an example of an 'early control system'. This usually involves the author leafing through old books in order to find the said early control system and the publisher getting permission to reprint a barely recognizable illustration. We have decided to eliminate the middle-man and to ask you to find your own early example of an adaptive or self-tuning control system. Use this as an opportunity to reflect upon the distinction between fixed coefficient feedback and adaptive feedback (the two can easily be confused). You may use the book *A History of Mechanical Inventions* by A.P. Usher (Dover, 1988) as a starting point.

Problem 1.2

The Eurotherm expert self-tuning controller described in Section 1.4.2 is based on a set of observations of system behaviour and a corresponding set of *ad-hoc* rules for adjusting the parameters of a three-term controller. We invite you to develop your own rule-based, expert self-tuner to fit the following situation. The unknown system is second order with no time delay. It is permitted to inject a square wave test signal and your self-tuning computer can measure rise times, overshoots etc. Propose a set of 'tuning rules' which will adjust the gain of a proportional controller that will double the open loop bandwidth. Then try adding a derivative term into the controller and developing additional rules to tune it.

Problem 1.3

The ideas of feedback control theory can be used to explain biological processes. Read *The Selfish Gene* by R. Dawkins (Penguin, 1989) and discuss the idea of the gene as an adaptive entity.

Problem 1.4

Adaptive systems are essentially time varying and we must be careful not to carry over standard results from time invariant systems theory without checking their validity.

For example, the time varying discrete-time linear system

$$\mathbf{x}(t+1) = \mathbf{A}(t)\mathbf{x}(t)$$

may be unstable even though all the eigenvalues of $\mathbf{A}(t)$ are less than unity in size for all t. Find an example of this behaviour when the state dimension is two.

1.8   NOTES AND REFERENCES

1.8.1   Useful books

There are a number of books which deal with self-tuning systems. Specifically recommended are:

Goodwin, G.C. and Sin, K.S.
  Adaptive Filtering, Prediction and Control, Prentice Hall, 1984.
This is an excellent theoretically oriented book.

Astrom, K.J. and Wittenmark, B.
  Adaptive Control, Addison Wesley, 1988.
The book covers self-adaptive feedback control with the breadth and readable style associated with these distinguished authors.

Widrow, B. and Stearns, S.
  Adaptive Filters, Prentice Hall, 1985.
This book focuses upon adaptive filtering and in particular the methods associated with Widrow and his co-workers.

Treichler, J.R., Johnson, C.R. and Larimore, M.G.
  Theory and Design of Adaptive Filters, Wiley, 1987.
This book contains useful implementation notes for commercial signal processing chips.

Peterka, V.
  Control of uncertain processes: applied theory and algorithms, *Kybernetika* (Special Supplement), **22**, 1986.
This textbook length appendix to *Kybernetika* is difficult to obtain. However, it is well worth seeking out because of its scholarly and meticulous Bayesian exposition.

The theoretical questions arising from the stability and convergence of adaptive algorithms involve many technical difficulties. Some books which treat this area are:

Ljung, L. and Soderstrom, T.
  Theory and Practice of Recursive Identification, McGraw Hill, 1983.
which concentrates on the ODE approach to the stochastic convergence problem.

Davis, M.H.A. and Vinter, R.B.
Stochastic Modelling and Control, Chapman and Hall, 1985.
This is an excellent concise text on the background theory of stochastic systems. A more substantial treatment is given in:

Caines, P.E.
Linear Stochastic Systems, Wiley, 1988.

Also see:

Anderson, B.D.O., Bitmead, R.R., Johnson, C.R. Jr., Kokotovic, P.V., Kosut, R.L., Mareels, I.M.Y., Praly, L. and Riedle, B.D.
Stability of Adaptive Systems – Passivity and Averaging Analysis, MIT Press, 1986.
This multi-author book covers more recent convergence material based upon averaging methods not treated by Ljung and Soderstrom.

Sastry, K. and Bodson, M.
Adaptive Control – Stability, Convergence and Robustness, Prentice Hall, 1989.

1.8.2   Survey articles

There are numerous technical articles which provide surveys of adaptive systems. Some useful articles are:

Astrom, K.J.
Theory and applications of adaptive control – a survey, *Automatica*, **19**(5), 471–86, 1983.
Wellstead, P.E. and Zanker, P.
Techniques of self-tuning, *Optimal Control Applications and Methods*, **3**, 305–22, 1982.
Seborg, D.E., Edgar, T.F. and Shah, S.L.
Adaptive control strategies for process control: a survey, *AICheE Journal*, **32**(6), 881–913, 1986.

1.8.3   Historical references

A historian's platitude states that it is not possible to understand today without a knowledge of yesterday. As an implicit endorsement of this sentiment, we offer the following suggestions for reading:

Bennett, S.
A History of Control Engineering, Peter Peregrinus, 1979.
Nahin, P.J.
Oliver Heaviside: Sage in Solitude, IEEE Press, 1988.
Hyman, A.
Charles Babbage: Pioneer of the Computer, Oxford University Press, 1984.
Mishkin, E. and Braun, L.
Adaptive Control Systems, McGraw Hill, 1961.

1.8.4   General references

The physical modelling of systems as a method of determining dynamical behaviour is treated in numerous books on system dynamics. The following three are typical:

Shearer, J.L., Murphy, A.T. and Richardson, H.H.
Introduction to System Dynamics, Addison Wesley, 1971.
Macfarlane, A.G.J.
Dynamical System Models, Harrap, 1970.
Wellstead, P.E.
Introduction to System Modelling, Academic Press, 1979.

System identification methods for estimating system parameters are covered in the following texts:

Norton, J.
System Identification, Academic Press, 1986.
Ljung, L.
System Identification: Theory for the User, Prentice Hall, 1987.
Soderstrom, T. and Stoica, P.
System Identification, Prentice Hall, 1989.

The general areas of fixed coefficient control and signal processing are covered in such texts as:

Franklin, G.G. and Powell, J.D.
Digital Control of Dynamic Systems, Addison Wesley, 1980.
Oppenheim, A.V. and Schafer, R.W.
Digital Signal Processing, Prentice Hall, 1975.

The early self-tuning control system from Kalman is described in:

Kalman, R.W.
Design of a self-optimising control system, *Trans. ASME* **80** (Series D), 468, 1958.

The paper by Peterka which, in the opinion of many, started the modern interest in self-tuning is:

Peterka, V.
Adaptive digital regulation of noisy systems, *Second IFAC Symposium on Identification and Process Parameter Estimation*, paper 6.2, Academia, Prague, 1970.

There are numerous books and papers on Model Reference Adaptive Control. An early paper of great significance is:

Parks, P.C.
Liapunov Redesign of Model Reference Adaptive Control Systems, *IEEE Trans.* **AC-11**, 362, 1966.

A useful book on various MRAC methods is:

Landau, Y.D.
Adaptive Control – The Model Reference Approach, Marcel Dekker, 1979.

A paper which summarizes adaptive signal processing methods associated with Widrow and his co-workers is:

Widrow, B., Glover, J.R., McCool, J.M., Kannitz, J., Williams, C.S., Hearn, R.H., Zeidler, J.R., Dong, E. Jr and Goodlin, R.C.
Adaptive noise cancelling: principles and applications, *Proc. IEEE*, **63**, 1692, 1975.

1.8.5   Application references

The application of self-tuning to engine control is based upon material supplied by Lucas Automotive Ltd. The general technique used – extremum control – is discussed in detail in Chapter 13. An early reference to the procedure is:

Draper, C.S. and Li, Y.
Principles of Optimising Control Systems, ASME Publication, 1951.

The specific algorithms used in the Lucas system are described in:

Holmes, M. and Cokerham, K.
Adaptive ignition control, *Sixth International Conference on Automotive Electronics*, IEE, London, 12–15 October 1987.

The description of the Eurotherm self-tuning controller is based upon information supplied by Eurotherm Ltd. A further description is given in:

Tinham, B.
A new era of multi algorithm control, *Control and Instrumentation*, **22**, 127–30, 1990.

The ABB Novatune/MasterPiece 200 material presented here was supplied by ABB Automation. The Novatune instrument is more fully described in:

Bengtsson, G. and Egardt, B.
Experiences with self-tuning control in the process industry, *Proc. Ninth IFAC World Congress, Budapest*, Pergamon Press, 1984.

The CEN 231 adaptive echo canceller description was supplied courtesy of TRT Ltd. A more complete technical description is given in:

Erdreich, M., Lassaux, J. and Mamann, J.M.
30-Channel Digital Echo Canceller for Telephone Circuits, *Commutation and Transmission*, **4**, 53–66, 1985.

The self-tuning vibration control description included material supplied by the National Laboratory for Civil Engineering, LNEC, Lisbon and Schlumberger Instruments. Further descriptions of the LNEC work is given in:

Emilio, F.T., Duarte, R.T., Carvalhal, F.J. and Pereira, J.J.
A new type of three-degrees-of-freedom shaking table, *Proc. Eighth European Conference on Earthquake Engineering*, LNEC, Lisbon, Portugal, September 1986.

See also:

Pereira, J.J. and Carvalhal, F.J.
Adaptive control technique for the simulation of seismic actions, *Proc. Conference Design of Concrete Structures*, Watford, UK, November, 1983. Elsevier, 1984.

The Schlumberger SI1210 is described further in:

Controlled Random Vibration Testing, Schlumberger technical report 117/83.

This document is available from the following address: Schlumberger Technologies, Instruments Division, Victoria Road, Farnborough, Hampshire GU14 7PW.

Part 1

System Identification for Self-tuning

In the introductory chapter a self-tuning system framework was proposed comprising three parts: an identifier, an algorithm synthesizer and an implementation module. Here, in the first part of the main text, the discussion concerns the identification algorithms associated with self-tuning. This discussion necessarily involves questions of system representation (Chapter 2), recursive estimation (Chapter 3), the practical use of recursive estimators (Chapter 4), numerical aspects of estimation (Chapter 5) and the convergence behaviour of recursive algorithms (Chapter 6).

2 System and Signal Models

2.1 OUTLINE AND LEARNING OBJECTIVES

In this chapter we consider the representation of systems and signals in a form suitable for self-tuning. In particular, since self-tuning is a digital computer-based technique, we consider digital parametric models of systems and signals. Moreover, these models are developed in a form suited to the identification algorithms to be used in self-tuning and described in Chapters 3, 4 and 5.

Because we want to represent systems for the purposes of both control and signal processing, the chapter will consider a general representation (Figure 2.1) in which both a system and a signal component are present. This general representation can then be specialized to meet the requirements of either control or signal processing.

The learning objective is to understand the way in which systems and signals are modelled for self-tuning. The system models are related to physical systems via linearization and discretization. The signal modelling treatment is intended to present the basic types of signals which can be encountered in control and signal processor engineering. Accordingly, it is assumed that the reader has a knowledge of z-transform theory. Thus the treatment is restricted to those aspects of discrete transform methods which are appropriate to self-tuning control and signal processing.

2.2 SYSTEM MODELS

The real dynamic processes discussed in Chapter 1 evolve continuously in time and are nonlinear in nature. No mathematical model can ever display every nuance of behaviour of the real system that it is constructed to represent, but (luckily!) this is never required. What we demand is a model that is satisfactory under given experimental conditions in pursuing a chosen experimental goal. Any chosen model is therefore only conditionally valid and the preservation of the relevant conditions is crucial (see Chapter 4).

Figure 2.1
General
representation of
a combined
signal and
system source.

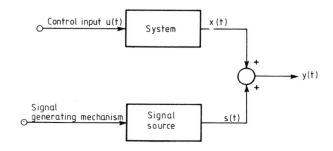

The most commonly used parametric models associated with the self-tuning area are *linear* both in the data and the parameters. As we will see, this eases considerably the problems of design, estimation and analysis. Secondly, the recursive nature of self-tuning algorithms suggest that *discrete time* modelling is appropriate.

Consider the situations shown in Figure 2.2. The input signal $u(t)$ is a control variable and $x(t)$ is the system output. The assumption is that the system model is a digital or discrete time model and that the signals $u(t)$ and $x(t)$ are discrete time sequences. The time index t takes integer values -3, -2, -1, 0, 1, 2, 3, ..., as shown in Figure 2.3(a). Usually, the discrete time sequences will correspond to sampled data values obtained from a continuous time signal $u_c(\tau)$ as indicated in Figure 2.3(b). Throughout this text (unless otherwise stated) we will write continuous time as τ (seconds) and note continuous time signals by subscript c. When $u(t)$ is a discrete sequence obtained from a continuous signal $u_c(\tau)$ the discrete values of $u(t)$ correspond to $u_c(\tau)$ at equispaced time intervals (see again Figure 2.3(b)):

Figure 2.2
System model
used in control
studies.

Figure 2.3
Illustrating (a) a
discrete signal
$u(t)$ and (b) its
continuous time
equivalent $u_c(\tau)$.
Note that τ is
continuous time
and the index t
corresponds to
the value of τ at
$t\tau_s$ s.

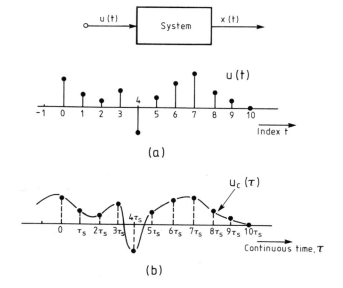

$$u(t) = u_c(t\tau_s)$$

where τ_s is the sampling interval.

The system in Figure 2.2 is assumed to be a discrete linear model of an underlying continuous time system in a digital computer control configuration. In this situation, the control input $u(t)$ can be related to the noise-free response $x(t)$ through the linear difference equation

$$x(t)+a_1 x(t-1)+ \ldots +a_{n_a} x(t-n_a) = b_0 u(t)+b_1 u(t-1)+ \ldots + b_{n_b} u(t-n_b). \quad (2.1)$$

Introducing the unit backward shift operator z^{-1} defined by

$$z^{-i} x(t) = x(t-i) \quad (2.2)$$

the model (2.1) can be expressed in the discrete time transfer function form

$$x(t) = \left[\frac{B}{A}\right] u(t) \quad (2.3)$$

where the objects B and A are polynomials in the shift operator z^{-1}, given by

$$\left. \begin{array}{l} A(z^{-1}) = 1+a_1 z^{-1}+ \ldots +a_{n_a} z^{-n_a} \\[2mm] B(z^{-1}) = b_0+b_1 z^{-1}+ \ldots +b_{n_b} z^{-n_b} \end{array} \right\} \quad (2.4)$$

In subsequent paragraphs we will discuss the use of this model in a computer. For the moment we briefly examine its relationship to the dynamics of the underlying continuous time system. This is done by linking the discrete time transfer function via the z-transform of sampled data theory to the Laplace transfer function which characterizes the (linearized) continuous time system dynamics.

Consider a continuous time system linked to a digital control system as illustrated in Figure 2.4. In this figure it is assumed that the noise-free system output $x_c(\tau)$ is sampled by an analogue-to-digital (A/D) converter to obtain the output sequence $x(t)$. Likewise the digital controller output $u(t)$ is passed through a digital-to-analogue converter (D/A) to form the continuous time signal $u_c(\tau)$. If the digital controller sampling interval is τ_s s, then the A/D converter can be modelled as a sampler with the same sampling interval. The D/A converter functions as follows. At sample interval t the D/A converter outputs the control sequence value $u(t)$ and holds the value at the D/A output. When sample interval t has elapsed the sequence value $u(t+1)$ is output and

Figure 2.4
Computer
control system
in which a
continuous time
system with
transfer function
$G(s)$ is controlled
by a digital
computer via a
digital-to-
analogue and an
analogue-to-
digital interface.

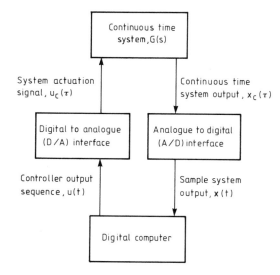

held at the D/A output. This sequence of 'sample and hold' continues and is illustrated in Figure 2.5. The sample and hold action is associated in digital control with the zero order hold (ZOH) mechanism. Essentially, the ZOH, when programmed to operate at a sample rate $1/\tau_s$ Hz, takes as its input a member of the control sequence $u(t)$. It then outputs this value and holds it constant for the duration of the sample interval τ_s s as shown in Figure 2.5. The ZOH is a linear time invariant system and has a transfer function given by

Figure 2.5 The
sample and hold
action of a
digital-to-
analogue (D/A)
converter. (a)
The control
sequence $u(t)$.
The original
control sequence
$u(t)$ (b) is
converted to a
continuous time
signal $u_c(\tau)$ by
feeding $u(t)$
through a
sample and hold
mechanism.

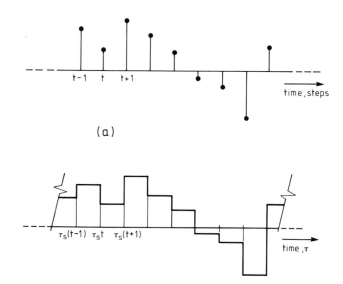

$$H(s) = \frac{1 - \exp(-s\tau_s)}{s} \tag{2.5}$$

where s is the conventional Laplace transform operator. As noted above, at the output the system response is turned into a discrete time sequence by a sampling device. This sampling action can be modelled as a switch which closes instantaneously every τ_s s. Thus the D/A can be represented as shown in Figure 2.6 by a sampler and a ZOH mechanism and the A/D by a sampler. The control computer itself can be represented by the controller algorithm used to regulate the process. The nature and design of these algorithms is considered in Part 2.

The representation of D/A and A/D interfaces shown in Figure 2.6 is important because it allows us to develop the form of system models which we will use in self-tuning control. This is done by considering the sampled data transfer function of the system as seen by the controller. In particular, we are interested in the sampled data representation between points P and P′ in Figure 2.6. Thus the sampled data transfer function of the system between the control sequence $\{u(t)\}$ and sampled output sequence $\{x(t)\}$ is given by the z-transform of the hold mechanism transfer function $H(s)$ and the system transfer function $G(s)$. Thus

$$\frac{x(z)}{u(z)} = \mathbf{Z}\left\{H(s) \cdot G(s)\right\} \tag{2.6}$$

In this expression $x(z)$, $u(z)$ are the z-transforms of the sequences $\{x(t)\}$, $\{u(t)\}$ respectively and $\mathbf{Z}[\cdot]$ denotes the z-transform operator which maps a

Figure 2.6
Representation
of a computer
control system.

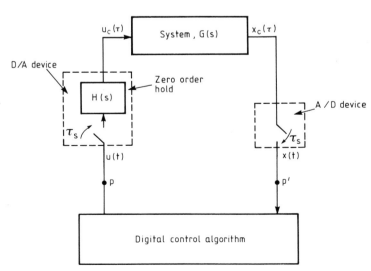

continuous time transfer function into its discrete time counterpart. Normally, this transformation may be most easily carried out by making a partial fraction expansion of the continuous time transfer function and using tables to convert each term in the expansion.

Note that the z-transform of the hold and system transfer functions must be taken in the product form of equation (2.6). It is *not* true to say that:

$$\mathbf{Z}\{H(s) \cdot G(s)\} = \mathbf{Z}\{H(s)\}\, \mathbf{Z}\{G(s)\}$$

Using (2.5) for the hold transfer function, equation (2.6) can be written as

$$\frac{x(z)}{u(z)} = \mathbf{Z}\left\{\frac{(1-\exp(-sT_s))}{s} \cdot G(s)\right\}$$

$$= (1-z^{-1})\,\mathbf{Z}\left\{\frac{G(s)}{s}\right\} \tag{2.7}$$

Using z-transform tables, we can obtain from the transfer function of a continuous time system the equivalent discrete time system transfer function, as perceived by the digital controller via the hold unit and output sampler. Table 2.1 gives a selection of commonly used z-transforms which can be used for this purpose. For example, consider the case of a first order system.

Example 2.1

Given

$$G(s) = \frac{a}{s+a}$$

then

$$\frac{G(s)}{s} = \frac{a}{s(s+a)} = \frac{1}{s} - \frac{1}{s+a}$$

and

$$\mathbf{Z}\left[\frac{G(s)}{s}\right] = \mathbf{Z}\left[\frac{1}{s}\right] - \mathbf{Z}\left[\frac{1}{s+a}\right]$$

Table 2.1 Table of commonly used z-transforms

Laplace transform $G(s)$	z-transform $G(z)$
$\dfrac{1}{s}$	$\dfrac{z}{z-1}$
$\dfrac{1}{s^2}$	$\dfrac{\tau_s z}{(z-1)^2}$
$\dfrac{1}{s+a}$	$\dfrac{z}{z-\exp(-a\tau_s)}$
$\dfrac{a}{s(s+a)}$	$\dfrac{z(1-\exp(-a\tau_s))}{(z-1)\,(z-\exp(-a\tau_s))}$
$\dfrac{1}{(s+a)^2}$	$\dfrac{\tau_s z\exp(-a\tau_s)}{(z-\exp(-a\tau_s))^2}$
$\dfrac{a}{s^2(s+a)}$	$\dfrac{\tau_s z}{(z-1)^2} - \dfrac{(1-\exp(-a\tau_s))z}{a(z-1)\,(z-\exp(-a\tau_s))}$
$\dfrac{a}{s^2 + a^2}$	$\dfrac{z\sin(a\tau_s)}{z^2 - 2z\cos(a\tau_s) + 1}$
$\dfrac{s}{s^2 + a^2}$	$\dfrac{z(z-\cos(a\tau_s))}{z^2 - 2z\cos a\tau_s + 1}$
$\dfrac{1}{(s+a)^2 + b^2}$	$\dfrac{1}{b}\left[\dfrac{z\exp(-a\tau_s)\sin b\tau_s}{z^2 - 2z\exp(-a\tau_s)\cos(b\tau_s) + \exp(-2a\tau_s)}\right]$
$\dfrac{s+a}{(s+a)^2 + b^2}$	$\dfrac{z^2 - z\exp(-a\tau_s)\cos b\tau_s}{z^2 - 2z\exp(-a\tau_s)\cos b\tau_s + \exp(-2a\tau_s)}$

From Table 2.1:

$$\mathbf{Z}\left[\frac{1}{s}\right] = \sum_{i=0}^{\infty} z^{-i} = \frac{1}{1-z^{-1}}$$

and

$$\mathbf{Z}\left[\frac{1}{s+a}\right] = \sum_{i=0}^{\infty} z^{-i}\exp(-ai\tau_s) = \frac{1}{1-\exp(-a\tau_s)z^{-1}}$$

Hence

$$\frac{x(z)}{u(z)} = (1-z^{-1})\left\{\frac{1}{1-z^{-1}} - \frac{1}{1-\exp(-a\tau_s)z^{-1}}\right\}$$

$$= \frac{z^{-1}(1-\exp(-a\tau_s))}{1-\exp(-a\tau_s)z^{-1}}.$$

□

The above procedure can be repeated on the entries in Table 2.1 and a table of z-transforms constructed for systems with ZOHs in series with them. Such a table is given in Table 2.2.

In general the z-transform models obtained in this way can be written as the ratio of two polynomials in z^{-1}. Thus:

$$\mathbf{Z}\{H(s)G(s)\} = z^{-1}\frac{B(z^{-1})}{A(z^{-1})} \tag{2.8}$$

where

$$B(z^{-1}) = b_0 + \ldots + b_{n_b}z^{-n_b}$$

$$A(z^{-1}) = 1 + a_1z^{-1} + \ldots + a_{n_a}z^{-n_a}$$

(cf. equation (2.4)). Note that the structure (2.8) embodies the delay of one sampling interval generated by the ZOH, even in the absence of a pure time delay in the system (see Section 2.2.1).

In this notation, Example 2.1 corresponds to:

$$B(z^{-1}) = b_0 \qquad \text{with } b_0 = 1 - \exp(-a\tau_s)$$

$$A(z^{-1}) = 1 + a_1z^{-1} \qquad \text{with } a_1 = -\exp(-a\tau_s)$$

Table 2.2 Some z-transforms of systems incorporating ZOHs

Laplace transform $G(s)$	z-transform $G(z)$
$\dfrac{1}{s}$	$\dfrac{\tau_s}{z-1}$
$\dfrac{1}{s^2}$	$\dfrac{z\tau_s^2}{(z-1)^2}$
$\dfrac{a}{s+a}$	$\dfrac{1-\exp(-a\tau_s)}{z-\exp(-a\tau_s)}$
$\dfrac{a^2}{(s+a)^2}$	$\dfrac{z(1-(1+a\tau_s)\exp(-a\tau_s)) + \exp(-a\tau_s)\,(a\tau_s-1)}{(z-\exp(-a\tau_s))^2}$
$\dfrac{sa^2}{(s+a)^2}$	$\dfrac{\tau_s a^2 \exp(-a\tau_s)\,z - \tau_s}{(z-\exp(-a\tau_s))^2}$

In the following (as in equation (2.3)) we will often drop polynomial arguments for simplicity.

In self-tuning estimation the system model is usually represented in the form of a difference equation in which the output $x(t)$ is written as a linear function of past values of itself and the actuation signal $u(t)$ (see equation (2.1)). This is achieved by noting that z^{-1} in equation (2.8) can be interpreted as the unit backward shift operator (see equation (2.2)). Using the shift interpretation, (2.8) can be rewritten in the difference equation form (2.1).

Example 2.2

Given a z-transform

$$\frac{B}{A} = \frac{b_0 z^3 + b_1 z^2 + b_2 z + b_3}{z^3 + a_1 z^2 + a_2 z + a_3}$$

divide through top and bottom by z^3 to give

$$\frac{B}{A} = \frac{b_0 + b_1 z^{-1} + b_2 z^{-2} + b_3 z^{-3}}{1 + a_1 z^{-1} + a_2 z^{-2} + a_3 z^{-3}}$$

Using this as a discrete transfer between sequences $\{u(t)\}$ and $\{x(t)\}$ we have

$$x(t) = \left[\frac{b_0 + b_1 z^{-1} + b_2 z^{-2} + b_3 z^{-3}}{1 + a_1 z^{-1} + a_2 z^{-2} + a_3 z^{-3}} \right] u(t)$$

This can now be rearranged and the shift operator interpretation used to form the following difference equation relating $\{u(t)\}$ to $\{x(t)\}$:

$$x(t) = -a_1 x(t-1) - \dot{a}_2 x(t-2) - a_3 x(t-3) + b_0 u(t)$$
$$+ b_1 u(t-1) + b_2 u(t-2) + b_3 u(t-3)$$

$\square$

In general, for a system given by a z-transform representation of the form (2.8), the equivalent difference equation can be written as

$$x(t) = - \sum_{i=1}^{n_a} a_i x(t-i) + \sum_{j-0}^{n_b} b_j u(t-j-1) \tag{2.9}$$

Moreover, recalling the equivalence between the z-transform and the difference equation model, it is possible to say that the degrees n_a, n_b of the polynomials A,B are given by

n_a = number of poles of $G(s)$, the underlying continuous time system transfer function;

$n_b \leq n_a$.

2.2.1 Time delays

The systems considered so far have not included pure time delays. In practice, however, many processes involve transport times and computational delays which jointly mean that the system transfer function will also include a pure time delay. If the pure time delay τ_d is an exact multiple of the sample interval τ_s, e.g.

$$\tau_d = k\tau_s \tag{2.10}$$

for some positive integer k, then this can be represented in discrete time as a backward shift of k steps operating upon the control sequence $u(t)$. In turn these k backward shifts are represented by a factor z^{-k} in the discrete transfer function (Figure 2.7). In this way a linear system involving a *total* time delay

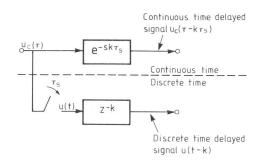

of k sample intervals (including a possible contribution of z^{-1} from a ZOH)
can be modelled by a discrete time transfer function

$$x(t) = \frac{z^{-k} B}{A} u(t) \tag{2.11}$$

where B and A are given by (2.4).

In the real world, the time delay in a system is never an exact multiple of
the sample interval. It is always possible, however, to write a time delay τ_d
in terms of an integer multiple k of the sample interval τ_s and a partial time
delay $m\tau_s$ (where $0 \leqslant m < 1$). In this way it is possible to express the general
delay τ_d as

$$\tau_d = k\tau_s - m\tau_s \tag{2.12}$$

The influence of the partial time delay m is to introduce an extra zero in
the numerator of the z-transform of the process. The position of this extra
zero will vary with m and for certain values of m the extra zero will be outside
the unit circle in the z-plane. Thus the partial time delay can give rise to a
nonminimum phase z-transfer function even if the original (continuous time)
system has all its zeroes in the left half-plane.

The following example illustrates this:

Example 2.3

Consider the first order system of Example 2.1 with an additional pure time
delay τ_d:

$$G(s) = \frac{a\exp(-s\tau_d)}{s+a}$$

where $\tau_d = k\tau_s - m\tau_s$ and in which $0 \leqslant m < 1$. Figure 2.8 illustrates this situation. Here m is the partial time delay in the system $G(s)$ (see Figure 2.4). The z-transform of this transfer function, together with a ZOH, is

$$\mathbf{Z}\{G(s)\} = \mathbf{Z}\left\{\frac{(1-\exp(-s\tau_s))}{s} \frac{a\exp(-s\tau_d)}{s+a}\right\}$$

$$= z^{-k}(1-z^{-1})\mathbf{Z}\left\{\frac{a\exp(ms\tau_s)}{s(s+a)}\right\}$$

$$= z^{-k}(1-z^{-1})\,\mathbf{Z}\left\{\frac{\exp(ms\tau_s)}{s} - \frac{\exp(ms\tau_s)}{s+a}\right\}$$

Using the time shift properties of Laplace transforms

$$\mathbf{L}^{-1}\left\{\frac{\exp(ms\tau_s)}{s}\right\} \rightarrow U(\tau+m\tau_s)$$

$$\mathbf{L}^{-1}\left\{\frac{\exp(ms\tau_s)}{s+a}\right\} \rightarrow U(\tau+m\tau_s)\left\{\exp(-a(\tau+m\tau_s))\right\}$$

where $\mathbf{L}^{-1}$ is the inverse Laplace transform operator and $U(\tau)$ is the unit step function.

Recalling that $0 \leqslant m < 1$, these time functions will have z-transforms

$$\mathbf{Z}\{U(\tau+m\tau_s)\} = \frac{1}{1-z^{-1}}$$

$$\mathbf{Z}\{U(\tau+m\tau_s)\exp(-a(\tau+m\tau_s))\} = \frac{\exp(-am\tau_s)}{1-\exp(-a\tau_s)z^{-1}}$$

Combining the individual z-transforms gives

$$\mathbf{Z}\{G(s)\} = z^{-k}\left\{\frac{1-z^{-1}\exp(-a\tau_s) - (1-z^{-1})\exp(-am\tau_s)}{(1-z^{-1}\exp(-a\tau_s))}\right\}$$

Figure 2.8
Illustrating a partial time delay $m\tau_s$ as the difference between the delay τ_d and the integer multiple $k\tau_s$.

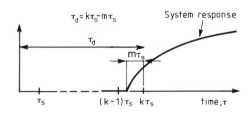

$$= z^{-k} \left\{ \frac{(1-\exp(-am\tau_s)) - (\exp(-a\tau_s) - \exp(-am\tau_s))\, z^{-1}}{(1-z^{-1}\exp(-a\tau_s))} \right\} \quad (2.13)$$

The important point about equation (2.13) is the way in which the numerator of the z-transfer function is influenced by the partial time delay m. In particular, it introduces additional terms which contribute a zero in the transfer function at location z_m given by

$$z_m = \frac{\exp(-a\tau_s) - \exp(am\tau_s)}{(1-\exp(-am\tau_s))}$$

When $m = 1$, then $z_m = 0$ (i.e. a zero at the origin) and the z-transform becomes

$$\mathbf{Z}\{G(s)\} = \frac{z^{-k}(1-\exp(-a\tau_s))}{(1-z^{-1}\exp(-a\tau_s))}$$

When m tends to zero, $z_m \to -\infty$ (i.e. an additional integer time delay) and the z-transform becomes

$$\mathbf{Z}\{G(s)\} = \frac{z^{-k-1}(1-\exp(-a\tau_s))}{(1-z^{-1}\exp(-a\tau_s))}$$

From the above it can be seen that the zero due to the partial time delay moves along the negative real axis of the z-plane as m varies. For small values of m it will be outside the unit circle. The occurrence of nonminimum phase zeroes associated with small delays can cause problems with minimum variance controllers. More of this will be said in Chapter 8.

□

The discrete transfer function of any system involving time delays may be obtained using the method followed in Example 2.3. A table of transforms including partial time delays is given in Table 2.3. Note that in this case the transforms are written in terms of z^{-1} so that they may be directly used in difference equation models.

2.3 SIGNAL MODELS

In this section we consider various kinds of signals which we may encounter in a self-tuning system. The signal $s(t)$ may be part of a control system, usually

Table 2.3   Transforms for systems involving fractional time delays

Laplace transform $G(s)$	z-transform $G(z)$
$s\exp(-ms\tau_s)$	$\dfrac{1-m}{\tau_s}z^{-1} + \dfrac{2m-1}{\tau_s}z^{-2} - \dfrac{m}{\tau_s}z^{-3}$
$\exp(-ms\tau_s)$	$(1-m)z^{-1} + mz^{-2}$
$\dfrac{\exp(-ms\tau_s)}{s}$	$\dfrac{\tau_s\{(1-m)z^{-1} + mz^{-2}\}}{1-z^{-1}}$
$\dfrac{a\exp(-ms\tau_s)}{s+a}$	$\dfrac{(1-\exp((m-1)a\tau_s))z^{-1} + \exp(-a\tau_s)(\exp(am\tau_s)-1)z^{-2}}{1-\exp(-a\tau_s)z^{-1}}$
$\dfrac{\exp(-ms\tau_s)}{s^2}$	$\dfrac{\tau_s^2((1-m)^2z^{-1} + (1+2m-2m^2)z^{-2} + m^2z^{-3})}{(1-z^{-1})^2}$
$\dfrac{sa^2\exp(-ms\tau_s)}{(s+a)^2}$	$\dfrac{\tau_s a^2\exp(-a\tau_s)\exp(am\tau_s)[(1-m)z^{-1} + (m\exp(-a\tau_s)-1+m)z^{-2} - m\exp(-a\tau_s)z^{-3}]}{(1-z^{-1}\exp(-a\tau_s))^2}$

In this table, $m\tau_s$ is a transport delay such that $0 \leqslant m \leqslant 1$, and τ_s is the sample interval. A ZOH is assumed present.

considered as an additive disturbance at the output (as shown in Figure 2.1). In this case the self-tuning controllers of Part 2 will attempt to remove the influence of the disturbance signal $s(t)$. In a self-tuning signal processing application the signal $s(t)$ is available (Figure 2.9). The need to represent the signal in this situation comes from a requirement to extract information from the signal. The signals which we need to consider can be classified into deterministic and random sources. We deal with deterministic signals first.

2.3.1 Deterministic signals

Offset

The simplest kind of signal is a constant value d. A system with offset d can be represented by the equation

Figure 2.9
Signal source
used in signal
processing.

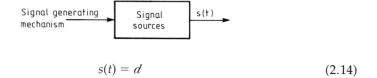

$$s(t) = d \qquad (2.14)$$

When present in a control system, the offset is removed by integral action in a feedback controller, but some self-tuning schemes estimate d and subtract out its effect by modifying the control signal. In a signal processing application, a constant signal level is often present due to the way in which the sensing instrumentation is designed. In addition, a constant signal is often used as a test of system functionality when no other signal is present.

Drift

A generalization of the constant offset is the drift signal where the offset becomes a function of time. In many situations drift can be modelled by a polynomial function of time $\mathscr{D}(t)$:

$$s(t) = \mathscr{D}(t) = d_0 + d_1 t + \ldots + d_{n_d} t^{n_d} \qquad (2.15)$$

A first order drift polynomial ($n_d = 1$) will cover many situations, especially the common case where the output offset is slowly changing at a constant rate. Also it is convenient to assume that $\mathscr{D}(t)$ is filtered by the A object of equation (2.4), so that the effective drift is $\mathscr{D}(t)/A$.

A further form of time-dependent drift occurs when the offset is periodic. Periodic drift can be represented in a model by the appropriate periodic function (e.g. sinusoidal, square wave, triangular wave). For example, a sine wave drift can be modelled as

$$\mathscr{D}(t) = d_s \sin (\Omega T t) + d_c \cos (\Omega T t) \qquad (2.16)$$

The parameter Ω determines the period of the oscillation and the quantities d_s and d_c jointly determine the amplitude and phase. In self-tuning schemes d_s, d_c are usually unknown but the period Ω is known. The self-tuning algorithm must then determine d_s, d_c. The influence of $\mathscr{D}(t)$ can then be measured and removed by modulating the control signal $u(t)$ in a control situation. In a signal processor a model of $\mathscr{D}(t)$ can be subtracted from the signal $s(t)$. The situation is more difficult if the parameter Ω is unknown.

A deterministic signal may not be an explicit function of the time index (although this is usual), but may be expressed in terms of some other independent variable. For example, if the offset is known to vary as a linear function of a variable w (which might be the operating temperature of a process or altitude of an aircraft), then an appropriate offset model is

$$s(w) = \mathcal{D}(w) = d_0 + d_1 w + \ldots + d_{n_d} w^{n_d} \tag{2.17}$$

Measurable signal sources

Often a system output or signal source is influenced by a known source $v_c(\tau)$ via a disturbance transfer function. The disturbance can be represented by a discrete time transfer function acting upon a sequence $v(t)$ representing a sampled version of the known source. Thus a measurable disturbance is represented by

$$s(t) = \frac{D}{A} v(t) \tag{2.18}$$

where

$$D = d_0 + d_1 z^{-1} + \ldots + d_{n_d} z^{-n_d}$$

$$A = 1 + a_1 z^{-1} + \ldots + a_{n_a} z^{-n_a}$$

Note the following points concerning D:

(a) Some of the leading coefficients d_0, d_1, . . ., may be zero, corresponding to a time delay in the disturbance transfer function.

(b) Unless the continuous time signal $v_c(\tau)$ is z-transformable, then the disturbance representation (2.18) is theoretically incorrect. In practice, however, the errors caused are small compared with the improved performance obtained by modelling the disturbance.

A known disturbance is usually removed in a conventional control scheme by feedforward control. This is also the case in self-tuning with the added feature that the coefficients of D are estimated and may be used to subtract out the influence of $v(t)$ by modifying $u(t)$. A similar subtraction procedure is used in signal processors for interference cancellation. In this situation the signal $v(t)$ represents some measurable function of a process which is interfering with a desired signal (see Chapter 11).

2.3.2   Random signals

An important class of random signals is that associated with small unpredictable changes in a system and unobservable noise-like disturbances. Such disturbances can be aggregated and modelled by a single noise source which is often assumed to be a stationary, gaussian noise sequence. Such a sequence can be represented by a white gaussian noise sequence $e(t)$ with zero mean and variance σ_e^2 which has been passed through a stable linear time invariant filter $N(z^{-1})$ (see Figure 2.10 and Appendix).

If $s(t)$ is the noise sequence as measured at the output of N then $s(t)$ will have a power spectrum (or spectral density function) given by

$$S_{ss}(\omega) = N(\exp(j\omega))N(\exp(-j\omega))\sigma_e^2 \qquad (2.19)$$

Note that z is evaluated on the unit circle ($z = \exp(j\omega)$).

Since $S_{ss}(\omega)$ is the only quantity we can measure which will give us information about N, we are free to construct any N which will match the power spectrum. If the power spectrum can be approximated by a rational polynomial function of $\cos \omega$, then it is always possible to obtain a noise transfer function N which is stable (all poles inside the unit circle) and has all its zeroes on or inside the unit circle. Zeroes on the unit circle are not of interest here and in their absence the noise transfer function will be inverse stable (or minimum phase by analogy with the corresponding term for continuous time systems), with all zeroes in the left half-plane.

Example 2.4

Consider a zero mean white noise sequence $\{e(t)\}$ with variance 2 which is fed through a discrete time transfer function N given by:

$$N(z^{-1}) = \frac{1 + 0.5z^{-1}}{1 + 0.6z^{-1}}$$

to form the noise signal $s(t)$.

Setting $z = \exp(j\omega)$ the frequence response of N is given by

$$N(\exp(j\omega)) = \frac{1 + 0.5\exp(-j\omega)}{1 + 0.6\exp(-j\omega)} \qquad -\pi < \omega < \pi$$

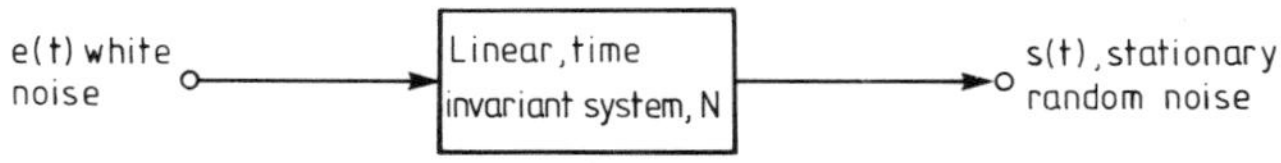

Figure 2.10   Representation of a stationary random signal as filtered white noise.

The power spectrum $S_{ss}(\omega)$ of the sequence $s(t)$ is

$$S_{ss}(\omega) = 2\,\frac{(1+0.5\exp(-j\omega))\,(1+0.5\exp(j\omega))}{(1+0.6\exp(-j\omega))\,(1+0.6\exp(j\omega))}$$

$$= \frac{2.5 + 2\cos\omega}{1.36 + 1.2\cos\omega}$$

Given $S_{ss}(\omega)$ it is possible to reverse this analysis (by substituting $\cos\omega = \frac{1}{2}(z+z^{-1})$) and to obtain a number of filters with the correct power spectrum. Two of these filters, however, will be unstable with a pole at $-(0.6)^{-1}$ and of the remaining two only one (our original N) will be inverse stable (minimum phase). Figure 2.11 illustrates the pole/zero configurations for the possible filters. □

In general, a stationary random signal source is represented by the stable transfer function model:

$$s(t) = \frac{C}{A}\,e(t) \tag{2.20}$$

where

$$C = 1 + c_1 z^{-1} + c_2 z^{-2} + \ldots + c_{n_c} z^{-n_c}$$

$$A = 1 + a_1 z^{-1} + a_2 z^{-2} + \ldots + a_{n_a} z^{-n_a}$$

Figure 2.11 The possible pole/zero configurations associated with the power spectrum $S_{ss}(\omega)$ of Example 2.4. The configurations are: (a) stable and minimum phase; (b) stable and nonminimum phase; (c) unstable and minimum phase; (d) unstable and nonminimum phase.

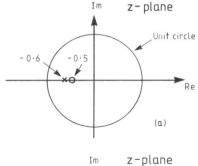

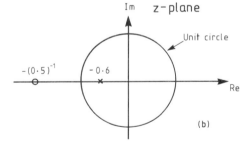

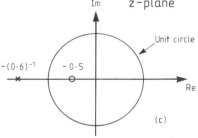

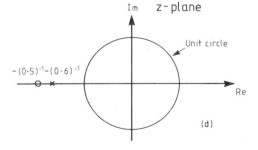

The above model for a random process is termed an *AutoRegressive Moving Average* (ARMA) model.

The stationary noise model of equation (2.20) is not always sufficient to represent random drift disturbances which influence some systems. In such circumstances, a random noise model is postulated in the form of a stationary *increment* to $s(t)$:

$$s(t) = s(t-1) + \frac{C}{A} e(t) \qquad (2.21a)$$

or

$$s(t) = \frac{C}{\Delta A} e(t) \qquad (2.21b)$$

where $\Delta = 1-z^{-1}$, the first difference operator. The incremental assumption builds an integrator into the noise model and consequently this form is termed an *AutoRegressive Integrated Moving Average* (ARIMA) representation. In the special case $C=1=A$, the signal is an integral of white noise and is often referred to as a *random walk* or 'Brownian motion'.

2.3.3 Overall signal model

The representations of deterministic, random and measurable disturbances can be drawn together to form a composite signal model which will meet most signal representation situations. Specifically, by combining the equations (2.15), (2.18) and (2.20) we can write:

$$s(t) = \frac{\mathcal{D}(t)}{A} + \frac{D}{A} v(t) + \frac{C}{A} e(t) \qquad (2.22)$$

Using this model it is possible to represent almost any signal form which might be encountered in self-tuning studies.

2.4 COMBINED SYSTEM AND SIGNAL MODELS

The system model given by equation (2.3) is sufficient for many control problems. Numerous practical applications occur, however, in which the system output is influenced by additional disturbances. Such additional disturbances can be represented by a combination of the system model (equation (2.3)) and the signal model (equation (2.22)) so that the overall process output is as indicated in Figure 2.1:

$$y(t) = x(t) + s(t)$$

This may be written out in full as

$$y(t) = \frac{B}{A} u(t-1) + \frac{\mathcal{D}(t)}{A} + \frac{D}{A} v(t) + \frac{C}{A} e(t) \qquad (2.23)$$

In any specific situation it is unlikely that all the signal components included in equation (2.23) will be present. Typical situations which occur are:

(i) $\mathcal{D}(t) = d_0$ This corresponds to the common situation when a process output is subject to a constant offset. Such an offset may be an intrinsic part of the process or be due to sensor instrumentation.

(ii) $v(t)$ *present* This corresponds to the situation in which a measurable but uncontrollable signal is influencing the system output $y(t)$. Feedforward control compensates for this.

(iii) $e(t)$ *present* This corresponds to the case in which aggregated random disturbances are corrupting the measurable output of the system.

For this last case, models of the form

$$y(t) = \frac{B}{A} u(t-1) + \frac{C}{A} e(t) \qquad (2.24)$$

are of the ARMA type with an additional Control (or eXogenous) input and model (2.24) is therefore termed CARMA (or ARMAX). In certain cases a better noise description is achieved if we use integrated noise in which case the CARIMA or (ARIMAX) representation arises:

$$y(t) = \frac{B}{A} u(t-1) + \frac{C}{\Delta A} e(t) \qquad (2.25)$$

The CARIMA form is widely used when the system disturbances are dominated by a random drift for which integrated white noise is an appropriate representation. It is also frequently used to justify the use of incremental control in a stochastic setting. More of this will be said in Part 2.

2.5   THE δ-OPERATOR REPRESENTATION

A potential problem in sampled data systems is that the z-transform representation may become ill-conditioned. Typically, if the sampling is fast, the z-plane poles are all approximately unity. This leads to problems of numerical ill-conditioning. Some of these problems can be avoided by using the δ-operator, where

$$\delta = \tau_s^{-1}(z-1) \tag{2.26}$$

This is a scaled shift in the origin such that the stability region in the δ-plane is a circle origin $(\tau_s^{-1},0)$ and radius τ_s^{-1} (Figure 2.12). As previously noted, problems of numerical stability arise with z-plane models when τ_s decreases. From Figure 2.12 it is clear that for the δ-transform the problem is less severe, since as τ_s decreases the stability region tends to that for continuous time (the left half-plane). In computational terms, what happens is that, as τ_s decreases, the δ-plane poles tend to bunch up around the origin instead of bunching around the (1,0) point in the z-plane. Numerically, it is easier to deal with small numbers (the δ-plane case) than small differences between large numbers (the z-plane case).

The δ-operator is in fact a way of approximating the derivatives of the data $u(t)$ and $y(t)$ and in this way it comes much closer to yielding a description of the differential equation which describes the physical system which we are modelling. To see this, consider the following example in which we apply the δ-operator to the sequence $\{y(t)\}$. From equation (2.26):

Figure 2.12
Stability regions
for z- and δ-
planes.

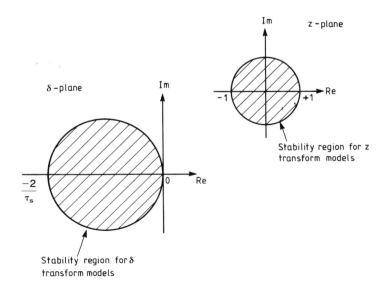

$$\delta y(t) = \tau_s^{-1}\,(y(t+1) - y(t))$$

which approximates the first derivative of $y(t)$ with respect to t.
  Likewise, we can write

$$\delta^2 y(t) = \tau_s^{-2}\,(z-1)^2\,y(t)$$

$$= \tau_s^{-2}((y(t+2) - y(t+1)) - (y(t+1) - y(t)))$$

$$= \tau_s^{-1}(\delta y(t+1) - \delta y(t))$$

which approximates the second derivative of $y(t)$ with respect to t.
  We have already seen that the z-transform-based discrete time models can cause problems because of the mapping from the s to z-plane (see Section 2.2 and the discussion of nonminimum phase B created by sampling a minimum phase continuous time system). The δ-operator approach should give good approximations to the continuous time parameters. Thus we can (approximately at least) associate a δ-transfer function with the transfer function of the physical system which we are studying.
  In the δ-operator framework the transfer function takes the form

$$y(t) = \frac{B(\delta)}{A(\delta)}\,u(t) \tag{2.27a}$$

where all time delays have been omitted and

$$A = \delta^n + a'_{n-1}\,\delta^{n-1} + \ldots + a'_i\delta^i + \ldots + a'_0$$

$$B = b'_n\,\delta^n + \ldots + b'_i\delta^i + \ldots + b'_0$$

When simulating the system defined by equation (2.27a) we require to advance the time sequence from time t to $t+1$. Thus

$$y(t+1) = \frac{B(\delta)}{A(\delta)}\,u(t+1) \tag{2.27b}$$

In order to compute $y(t+1)$ in the above expression it is necessary to do so using operators of the form $\delta^i y(t)$, $\delta^i u(t)$ from equation (2.27a) at the previous time step t. This shift from one time interval to the next is achieved using the fact that by multiplying equation (2.26) by δ^i and rearranging we obtain:

$$z\delta^i = \tau_s\delta^{i+1} + \delta^i \tag{2.28}$$

so that

$$\delta^i u(t+1) = \tau_s \delta^{i+1} u(t) + \delta^i u(t) \qquad (2.29a)$$

$$\delta^i y(t+1) = \tau_s \delta^{i+1} y(t) + \delta^i y(t) \qquad (2.29b)$$

This equation can be used to generate all the values of $\delta^i u(t+1)$ and $\delta^i y(t+1)$ except the highest powers $\delta^n u(t+1)$ and $\delta^n y(t+1)$. These can be obtained from the defining relation (2.26) by noting that

$$z^{n+1} = z(z^n) = z(\tau_s \delta + 1)^n \qquad (2.30)$$

Using this expression it is possible to write

$$z^{n+1} u(t) = u(t+n+1) = (\tau_s \delta + 1)^n u(t+1) \qquad (2.31)$$

or, expanding the right-hand side,

$$u(t+n+1) = \left[(\tau_s \delta)^n + \sum_{i=1}^{n} \begin{bmatrix} n \\ i \end{bmatrix} (\tau_s \delta)^{n-i} \right] u(t+1) \qquad (2.32)$$

Rearranging this expression, the highest power $\delta^n u(t+1)$ can be written thus:

$$(\tau_s \delta)^n u(t+1) = u(t+n+1) - \sum_{i=1}^{n} \begin{bmatrix} n \\ i \end{bmatrix} (\tau_s \delta)^{n-i} u(t+1) \qquad (2.33)$$

Equation (2.33) can be used to compute $\delta^n u(t+1)$ for use in equation (2.27b), since all the terms on the right-hand side are known, *provided* we have the value of input up to time $t+n+1$. In this connection it is important to note that in δ models, the object $\delta^i u(t)$ is a function of the time shift object $u(t+i)$. Thus in δ model terms we are advancing from $\delta^n u(t)$, $\delta^n y(t)$ to $\delta^n u(t+1)$, $\delta^n y(t+1)$ while the actual time sequence is advancing from $t+n$ to $t+n+1$.

This difference between 'δ time' and 'shift time' can be seen again when we compute $y(t+n+1)$ from $\delta^n y(t+1)$. With the expressions (2.32) and (2.29) we have enough information to compute $\delta^n y(t+1)$ from equation (2.27b):

$$\delta^n y(t+1) = B(\delta)u(t+1) - \sum_{i=1}^{n} a_{n-i} \delta^{n-i} y(t+1) \qquad (2.34)$$

The last computation required is to obtain $y(t+n+1)$ from $\delta^n y(t+1)$. This is done by noting:

$$z^{n+1} y(t) = z^n y(t+1) = (\delta \tau_s + 1)^n y(t+1)$$

i.e.

$$y(t+n+1) = \sum_{i=0}^{n} \begin{bmatrix} n \\ i \end{bmatrix} (\tau_s \delta)^{n-i} y(t+1) \tag{2.35}$$

The following example illustrates how the output of a δ model may be computed using these equations.

Example 2.5

Consider a second order example as an illustration of how the above ideas work. In particular, suppose we have the following system:

$$y(t) = \left[\frac{b_0 + b_1 \delta + b_2 \delta^2}{a_0 + a_1 \delta + \delta^2} \right] u(t) \tag{2.36a}$$

so that at the next time step $t+1$:

$$y(t+1) = \left[\frac{b_0 + b_1 \delta + b_2 \delta^2}{a_0 + a_1 \delta + \delta^2} \right] u(t+1) \tag{2.36b}$$

In shifting the δ-model equation forward from t to $t+1$, we have the following relations, from equations (2.29) and assuming $\tau_s = 1$:

$$\delta^0 u(t+1) = u(t+1)$$

$$\delta u(t+1) = \delta^2 u(t) + \delta u(t) \tag{2.37}$$

$$\delta^0 y(t+1) = y(t+1)$$

$$\delta y(t+1) = \delta^2 y(t) + \delta y(t)$$

The value of $\delta^2 u(t+1)$ is computed from equation (2.33) as

$$\delta^2 u(t+1) = u(t+3) - (2\delta u(t+1) + u(t+1)) \tag{2.38}$$

The value of $\delta^2 y(t+1)$ is then computed from equation (2.36b) as

$$\delta^2 y(t+1) = b_0 u(t+1) + b_1 \delta u(t+1) + b_2 \delta^2 u(t+1) - a_0 y(t+1) - a_1 \delta y(t+1) \tag{2.39}$$

Equation (2.39) gives the output of the system in 'δ time'. The final step is to compute $y(t+3)$ in 'shift-time'. This uses equation (2.35), thus:

$$y(t+3) = (δ+1)^2 y(t+1)$$

$$= δ^2 y(t+1) + 2δy(t+1) + y(t+1) \qquad (2.40)$$

□

The above example shows how we can compute the output of a δ model recursively. This is useful for simulation and for data filtering. Indeed δ models are widely used in digital filtering where low round-off error and large sample rates are required.

Note that the δ-operator representation does not lend itself to handling large time delays. Small delays can be modelled by Padé approximation methods. However, significant transport lags of the kind found in process systems cannot be modelled in this manner.

2.6 SUMMARY

The chapter has dealt with the following issues:

- The distinction between systems and signals in self-tuning applications
- The representation of continuous time systems as discrete time transfer functions
- The kinds of signals encountered in control and signal processing and their discrete time representation
- The use of incremental models in noise representation
- The use of δ transform models in situations where fast sampling of slow dynamics is required

2.7 PROBLEMS

Problem 2.1

In a discrete time model $Ay(t) = Bu(t-k)$ of an underlying continuous system

$$y(s) = \frac{\exp(-τ_d s)B_c(s)}{A_c(s)}$$

determine the degree of A in terms of the degree of $A_c(s)$ and the value of k in terms of the time delay $τ_d$. Can you relate the degree of B to the degrees of B_c, A_c?

Problem 2.2

Establish the mapping of continuous time poles into discrete time poles. Consider distinct poles only. Explain why such a mapping does not exist for zeroes.

Problem 2.3

For the system shown in Figure P2.1 obtain the z-transfer function relating sequence $\{u(t)\}$ to $\{y(t)\}$. Derive the difference equation model for $\tau_s = 1$ s.

Problem 2.4

The system shown in Figure P2.2 is subject to output disturbances from a measurable source $v(t)$ and a constant offset $d_0 = 1$. Determine a discrete time transfer function for the system in the form

$$Ay(t) = Bu(t-1) + Dv(t) + d_0.$$

Problem 2.5

Given the (discrete time) power spectrum

$$S_{ss}(\omega) = \frac{1.04 - 0.4 \cos \omega}{1.81 - 1.8 \cos \omega}$$

find a difference equation representation of the corresponding discrete time stochastic process $s(t)$.

Figure P2.1

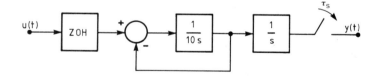

Figure P2.2

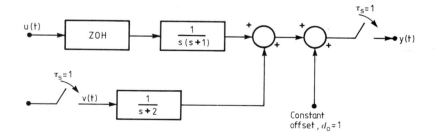

Problem 2.6

A random noise source $s(t)$ is generated by the difference equation

$$s(t) = e(t) + 2e(t-1)$$

If $e(t)$ is a zero mean, white noise process with unit variance, show that an equivalent noise source (i.e. with the same spectral density) can be generated by the equation

$$s(t) = w(t) + 0.5w(t+1)$$

where $w(t)$ is a zero mean white noise process. What is the variance of $w(t)$?
    If the process $s(t)$ is generated by the difference equation

$$s(t) = as(t-1) + e(t) - a^{-1}e(t-1) \qquad |a| < 1$$

show that it is uncorrelated and calculate its variance.

Problem 2.7

The output $y(t)$ of the system

$$y(t) = \frac{3z^{-1}}{1+2z^{-1}} u(t)$$

is corrupted by a zero mean stationary disturbance $w(t)$ with power spectrum

$$S_{ww}(\omega) = \frac{5-4\cos\omega}{1.49-1.4\cos\omega}$$

Derive a CARMA model in which $C(0) = 1$ and C has all its zeroes inside the unit circle.

Problem 2.8

Given a system description

$$(1-az^{-1})\,y(t) = bu(t-2) + e(t)$$

show that this can be rewritten as

$$y(t) = a^2 y(t-2) + b(1+az^{-1})u(t-2) + (1+az^{-1})e(t)$$

Comment on the usefulness of this form of representation for predicting $y(t+2)$ from current measurements.

Problem 2.9

The stationary random noise source $w_c(\tau)$ has (continuous time) power spectrum

$$S_{ww}(\omega) = \frac{b^2}{\omega^2 + a^2} \qquad (a > 0)$$

Set $s = j\omega$ and employ the spectral factorization technique indicated in Example 2.4 to derive a differential equation for $w_c(\tau)$.

If $w_c(\tau)$ is sampled at intervals of τ_s, show that the resulting discrete time noise source $w(t)$ satisfies the ARX model

$$\omega(t) = \exp(-a\tau_s)\,\omega(t-1) + be(t)$$

where $e(t)$ is zero mean white noise with variance

$$\sigma_e^2 = (1 - \exp(-2a\tau_s))/2a$$

Problem 2.10

Show that the random walk process:

$$y(t) = \frac{e(t)}{\Delta}$$

is nonstationary by deriving its variance as a function of time. What is $E[y(t_1)y(t_2)]$? (Assume that $e(t) = 0$ for $t \leq 0$.)

2.8   NOTES AND REFERENCES

Digital control and z-transform methods are discussed in any digital control text. However, the following book is particularly recommended:

Franklin, G.F. and Powell, J.D.
  Digital Control of Dynamical Systems, Addison Wesley, 1980.

A complementary treatment which takes a signal processing viewpoint is:

Oppenheim, A.V. and Schafer, R.W.
  Digital Signal Processing, Prentice Hall, 1975.

The representation of stochastic processes is discussed in:

Astrom, K.J.
  Introduction to Stochastic Control Theory, Academic Press, 1970.

The δ-operator treatment given here is taken from:

Edmunds, J.M.
  Sampled Data Systems using Difference Operator Models, CSC Report 627, UMIST,
  1983.

A detailed treatment of the accuracy implications of the delta model is given in:

Middleton, R.H. and Goodwin, G.C.
  Improved finite word length characteristics in digital control using delta operators,
  IEEE Trans. **AC-31**(11), 1015, 1986.

3 Recursive Estimation

3.1 OUTLINE AND LEARNING OBJECTIVES

We now turn to the techniques by which the parameters of a system model are obtained in self-tuning control and signal processing. This is usually accomplished by assuming a discrete time form for the system model and then using a recursive estimation algorithm to obtain estimates of the parameters of the model. Section 3.2 discusses how the system and signal representations for a computer control or signal processing scheme can be put into a form suitable for parameter estimation and how the unknown parameters are determined using linear least squares. Section 3.3 discusses the recursive least squares algorithm, the basic estimator used in many practically successful self-tuning identifiers. Subsequent sections discuss recursive algorithms which are related to least squares. The learning objective is to become familiar with the derivation, properties and use of the recursive least squares algorithm as applied to the estimation of system and signal parameters.

3.2 LEAST SQUARES ALGORITHM

Consider a discrete time transfer function model of a system (Figure 3.1) with control input sequence $u(t)$ and with output $y(t)$ subject to disturbances from a measurable source $v(t)$, drift and random noise. From Chapter 2, the model can be written in the form

$$Ay(t) = Bu(t-1) + Dv(t) + \mathcal{D}(t) + Ce(t) \tag{3.1}$$

where

$$A = 1 + a_1 z^{-1} + \ldots + a_{n_a} z^{-n_a}$$

$$B = b_0 + b_1 z^{-1} + \ldots + b_{n_b} z^{-n_b}$$

$$D = d_0 + d_1 z^{-1} + \ldots + d_{n_d} z^{-n_d}$$

Figure 3.1
Discrete time
system transfer
function of a
system including
all possible
input and
disturbance
components.

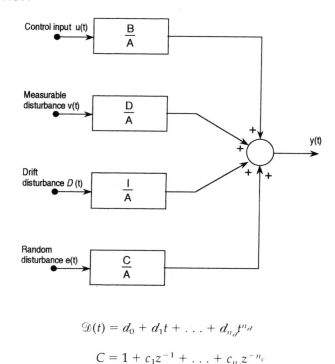

Figure 3.1
Discrete time
system transfer
function of a
system including
all possible
input and
disturbance
components.

$$\mathcal{D}(t) = d_0 + d_1 t + \ldots + d_{n_d} t^{n_d}$$

$$C = 1 + c_1 z^{-1} + \ldots + c_{n_c} z^{-n_c}$$

When the system to be controlled is known then the coefficients of polynomials $A, B, D, \mathcal{D}, C$ can be calculated from continuous system parameters. If the system is unknown, however, then the polynomial coefficients are treated as parameters to be determined by measurement or *estimation*. For estimation purposes, it is convenient to write equation (3.1) in a form which emphasizes the object to be estimated and the data available. This is achieved by using the backward shift interpretation of z^{-1} to cast equation (3.1) in the form

$$y(t) = \mathbf{x}^T(t)\boldsymbol{\theta} + e(t) \tag{3.2a}$$

where $\boldsymbol{\theta}$ is the vector of unknown parameters, defined by

$$\boldsymbol{\theta}^T = [-a_1, \ldots, -a_{n_a}, b_0, \ldots, b_{n_b}, d_0, \ldots, d_{n_d}, d_0, \ldots, d_{n_d}, c_1, \ldots, c_{n_c}] \tag{3.2b}$$

and $\mathbf{x}(t)$ is a *regression vector* partly consisting of measured input/output variables and defined by:

$$\mathbf{x}^T(t) = [y(t-1), \ldots, y(t-n_a), u(t-1), \ldots u(t-n_b-1), v(t), \ldots, v(t-n_d), \tag{3.2c}$$
$$1, t, \ldots, t^{n_d}, e(t-1), e(t-2), \ldots, e(t-n_c)]$$

Note that this vector also contains the values of $e(t-1)$, $e(t-2)$, ..., $e(t-n_c)$ which in general will be unknown, since they are the past values of the unobservable white noise disturbance $e(t)$. We will show how to include these unknown noise terms in an estimation algorithm in Section 3.6. For the moment we assume that $n_c=0$, in which case the C coefficients c_1, c_2, c_3, ..., are zero and the unknown noise no longer appears in $x(t)$. In this case $x(t)$ is often referred to as the data vector.

We now assume that (3.2) is an exact description of the system, i.e. it is the true data-generating mechanism, and that we wish to determine from available data the vector θ of true system parameters. In order to do this we further assume a model of the system of the correct structure:

$$y(t) = \mathbf{x}^T(t)\hat{\boldsymbol{\theta}} + \hat{e}(t) \tag{3.3}$$

where $\hat{\boldsymbol{\theta}}$ is a vector of adjustable model parameters and $\hat{e}(t)$ is the corresponding modelling (or fitting) error at time t. Our aim is to select $\hat{\boldsymbol{\theta}}$ so that overall modelling error is minimized in some sense. Note that (3.2a) and (3.3) imply:

$$\hat{e}(t) = e(t) + \mathbf{x}^T(t)\,(\boldsymbol{\theta}-\hat{\boldsymbol{\theta}})$$

so that $\hat{e}(t)$ depends on $\hat{\boldsymbol{\theta}}$ and, in some cases, the 'minimized' modelling errors will be equal to the white noise sequence corrupting the system output data.

Assume that the system described by equation (3.2) has been running for sufficient time to form N consecutive data vectors. The data obtained in this way allows the model (3.3) to be expressed in the vector/matrix form:

$$\begin{bmatrix} y(1) \\ y(2) \\ \cdot \\ \cdot \\ \cdot \\ y(N) \end{bmatrix} = \begin{bmatrix} \mathbf{x}^T(1) \\ \mathbf{x}^T(2) \\ \cdot \\ \cdot \\ \cdot \\ \mathbf{x}^T(N) \end{bmatrix} \hat{\boldsymbol{\theta}} + \begin{bmatrix} \hat{e}(1) \\ \hat{e}(2) \\ \cdot \\ \cdot \\ \cdot \\ \hat{e}(N) \end{bmatrix} \tag{3.4}$$

To be able to estimate the parameters uniquely the number N of equations in equation (3.4) must not be less than m, the number of unknown parameters in the vector $\boldsymbol{\theta}$. In the noise-free case ($e(t)=0$), the equation can be solved as a set of linear equations in $N = m$ unknowns, where $m = n_a + (n_b+1) + (n_d+1) + (n_d+1)$ (assuming $n_c=0$). The resulting modelling errors are identically zero. When noise is present (and, in practice, even in nominally noise-free systems) we must have N much larger than m and use an alternative procedure to reduce estimation errors induced by the noise. The technique most widely used in this connection is linear *least squares*, which we now introduce.

Rewrite equation (3.4) in the stacked notation thus:

$$\mathbf{y} = \mathbf{X}\hat{\boldsymbol{\theta}} + \hat{\mathbf{e}} \qquad (3.5)$$

in which

$$\mathbf{y}^T = [y(1), \ldots, y(N)]$$

$$\hat{\mathbf{e}}^T = [\hat{e}(1), \ldots, \hat{e}(N)]$$

and

$$\mathbf{X} = \begin{bmatrix} \mathbf{x}^T(1) \\ \mathbf{x}^T(2) \\ \cdot \\ \cdot \\ \cdot \\ \mathbf{x}^T(N) \end{bmatrix}$$

Rearrange (3.5) in terms of the error vector $\hat{\mathbf{e}}$:

$$\hat{\mathbf{e}} = \mathbf{y} - \mathbf{X}\hat{\boldsymbol{\theta}} \qquad (3.6)$$

and select an estimate $\hat{\boldsymbol{\theta}}$ of the true vector of parameters which minimizes J, the sum of squares of errors:

$$J = \sum_{t=1}^{N} \hat{e}^2(t) = \hat{\mathbf{e}}^T\hat{\mathbf{e}} \qquad (3.7)$$

To find the least squares estimate, rewrite equation (3.7) in terms of the data vectors and parameter vector:

$$J = (\mathbf{y} - \mathbf{X}\hat{\boldsymbol{\theta}})^T (\mathbf{y} - \mathbf{X}\hat{\boldsymbol{\theta}})$$

$$= \mathbf{y}^T\mathbf{y} - \hat{\boldsymbol{\theta}}^T\mathbf{X}^T\mathbf{y} - \mathbf{y}^T\mathbf{X}\hat{\boldsymbol{\theta}} + \hat{\boldsymbol{\theta}}^T\mathbf{X}^T\mathbf{X}\hat{\boldsymbol{\theta}}$$

Setting to zero the derivative of J with respect to $\hat{\boldsymbol{\theta}}$ for a stationary point:

$$\frac{\partial J}{\partial \hat{\boldsymbol{\theta}}} = -2\mathbf{X}^T\mathbf{y} + 2\mathbf{X}^T\mathbf{X}\,\hat{\boldsymbol{\theta}} = 0$$

yields the *normal equations*:

$$\chi^T\chi\hat{\theta} = \chi^Ty$$

and these solve for a unique minimum if the second derivative matrix

$$\frac{\partial J^2}{\partial \hat{\theta}^2} = 2(\mathbf{X}^T\mathbf{X})$$

is positive definite.

Hence the least squares estimator for the parameter vector is

$$\hat{\theta} = [\mathbf{X}^T\mathbf{X}]^{-1}[\mathbf{X}^T\mathbf{y}] \qquad (3.8)$$

The resulting modelling error $\hat{\mathbf{e}}$ is denoted by

$$\boldsymbol{\eta}^T = [\eta(1), \ldots, \eta(N)]$$

whose components are called *residuals*.

Example 3.1

Consider a simple example in which we are seeking a relationship of the form $y = f(u)$ between three points whose cartesian (u,y) coordinates are $(0,0)$, $(1,0.9)$ and $(2,2.1)$. We can assume that these correspond to measurements at times $t = 1,2,3$, respectively, so that

$$u(1) = 0, \quad u(2) = 1, \quad u(3) = 2$$

The corresponding output data are

$$y(1) = 0, \quad y(2) = 0.9, \quad y(3) = 2.1$$

and therefore

$$\mathbf{y} = [0, 0.9, 2.1]^T$$

It is clear that y and u are almost linearly related but let us first assume a model

$$y(t) = \text{constant} = d_0$$

so that $\mathbf{x}(t) = [1]$ and $\mathbf{X} = [1\ 1\ 1]^T$. Then (3.8) yields $\hat{d}_0 = 1$ and

$$\boldsymbol{\eta} = \begin{bmatrix} -1 \\ -0.1 \\ 1.1 \end{bmatrix}$$

Corresponding to the linear model

$$y(t) = d_0 + d_1 u(t)$$

then

$$\mathbf{x}(t) = \begin{bmatrix} 1 \\ u(t) \end{bmatrix} \qquad \text{and} \qquad \mathbf{X} = \begin{bmatrix} 1 & 0 \\ 1 & 1 \\ 1 & 2 \end{bmatrix}$$

leading to

$$\begin{bmatrix} \hat{d}_0 \\ \hat{d}_1 \end{bmatrix} = \begin{bmatrix} -0.05 \\ 1.05 \end{bmatrix}$$

$$\boldsymbol{\eta} = \begin{bmatrix} 0.05 \\ -0.1 \\ 0.05 \end{bmatrix}$$

This seems a 'good fit' (close to $y = u$), but is it? We can (of course) get a *perfect* fit ($\boldsymbol{\eta} = 0$) by fitting the quadratic model

$$y(t) = d_0 + d_1 u(t) + d_2 u^2(t)$$

so that $\mathbf{x}(t) = [1, u(t), u^2(t)]^T$ and

$$\mathbf{X} = \begin{bmatrix} 1 & 0 & 0 \\ 1 & 1 & 1 \\ 1 & 2 & 4 \end{bmatrix}$$

This $\mathbf{X}$ is invertible so that (3.8) reduces to

$$\hat{\boldsymbol{\theta}} = \mathbf{X}^{-1} \mathbf{y}$$

which is equivalent to solving (in this case) three equations for three unknowns, yielding

$$\begin{bmatrix} \hat{d}_0 \\ \hat{d}_1 \\ \hat{d}_2 \end{bmatrix} = \begin{bmatrix} 0 \\ 0.75 \\ 0.15 \end{bmatrix}$$

Zero residuals are not expected whenever the number of estimated parameters equals the number of data points. Indeed, as we increase the number of parameters we expect J (defined by equation (3.7)) to decrease monotonically to zero because we must always get a smaller minimum by increasing the number of degrees of freedom. In our case the minimum value of J takes the sequence of values

$$2.22, \qquad 0.015, \qquad 0$$

and it is clear that in order to choose a best fit that does not coincide with the (usually useless) perfect fit, we need to penalize the number of parameters used. If, for example, we add to each term of the sequence the natural logarithm of the number of estimated parameters, we arrive at the transformed sequence

$$2.22, \qquad 0.708, \qquad 1.099$$

suggesting that the linear model is best. This approach can be placed on a firmer theoretical basis and used for more complex problems of model structure determination (see Section 3.9 Notes and references).

$\square$

3.2.1 Properties of the least squares estimator

The least squares estimator $\hat{\boldsymbol{\theta}}$ is a random variable whose properties can be analysed using equation (3.3) which defines the actual system and disturbances. Two properties are important in this respect: *bias* and *covariance*. The term bias refers to the systematic error which can occur in the parameter estimate. The term covariance is related to the spread of estimates arising from the random errors.

Bias

Stacking equation (3.2a) for $t = 1, \ldots, N$ (cf. equation (3.5)) and substituting in equation (3.8) yields:

$$\hat{\boldsymbol{\theta}} = [\mathbf{X}^\mathrm{T}\mathbf{X}]^{-1} [\mathbf{X}^\mathrm{T}\mathbf{X}\boldsymbol{\theta} + \mathbf{X}^\mathrm{T}\mathbf{e}]$$

$$= \boldsymbol{\theta} + [\mathbf{X}^\mathrm{T}\mathbf{X}]^{-1} \mathbf{X}^\mathrm{T}\mathbf{e} \tag{3.9}$$

where $\mathbf{e}$ is a vector of actual white noise values and is defined by

$$\mathbf{e}^\mathrm{T} = [e(1), \ldots, e(N)]$$

The average deviation of the estimate of the parameter from its true value (termed the bias in the estimator) is given by a rearrangement of equation (3.9):

$$\hat{\boldsymbol{\theta}} - \boldsymbol{\theta} = [\mathbf{X}^T\mathbf{X}]^{-1}\mathbf{X}^T\mathbf{e} \tag{3.10}$$

When the data which makes up $\mathbf{X}$ is deterministic, the expected value of $\hat{\boldsymbol{\theta}} - \boldsymbol{\theta}$ can be written as

$$E[\hat{\boldsymbol{\theta}}-\boldsymbol{\theta}] = [\mathbf{X}^T\mathbf{X}]^{-1}\mathbf{X}^T E[\mathbf{e}]$$

$$= 0 \qquad \text{if } e(t) \text{ is zero mean.}$$

If the elements of $\mathbf{x}(t)$ are random but independent of $e(t)$, then we can write the expected value of the parameter error vector $\hat{\boldsymbol{\theta}} - \boldsymbol{\theta}$ as

$$E[\hat{\boldsymbol{\theta}}-\boldsymbol{\theta}] = E_x\{[\mathbf{X}^T\mathbf{X}]^{-1}\mathbf{X}^T\}\, E_e[\mathbf{e}]$$

$$= 0 \quad \text{if } e(t) \text{ is zero mean.}$$

where the subscripts x,e of the expectation operator $E[\cdot]$ denote expectations over the $\mathbf{X}$ and the $\mathbf{e}$ objects respectively. It is often the case in the identification of engineering systems that the data $\mathbf{x}(t)$ are correlated with the noise $\{e(s), s \leq t-1\}$, and in these cases the above arguments are inappropriate. When this occurs it is convenient to use the concept of the *probability limit* (see Appendix). When the data $\mathbf{x}(t)$ and the noise $e(t)$ are correlated this concept can be used to show that the estimates are unbiased if we have:

$$\operatorname*{plim}_{N\to\infty} (N^{-1}\mathbf{X}^T\mathbf{e}) = 0$$

Note that the entries in the vector $N^{-1}\mathbf{X}^T\mathbf{e}$ are essentially the cross correlations between the data and noise sequences, so that the asymptotic uncorrelatedness of the data and the noise is necessary for unbiased estimates (in the plim sense). The following example illustrates this.

Example 3.2

Consider the system

$$y(t) = a\, y(t-1) + e(t) \qquad |a| < 1$$

so that $x(t) = y(t-1)$.

Equation (3.10) takes the form

$$\hat{\boldsymbol{\theta}} - \boldsymbol{\theta} = \left[\sum_{t=1}^{N} \mathbf{x}(t)\mathbf{x}^{\mathsf{T}}(t) \right]^{-1} \sum_{t=1}^{N} \mathbf{x}(t)e(t)$$

i.e.

$$\hat{a} - a = \left[\frac{1}{N} \sum_{t=1}^{N} y^2(t-1) \right]^{-1} \left[\frac{1}{N} \sum_{t=1}^{N} y(t-1)e(t) \right]$$

Denoting the second bracketed term by $\xi(N)$, we wish to show that plim $\xi(N)$ is zero. This can be achieved by use of the Chebyshev inequality (see Appendix), which implies that

$$\text{prob}\,[|X - m_x| > \epsilon] \leq (\text{var } X)/\epsilon^2$$

for any positive ϵ, where m_x, var X denote the mean and variance respectively of the random variable X (provided that these moments exist).
   Then

$$E\,\xi(N) = \frac{1}{N} \sum_{t-1}^{N} E[y(t-1)e(t)] = 0$$

and therefore

$$\text{var } \xi(N) = E\left[\frac{1}{N} \sum_{t=1}^{N} y(t-1)e(t) \right]^2$$

$$= \frac{1}{N^2} \sum_{s,\,t=1}^{N\,\,N} E[y(s-1)y(t-1)e(s)e(t)]$$

If the sequence $\{e(t)\}$ is *independent* white noise with zero mean, then the only nonzero terms in the double sum correspond to $s = t$ only, leading to

$$\text{var } \xi(N) = \frac{\sigma_e^2}{N^2} \sum_{t=1}^{N} Ey^2(t-1)$$

$$= \frac{\sigma_e^2}{N} Ey^2(t)$$

since y is stationary because $|a| < 1$.

Ignoring initial conditions,

$$y(t) = e(t) + ae(t-1) + a^2 e(t-2) + \ldots$$

Therefore

$$E\, y^2(t) = \sigma_e^2 \,(1 + a^2 + a^4 + \ldots) = \frac{\sigma_e^2}{1 - a^2}$$

and finally

$$\mathrm{prob}[|\xi(N)| > \epsilon] \leqslant \frac{1}{N}\left[\frac{1}{\epsilon^2}\frac{\sigma_e^4}{1-a^2}\right]$$

Clearly the right-hand side goes to zero as N increases to infinity and therefore

$$\mathrm{plim}\ \xi(N) = 0. \qquad\qquad \square$$

Covariance

When the data $\mathbf{X}$ and $\mathbf{e}$ are uncorrelated and $\mathbf{e}$ is zero mean, then we can compute the covariance of the least squares estimates as:

$$\mathrm{cov}(\hat{\boldsymbol{\theta}}) = E[(\hat{\boldsymbol{\theta}} - \boldsymbol{\theta})\,(\hat{\boldsymbol{\theta}} - \boldsymbol{\theta})^\mathrm{T}]$$

$$= E[((\mathbf{X}^\mathrm{T}\mathbf{X})^{-1}\,\mathbf{X}^\mathrm{T}\mathbf{e})\,((\mathbf{X}^\mathrm{T}\mathbf{X})^{-1}\,\mathbf{X}^\mathrm{T}\mathbf{e})^\mathrm{T}]$$

$$= E_x[(\mathbf{X}^\mathrm{T}\mathbf{X})^{-1}\,\mathbf{X}^\mathrm{T}\,E_e[\mathbf{e}\mathbf{e}^\mathrm{T}]\,\mathbf{X}(\mathbf{X}^\mathrm{T}\mathbf{X})^{-1}] \qquad (3.11)$$

Again we have used the rules of expectation to separate the $\mathbf{X}$ and $\mathbf{e}$ objects. Further, in our case the errors are a white noise sequence implying that they are mutually uncorrelated so that

$$E[\mathbf{e}\mathbf{e}^\mathrm{T}] = \sigma_e^2\,\mathbf{I}_m \qquad (3.12)$$

Hence the covariance of errors in the least squares parameter estimates is given by

$$\mathrm{cov}(\hat{\boldsymbol{\theta}}) = \sigma_e^2\,E_x[(\mathbf{X}^\mathrm{T}\mathbf{X})^{-1}] \qquad (3.13)$$

If the data is deterministic the expectation operation can be dropped.

Equation (3.13) indicates that the matrix $(X^TX)^{-1}$ is of great importance in self-tuning systems since it gives a direct measure of the variability and covariability of the parameter estimates. This is useful since it is common for the accuracy of estimates to be linked to the size of the covariance matrix elements.

In order to give a feeling for the concept of covariance, consider the following example.

Example 3.3

For a system

$$y(t) = b_0 u(t) + b_1 u(t-1) + e(t)$$

with no time delay, the matrix X is given by

$$X = \begin{bmatrix} u(1), u(0) \\ u(2), u(1) \\ \cdot \\ \cdot \\ \cdot \\ u(N), u(N-1) \end{bmatrix}$$

The matrix X^TX is then computed from

$$X^TX = \begin{bmatrix} u(1), u(2), \ldots, u(N) \\ u(0), u(1), \ldots, u(N-1) \end{bmatrix} \begin{bmatrix} u(1), u(0) \\ u(2), u(1) \\ \cdot \\ \cdot \\ \cdot \\ u(N), u(N-1) \end{bmatrix}$$

$$= \begin{bmatrix} \sum_{t=1}^{N} u^2(t) & \sum_{t=1}^{N} u(t)u(t-1) \\ \sum_{t=1}^{N} u(t-1)u(t) & \sum_{t=1}^{N} u^2(t-1) \end{bmatrix}$$

Note the following points:

(i) The dimensions of X^TX depend upon the number of unknown parameters, *not* the number of data samples. In this case the matrix is two by two; in general for m unknown parameters the matrix is $m \times m$.

(ii) The entries in $\mathbf{X}^T\mathbf{X}$ are correlation-like sums over the available N data points and the matrix is symmetric, so that only the upper triangular half needs to be computed.

(iii) If $u(t)$ is constant and equal to unity then

$$y(t) = (b_0 + b_1) + e(t)$$

and

$$\frac{1}{N}[\mathbf{X}^T\mathbf{X}] = \begin{bmatrix} 1 & 1 \\ 1 & 1 \end{bmatrix}$$

This matrix is singular and hence a unique least squares solution cannot be obtained because $\mathbf{X}^T\mathbf{X}$ is noninvertible. This corresponds to a set of simultaneous equations where the number of equations is less than the number of unknowns. Clearly, the two parameters appear only as a sum and therefore cannot be estimated separately. For a unique solution $u(t)$ must vary sufficiently to ensure that

$$\det \frac{1}{N}\left[[\mathbf{X}^T\mathbf{X}]\right] \neq 0$$

as N increases. Such conditions are usually associated with the term 'sufficiently exciting' or *persistently exciting*. In self-tuning, it is important that $u(t)$ changes sufficiently to avoid a rank deficient $\mathbf{X}^T\mathbf{X}$ matrix. If the system is noise corrupted then fed back noise may provide sufficient excitation, otherwise excitation must be provided by varying the reference signal.

□

Self-tuning involves identification under closed loop control and this can cause problems of *identifiability*. Again this involves $\mathbf{X}^T\mathbf{X}$ having (possibly) a zero determinant caused by linear dependence among the rows and columns of $\mathbf{X}^T\mathbf{X}$.

Example 3.4

Consider a forward path system (Figure 3.2)

$$y(t) = a\,y(t-1) + b\,u(t-1) + e(t)$$

with proportional regulator

$$u(t) = g\,y(t).$$

Figure 3.2
Illustrating the
system
considered in
Example 3.3.

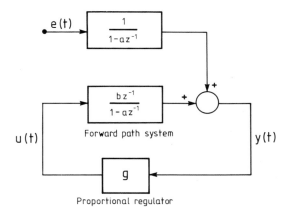

Figure 3.2
Illustrating the
system
considered in
Example 3.3.

It is possible to substitute the regulator equation into the forward path system equation to give either

$$y(t) = (a+gb)y(t-1) + e(t)$$

or

$$y(t) = (a/g + b)u(t-1) + e(t).$$

Note:

(i) It is not possible to solve for the system parameters a and b. We can only obtain an estimate of $\alpha = (a+gb)$.

(ii) There are *two* equivalent low order systems which have the same white driving noise.

Points (i) and (ii) indicate important facts about identifiability. First, we should be able to identify the true system parameters. Second, there should be a single model structure which can be identified. Structure here means the form of the model given by the degrees n_a, n_b, n_c, etc.

In this case the addition of a simple noise-free proportional control loop means we cannot identify the true system parameters a,b and that there are at least two first order systems (and in fact an *infinite* number) that fit the data.

□

In the example, identifiability of the true system is lost by the addition of the noise-free proportional regulator. This causes the data vector

$\mathbf{x}^T(t) = [y(t-1), u(t-1)]$ to have linearly dependent entries (specifically $u(t-1)$ is g times $y(t-1)$). This point is generally true — if $\mathbf{x}(t)$ has linearly dependent entries then $\mathbf{X}$ will have linearly dependent columns and $\mathbf{X}^T\mathbf{X}$ will be singular. Identifiability problems of this kind are cured by breaking up the linear dependence. Methods that achieve this (see Section 6.6.3) include:

(i) adding an independent noise component ('dither') to the feedback equation;

(ii) introducing longer lags in the feedback law; and

(iii) employing a time-varying or nonlinear feedback law.

Illustrative simulations are postponed until the end of Section 3.3 after we have encountered an on-line version of least squares.

3.2.2   Relationship between the residuals and data

Here we briefly discuss the relationship of the data to the residuals or fitting error sequence $\eta(t)$, ($t = 1, 2, \ldots$) obtained from least squares. This will be used in Chapter 7 in discussing the so-called self-tuning property.

The least squares parameter estimate is given by

$$\hat{\boldsymbol{\theta}} = [\mathbf{X}^T\mathbf{X}]^{-1}\mathbf{X}^T\mathbf{y} \qquad (3.8)\text{bis}$$

Noting that $\hat{\mathbf{e}} = \boldsymbol{\eta}$ in the least squares solution and premultiplying equation (3.5) by $\mathbf{X}^T$, yields

$$\mathbf{X}^T\mathbf{y} = [\mathbf{X}^T\mathbf{X}]\,\hat{\boldsymbol{\theta}} + \mathbf{X}^T\boldsymbol{\eta} \qquad (3.14)$$

These two equations imply that

$$\mathbf{X}^T\boldsymbol{\eta} = 0 \qquad (3.15)$$

Writing this out in full gives

$$[\mathbf{x}(1), \mathbf{x}(2), \ldots, \mathbf{x}(N)]\boldsymbol{\eta} = 0$$

and recalling the definition of $\mathbf{x}(t)$, this can be rewritten as

$$\sum_{t=1}^{N} y(t-i)\eta(t) = 0 \qquad \text{for } i=1, \ldots, n_a$$

$$\text{and}$$

$$\sum_{t=1}^{N} u(t-i)\eta(t) = 0 \qquad \text{for } i=1, \ldots, n_b+1$$

$$\left.\begin{array}{c} \\ \\ \\ \\ \\ \end{array}\right\} \qquad (3.16)$$

and so on for any other entries in $x(t)$, if present.

As N becomes large, and under the ergodic assumption (see Appendix) these equations imply that

$$E\{y(t-i)\eta(t)\} = 0 \qquad \text{for } i=1, \ldots, n_a$$

$$E\{u(t-i)\eta(t)\} = 0 \qquad \text{for } i=1, \ldots, n_b+1$$

$$\left.\begin{array}{c} \\ \\ \\ \end{array}\right\} \qquad (3.17)$$

Equations (3.16) and (3.17) are sometimes said to express the *orthogonality property* of least squares and are the basis of a number of other results, one being the self-tuning property discussed in Appendix 7.2.

3.3   RECURSIVE LEAST SQUARES

To be useful in self-tuning control the parameter estimation scheme should be iterative, allowing the estimated model of the system to be updated at each sample interval as new data become available.

In particular, it is useful to be able to visualize the estimation process in terms of Figure 3.3. In this scheme new input/output data become available at each sample interval. The model based on past information (summarized in $\hat{\theta}(t-1)$) is used to obtain an estimate $\hat{y}(t)$ of the current output. This is then compared with the observed output $y(t)$ to generate an error $\epsilon(t)$. This in turn generates an update to the model which corrects $\hat{\theta}(t-1)$ to the new value $\hat{\theta}(t)$. This recursive 'predictor–corrector' form allows significant saving in computation. Instead of recalculating the least squares estimate in its entirety, requiring the storage of *all* previous data, it is both efficient and elegant to merely store the 'old' estimate calculated at time t, denoted by $\hat{\theta}(t)$, and to obtain the 'new' estimates $\hat{\theta}(t+1)$ by an updating step involving the new observation only. To see how this is done, compare a least squares estimate based on data from time samples 1 to t with the estimate based on data from time samples 1 to $t+1$.

For the estimator using data from time 1 to t we have

Figure 3.3
Visualization of
recursive
estimation as an
iterative process
in which the
model
parameters are
adjusted by an
update mechan-
ism derived
from a measure
of the quality of
the model. In
this model
quality is
indicated by the
error $\epsilon(t)$.

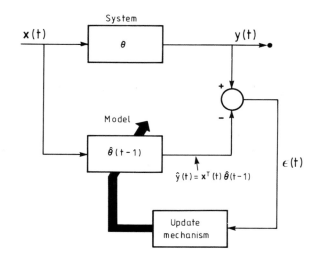

$$\hat{\boldsymbol{\theta}}(t) = (\mathbf{X}^{\mathrm{T}}(t)\mathbf{X}(t))^{-1}\,\mathbf{X}^{\mathrm{T}}(t)\,\mathbf{y}(t) \tag{3.18}$$

where we write $\boldsymbol{\chi}(t)$ as a function of time to indicate that it is based upon data from time steps t_0 to t. Thus:

$$\mathbf{X}(t) = \begin{bmatrix} \mathbf{x}^{\mathrm{T}}(1) \\ \mathbf{x}^{\mathrm{T}}(2) \\ \cdot \\ \cdot \\ \cdot \\ \mathbf{x}^{\mathrm{T}}(t) \end{bmatrix}, \qquad \text{likewise } \mathbf{y}(t) = \begin{bmatrix} y(1) \\ y(2) \\ \cdot \\ \cdot \\ \cdot \\ y(t) \end{bmatrix}$$

At time $t+1$ we obtain further measurements from the process which enable us to form

$$\mathbf{X}(t+1) = \begin{bmatrix} \mathbf{x}^{\mathrm{T}}(1) \\ \mathbf{x}^{\mathrm{T}}(2) \\ \cdot \\ \cdot \\ \cdot \\ \mathbf{x}^{\mathrm{T}}(t) \\ \cdots \\ \mathbf{x}^{\mathrm{T}}(t+1) \end{bmatrix} = \begin{bmatrix} \mathbf{X}(t) \\ \cdots \\ \mathbf{x}^{\mathrm{T}}(t+1) \end{bmatrix} \tag{3.19}$$

$$\mathbf{y}(t+1) = \begin{bmatrix} y(1) \\ y(2) \\ \cdot \\ \cdot \\ \cdot \\ y(t) \\ \cdots \\ y(t+1) \end{bmatrix} = \begin{bmatrix} \mathbf{y}(t) \\ \cdots \\ y(t+1) \end{bmatrix} \qquad (3.20)$$

The estimates at step $t+1$ are then given by

$$\hat{\boldsymbol{\theta}}(t+1) = [\mathbf{X}^T(t+1)\,\mathbf{X}(t+1)]^{-1}\,\mathbf{X}^T(t+1)\mathbf{y}(t+1) \qquad (3.21)$$

Now

$$\mathbf{X}^T(t+1)\mathbf{X}(t+1) = [\mathbf{X}^T(t)\mathbf{x}(t+1)] \begin{bmatrix} \mathbf{X}(t) \\ \cdots \\ \mathbf{x}^T(t+1) \end{bmatrix}$$

$$= \mathbf{X}^T(t)\mathbf{X}(t) + \mathbf{x}(t+1)\mathbf{x}^T(t+1) \qquad (3.22)$$

Thus given $\mathbf{x}(t+1)$ we can easily update the old matrix of correlations $\mathbf{X}^T(t)\mathbf{X}(t)$ to obtain the new matrix $\mathbf{X}^T(t+1)\,\mathbf{X}(t+1)$. However, we actually need to find a way to update the *inverse* of $\mathbf{X}(t)^T\mathbf{X}(t)$ directly without requiring a matrix inversion at each time step. In addition, we also need to update the term $\mathbf{X}^T(t+1)\mathbf{y}(t+1)$. Now, from equations (3.19) and (3.20):

$$\mathbf{X}^T(t+1)\mathbf{y}(t+1) = [\mathbf{X}^T(t)\,\mathbf{x}(t+1)] \begin{bmatrix} \mathbf{y}(t) \\ \cdots \\ y(t+1) \end{bmatrix}$$

$$\mathbf{X}^T(t+1)\mathbf{y}(t+1) = \mathbf{X}^T(t)\mathbf{y}(t) + \mathbf{x}(t+1)y(t+1) \qquad (3.23)$$

Introducing some shorthand by denoting

$$\left. \begin{aligned} \mathbf{P}(t) &= [\mathbf{X}^T(t)\,\mathbf{X}(t)]^{-1} \\[2mm] \mathbf{B}(t) &= \mathbf{X}^T(t)\,\mathbf{y}(t) \end{aligned} \right\} \qquad (3.24)$$

we have

$$\left. \begin{aligned} \hat{\boldsymbol{\theta}}(t+1) &= \mathbf{P}(t+1)\,\mathbf{B}(t+1) \\[2mm] \hat{\boldsymbol{\theta}}(t) &= \mathbf{P}(t)\,\mathbf{B}(t) \end{aligned} \right\} \qquad (3.25)$$

Also

$$\mathbf{P}^{-1}(t+1) = \mathbf{P}^{-1}(t) + \mathbf{x}(t+1)\mathbf{x}^\mathrm{T}(t+1) \tag{3.26}$$

and

$$\mathbf{B}(t+1) = \mathbf{B}(t) + \mathbf{x}(t+1)y(t+1) \tag{3.27}$$

Equation (3.27) gives a direct update from $\mathbf{B}(t)$ to $\mathbf{B}(t+1)$. The crucial step is to establish the same direct update from $\mathbf{P}(t)$ to $\mathbf{P}(t+1)$. The standard way to do this is by applying the Matrix Inversion Lemma:

$$(\mathbf{A}+\mathbf{BCD})^{-1} = \mathbf{A}^{-1} - \mathbf{A}^{-1}\mathbf{B}(\mathbf{C}^{-1} + \mathbf{DA}^{-1}\mathbf{B})^{-1}\mathbf{DA}^{-1}$$

to (3.26).

Assigning

$$\mathbf{A} = \mathbf{P}^{-1}(t), \qquad \mathbf{C} = 1, \qquad \mathbf{B} = \mathbf{x}(t+1), \qquad \mathbf{D} = \mathbf{x}^\mathrm{T}(t+1)$$

gives

$$\mathbf{P}(t+1) = \mathbf{P}(t)\left[\mathbf{I}_\mathrm{m} - \mathbf{x}(t+1)\left(1+\mathbf{x}^\mathrm{T}(t+1)\mathbf{P}(t)\mathbf{x}(t+1)\right)^{-1}\mathbf{x}^\mathrm{T}(t+1)\mathbf{P}(t)\right] \tag{3.28}$$

Equation (3.28) gives us the means to update $\mathbf{P}(t)$ to $\mathbf{P}(t+1)$ without inverting a matrix. In fact, the only inversion is of the scalar term $[1+\mathbf{x}^\mathrm{T}(t+1)\mathbf{P}(t)\mathbf{x}(t+1)]$.

The recursion for $\mathbf{P}(t+1)$ can be combined with the recursion for $\mathbf{B}(t+1)$ (equation (3.27)) in many ways to give a direct recursion for $\hat{\boldsymbol{\theta}}(t+1)$ from $\hat{\boldsymbol{\theta}}(t)$. The most common way is to define the error variable $\epsilon(t+1)$ (as indicated in Figure 3.3) by:

$$\epsilon(t+1) = y(t+1) - \mathbf{x}^\mathrm{T}(t+1)\hat{\boldsymbol{\theta}}(t) \tag{3.29}$$

and substitute for $y(t+1)$ in equation (3.27).

This gives:

$$\mathbf{B}(t+1) = \mathbf{B}(t) + \mathbf{x}(t+1)\mathbf{x}^\mathrm{T}(t+1)\hat{\boldsymbol{\theta}}(t) + \mathbf{x}(t+1)\epsilon(t+1)$$

Substituting for $\mathbf{B}(t)$, $\mathbf{B}(t+1)$ using equations (3.25) gives:

$$\hat{\boldsymbol{\theta}}(t+1) = \hat{\boldsymbol{\theta}}(t) + \mathbf{P}(t+1)\mathbf{x}(t+1)\epsilon(t+1) \tag{3.30}$$

In summary, the full recursive least squares (RLS) algorithm for updating $\hat{\boldsymbol{\theta}}(t)$ is as follows:

Algorithm: Matrix Inversion Lemma RLS, Version 1

At time step $t+1$:

(i) Form $\mathbf{x}(t+1)$ using the new data.

(ii) Form $\epsilon(t+1)$ using

$$\epsilon(t+1) = y(t+1) - \mathbf{x}^{\mathrm{T}}(t+1)\hat{\boldsymbol{\theta}}(t)$$

(iii) Form $\mathbf{P}(t+1)$ using

$$\mathbf{P}(t+1) = \mathbf{P}(t)\left[\mathbf{I_m} - \frac{\mathbf{x}(t+1)\mathbf{x}^{\mathrm{T}}(t+1)\,\mathbf{P}(t)}{1 + \mathbf{x}^{\mathrm{T}}(t+1)\mathbf{P}(t)\mathbf{x}(t+1)}\right]$$

(iv) Update $\hat{\boldsymbol{\theta}}(t)$

$$\hat{\boldsymbol{\theta}}(t+1) = \hat{\boldsymbol{\theta}}(t) + \mathbf{P}(t+1)\mathbf{x}(t+1)\epsilon(t+1)$$

(v) Wait for the next time step to elapse and loop back to step (i)

□

The following example demonstrates the performance of the RLS algorithm in a variety of experimental circumstances.

Example 3.5

Consider again the first order system of Example 3.4:

$$y(t) = ay(t-1) + bu(t-1) + e(t)$$

where the true values of the parameters are $a = 0.5$, $b = 1$ and the white noise variance is 0.1. Figures 3.4(a)–(g) show the evolution of the RLS estimates $\hat{a}(t)$, $\hat{b}(t)$ under the following experimental conditions (cf. Example 6.11):

(a) $u(t) = 0$

(b) $u(t) = 1$

(c) $u(t)$ zero mean unit variance white noise

(d) $u(t) = 0.1y(t)$

(e) $u(t) = 0.1y(t)+r(t)$ where $r(t)$ is zero mean unit variance white noise, uncorrelated with $e(t)$

(f) $u(t) = 0.1\ y(t-1)$

(g) $u(t) = 0.1\ \text{sgn}(y(t))$

The RLS recursion is initialized with the values

$$\mathbf{P}(0) = 10\mathbf{I}_2, \qquad \hat{a}(0) = 0, \qquad \hat{b}(0) = 1$$

A more detailed discussion on the choice of $\mathbf{P}(0)$, $\mathbf{x}(0)$ is postponed to Chapter 4.

The following remarks explain the behaviour of the estimates.

Figure 3.4(a) In this situation, because $u(t) = 0$, the parameter b is unidentifiable. Essentially, there is insufficient information to determine b and hence the estimate fails to converge to the correct value.

Figure 3.4(b) The parameters are now both identifiable. If, however, more complex B dynamics are present (as in Example 3.3) then parameter identifiability would be lost.

Figure 3.4(c) The excitation $u(t)$ is an independent noise signal and the parameters a,b are accordingly identifiable. The rapid convergence is typical of an open loop ARMAX identification scheme with relatively little system noise present.

Figure 3.4(d) The signal $u(t)$ is proportional to the output signal $y(t)$ (cf. Example 3.4). As a result the parameters a,b are not identifiable.

Figures 3.4(e,f,g) As noted in Example 3.4, the linear dependence caused by some feedback rules can be broken up by techniques such as:

(i) adding an independent noise component (Figure 3.4(e));

(ii) introducing longer lags in the feedback law (Figure 3.4(f)); and

(iii) employing a nonlinear feedback law (Figure 3.4(g)).

□

It will be useful to give a different formulation of the RLS algorithm.

Note that the term $\mathbf{P}(t+1)\mathbf{x}(t+1)$ in equation (3.30) is a column vector of adjustment gains. Designate this vector $\mathbf{L}(t+1)$, thus:

$$\mathbf{L}(t+1) = \mathbf{P}(t+1)\mathbf{x}(t+1) \tag{3.31}$$

Figure 3.4
Parameter
estimates (see
Example 3.5) for
an RLS estimator
applied to the
system
$y(t) = ay(t-1) + bu(t-1) + e(t)$
under the
different
experimental
conditions: (a)
$u(t) = 0$
(parameters
unidentifiable);
(b) $u(t) = 1$
(parameters
identifiable);

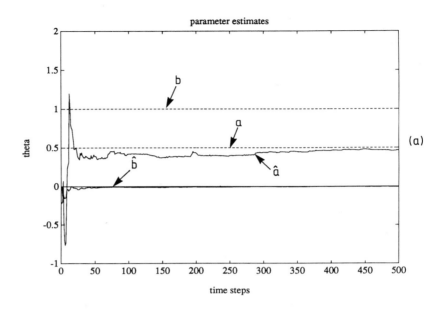

(a)

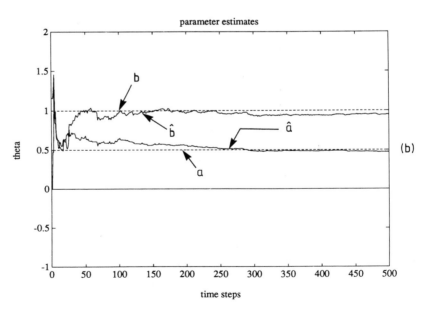

(b)

Figure 3.4 (*cont.*)
(c) *u(t)*
independent
noise
(parameters
identifiable); (d)
u(t) = 0.1*y(t)*
(parameters
unidentifiable);

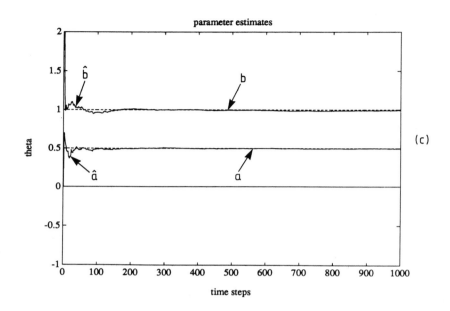

(c)

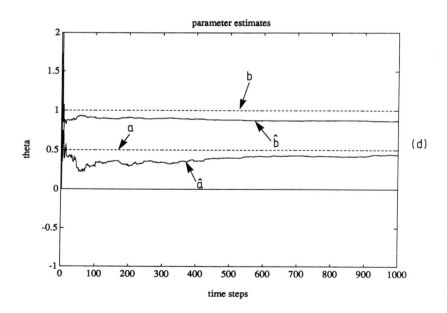

(d)

Figure 3.4 (*cont.*)
(e) as (d) but
with an
independent
noise source
added; (f)
$u(t) = 0.1y(t-1)$
(parameters
identifiable);

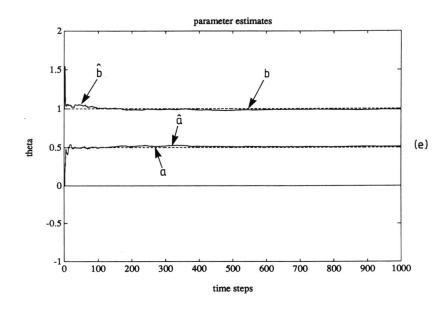

(e)

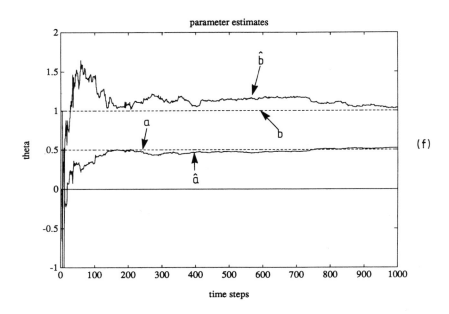

(f)

Figure 3.4 (*cont.*)
(g) $u(t) = 0.1$
sign($y(t)$)
(parameters
identifiable).

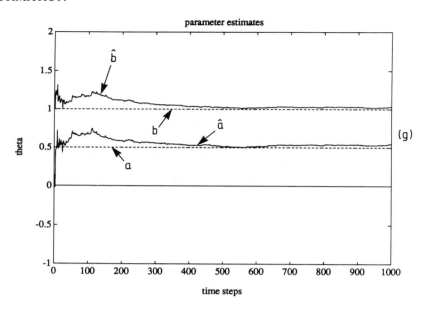

so that

$$\hat{\boldsymbol{\theta}}(t+1) = \hat{\boldsymbol{\theta}}(t) + \mathbf{L}(t+1)\epsilon(t+1) \tag{3.32}$$

By combining equation (3.31) and the expression for $\mathbf{P}(t+1)$ (equation (3.28)), we can obtain the following equation for $\mathbf{L}(t+1)$:

$$\mathbf{L}(t+1) = \frac{\mathbf{P}(t)\mathbf{x}(t+1)}{1+\mathbf{x}^{\mathrm{T}}(t+1)\mathbf{P}(t)\mathbf{x}(t+1)} \tag{3.33}$$

Similarly, $\mathbf{P}(t+1)$ can be re-expressed as

$$\mathbf{P}(t+1) = \mathbf{P}(t) - \mathbf{L}(t+1)[\mathbf{P}(t)\mathbf{x}(t+1)]^{\mathrm{T}} \tag{3.34}$$

The RLS in revised form is:

Algorithm: Matrix Inversion Lemma RLS, Version 2

At time step $t+1$:

(i) Form $\mathbf{x}(t+1)$ using the new data.

(ii) Form $\epsilon(t+1)$ as before:

$$\epsilon(t+1) = y(t+1) - \mathbf{x}^T(t+1)\hat{\boldsymbol{\theta}}(t)$$

(iii) Form $\mathbf{L}(t+1)$ using equation (3.33).

(iv) Update $\hat{\boldsymbol{\theta}}(t)$ to obtain $\hat{\boldsymbol{\theta}}(t+1)$:

$$\hat{\boldsymbol{\theta}}(t+1) = \hat{\boldsymbol{\theta}}(t) + \mathbf{L}(t+1)\epsilon(t+1)$$

(v) Form $\mathbf{P}(t+1)$ using equation (3.34).

(vi) Wait for the next time step to elapse and loop back to step (i).

$\square$

3.4 RESIDUALS AND PREDICTION ERRORS

A key variable in the RLS algorithm is the modelling error $\epsilon(t)$. We now relate it to the residual $\eta(t)$ associated with the least squares procedure.

Recall that $\epsilon(t)$ is defined as:

$$\epsilon(t) = y(t) - \mathbf{x}^T(t)\hat{\boldsymbol{\theta}}(t-1) \qquad (3.29)\text{bis}$$

In other words it is the error between the system output $y(t)$ and the *predicted* output using parameter estimates $\hat{\boldsymbol{\theta}}(t-1)$ from the *previous* time step t-1. For this reason it is usually called the (*a priori*) *output prediction error*.

The true modelling error (or *residual* or *a posteriori prediction error*) at time step t is

$$\eta(t) = y(t) - \mathbf{x}^T(t)\,\hat{\boldsymbol{\theta}}(t) \qquad (3.35)$$

where the only difference between $\epsilon(t)$ and $\eta(t)$ is that $\eta(t)$ is based upon the parameter estimates $\hat{\boldsymbol{\theta}}(t)$ at the *current* time step t.

As the estimation time t increases, the difference between $\eta(t)$ and $\epsilon(t)$ should become insignificant, but during the first few recursions the difference is significant and various estimation algorithms exploit this (see Section 3.6).

From (3.29) and (3.35)

$$\eta(t) = \epsilon(t) - \mathbf{x}^T(t)(\hat{\boldsymbol{\theta}}(t) - \hat{\boldsymbol{\theta}}(t-1)) \qquad (3.36)$$

Using equation (3.32), the difference $\hat{\boldsymbol{\theta}}(t) - \hat{\boldsymbol{\theta}}(t-1)$ can be substituted in equation (3.36) to yield

$$\eta(t) = \epsilon(t)(1 - \mathbf{x}^{\mathrm{T}}(t)\mathbf{L}(t)) \tag{3.37}$$

From the equation for $\mathbf{L}(t)$ (equation (3.33)), the above expression can be re-expressed as:

$$\eta(t) = \epsilon(t)/(1 + \mathbf{x}^{\mathrm{T}}(t)\mathbf{P}(t-1)\mathbf{x}(t)) \tag{3.38}$$

The denominator of this expression is a scalar and has already been calculated in the RLS step. Thus $\eta(t)$ can be obtained from $\epsilon(t)$ at any time step by a scalar division. This relationship will be exploited in Section 3.6.

3.5   GRADIENT METHODS

The RLS algorithm iteratively minimizes the least squares cost function (3.7). It is therefore illuminating to compare it with the standard gradient algorithms used in deterministic optimization (steepest descent, Newton's method, etc.). In this way, other potentially simpler forms of recursive estimation may suggest themselves.

The RLS update of the parameter vector $\boldsymbol{\theta}$ is

$$\hat{\boldsymbol{\theta}}(t+1) = \hat{\boldsymbol{\theta}}(t) + \mathbf{P}(t+1)\mathbf{x}(t+1)\epsilon(t+1) \tag{3.39}$$

where the matrix $\mathbf{P}(t+1)$, despite avoiding the inversion of a matrix, takes a relatively large number of operations to compute. We therefore seek to replace $\mathbf{P}(t+1)$ by an expression which is simpler to compute and (possibly) takes less storage space. We can interpret equation (3.39) as an iterative minimization step as follows.

Consider the error criterion, which for convenience scales $\epsilon(t)$ by $\sqrt{2}$.

$$\epsilon^2(t+1) = \frac{1}{2}\left[y(t+1) - \mathbf{x}^{\mathrm{T}}(t+1)\hat{\boldsymbol{\theta}}(t)) \right]^2 \tag{3.40}$$

The gradient with respect to $\hat{\boldsymbol{\theta}}(t)$ is

$$\frac{\partial(\epsilon^2(t+1))}{\partial\hat{\boldsymbol{\theta}}(t)} = -\left\{ \mathbf{x}(t+1)\epsilon(t+1) \right\} \tag{3.41}$$

Substituting in equation (3.39), a gradient descent-like iteration arises:

$$\hat{\boldsymbol{\theta}}(t+1) = \hat{\boldsymbol{\theta}}(t) - \mathbf{P}(t+1)\frac{\partial[\epsilon^2(t+1)]}{\partial\hat{\boldsymbol{\theta}}(t)} \tag{3.42}$$

where $\mathbf{P}(t+1)$ can be viewed as a positive definite matrix which modifies the direction of the gradient adjustments.

It is plausible, therefore, to replace $\mathbf{P}(t+1)$ by an adjustment gain $\mathbf{\gamma}(t+1)$ selected in some other (simpler) way.

Note:

(i) $\mathbf{\gamma}(t+1)$ can be a scalar or a square matrix.

(ii) $\mathbf{\gamma}(t+1)$ can be constant or time varying *provided* the convergence of $\hat{\mathbf{\theta}}(t)$ to $\mathbf{\theta}$ as t increases is assured.

In practice (ii) implies that $\mathbf{\gamma}(t+1)$ must be selected in a special manner because of the noise contribution $e(t)$ to the system equation

$$y(t) = \mathbf{x}^{\mathrm{T}}(t)\mathbf{\theta} + e(t) \tag{3.43}$$

If $e(t)$ is absent then equation (3.42) can be treated as a deterministic hill-climbing algorithm. In particular, $\mathbf{\gamma}(t)$ can be a constant. The presence of $e(t)$, however, means that $\mathbf{\gamma}(t)$ must decrease with time if the influence of the noise on the estimates is to be progressively reduced. However, $\mathbf{\gamma}(t)$ must not decrease *so fast* that the value of $\hat{\mathbf{\theta}}(t)$ does not reach $\mathbf{\theta}$. In summary, we must be able to descend to the minimum, yet squeeze out the influence of noise. For example, the sequence

$$\mathbf{\gamma}(t) = 1/t \tag{3.44}$$

satisfies this criterion of a decreasing gain, which does not decrease too fast (see Section 6.8). For scalar gain sequences satisfying these convergence conditions the algorithm

$$\hat{\mathbf{\theta}}(t+1) = \hat{\mathbf{\theta}}(t) + \mathbf{\gamma}(t+1)\mathbf{x}(t+1)\epsilon(t+1) \tag{3.45}$$

is termed a *stochastic approximation* (or stochastic gradient) algorithm.

The advantage of such gradient algorithms is simplicity and speed of computation. They are, however, characterized by slow convergence of the parameter estimate $\hat{\mathbf{\theta}}(t)$ and poor adaption to changes in system parameters. This is to be expected. Comparing (3.45) with (3.42), we can interpret the stochastic approximation algorithm in terms of 'steepest descent' optimization, which is well known to have a slow convergence rate in the neighbourhood of minima.

If the gain sequence does not converge to zero, the algorithm may be suitable for tracking slowly varying system parameters. Examples of such algorithms are as follows:

(a) *Sequential learning algorithm*

$$\boldsymbol{\gamma}(t) = \alpha/[\beta + \mathbf{x}^T(t)\mathbf{x}(t)] \qquad (\beta > 0; 0 < \alpha < 2) \qquad (3.46)$$

(b) *Constant scalar gain*

$$\boldsymbol{\gamma}(t) = \text{constant scalar} \qquad (3.47)$$

This algorithm is commonly used in adaptive filtering under the name of least-mean-square or LMS.

(c) *Constant matrix gain*

$$\boldsymbol{\gamma}(t) = \text{constant positive definite symmetric matrix} \qquad (3.48)$$

This algorithm looks like the RLS with $\mathbf{P}(t)$ held constant. Indeed, this effect can be achieved asymptotically if the RLS is modified by incorporating a forgetting factor or random walk to inject adaptivity to parameter drift (see Chapter 4).

Example 3.6

This example demonstrates the type of performance which can be expected from gradient based recursive estimators as compared to the RLS estimator.

Consider the AR system given by

$$y(t) = -a_1 y(t-1) - a_2 y(t-2) + e(t)$$

where the true values of the system parameters are $a_1 = -1.5$ and $a_2 = +0.56$. The white noise $e(t)$ is zero mean and has unit variance. Figure 3.5(a) shows the evolution of the RLS estimates $\hat{a}_1, \hat{a}_2$. The RLS recursion was initialized with the values

$$\mathbf{P}(0) = 10\mathbf{I}_2, \qquad \hat{a}_1(0) = 0, \qquad \hat{a}_2(0) = 0$$

Note the rapid convergence of the RLS estimates to the correct parameter values. This should be compared with the stochastic gradient results. Specifically:

Figure 3.5(b): Shows the evolution of the stochastic approximation algorithm with the gain sequence of equation (3.44). Note that even after 2000 time steps the estimates have not yet converged. Such slow convergence is typical of the stochastic approximation method.

Figure 3.5
Parameter
estimates for
various recursive
estimators
(described in
Example 3.6)
applied to the
system
$y(t) = -a_1y(t-1)$
$- a_2y(t-2) + e(t)$
Parameter
estimates are
shown
corresponding to
the following
estimators: (a)
RLS; (b)
stochastic
approximation;

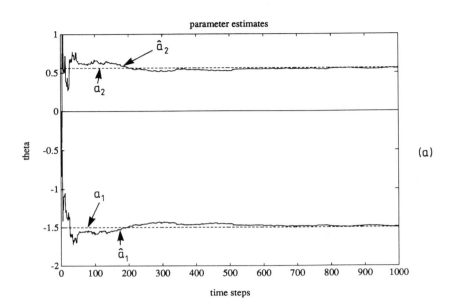

(a)

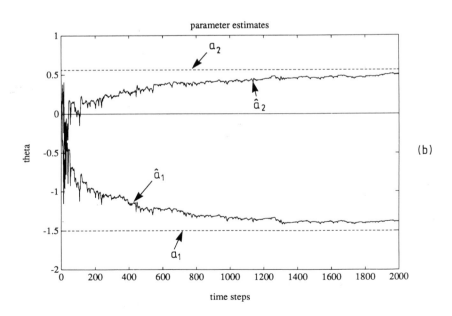

(b)

Figure 3.5 *(cont.)*
(c) sequential
learning $\alpha = 0.1$,
$\beta = 1$; (d)
sequential
learning
$\alpha = 0.01$, $\beta = 1$;

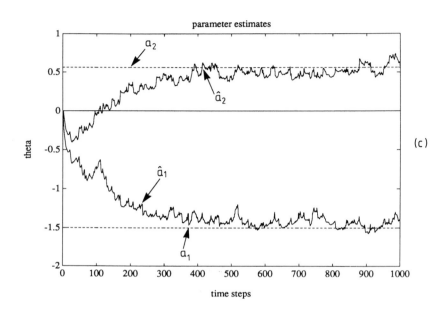

(c)

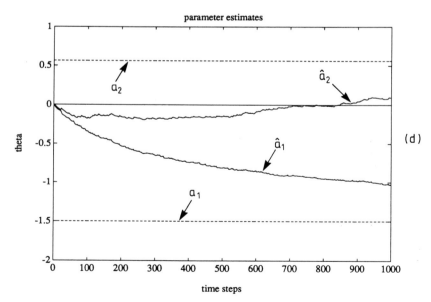

(d)

Figure 3.5 *(cont.)* (e) constant scalar gain $\gamma = 0.005$; (f) constant scalar gain $\gamma = 0.01$.

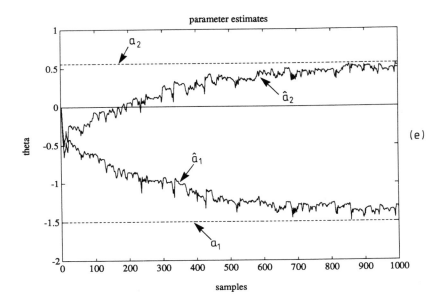

(e)

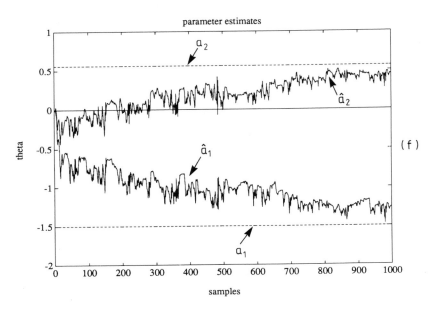

(f)

Figure 3.5(c,d): Shows the results of using the sequential learning gain (equation (3.46)) with $\alpha = 0.1$, $\beta = 1$ (Figure 3.5(c)) and $\alpha = 0.01$, $\beta = 1$ (Figure 3.5(d)). Note the trade-off between speed of convergence and variability of the estimates as α is changed.

Figure 3.5(e,f): Shows the results of using the constant scalar gain algorithm (equation (3.47)) with $\gamma = 0.005$ and 0.01 respectively. Note again the trade-off between parameter estimate variability and the size of the gain.

□

3.6   EXTENDED LEAST SQUARES AND APPROXIMATE MAXIMUM LIKELIHOOD

In least squares estimation the model assumed is

$$Ay(t) = Bu(t-1) + \hat{e}(t) \tag{3.49}$$

When the system output is corrupted by *coloured noise*, however, a more suitable model is

$$Ay(t) = Bu(t-1) + C\hat{e}(t) \tag{3.50}$$

where, in general, C cannot be assumed to be unity (see Section 3.2).

Of course, we can consider the term $C\hat{e}(t)$ as simply 'error' and estimate A,B in (3.50) using RLS estimation. In general, however, the estimates generated in this way will be biased as the following example shows.

Example 3.7

Consider the system

$$y(t) = ay(t-1) + \xi(t) \qquad |a| < 1 \tag{3.51}$$

where

$$\xi(t) = e(t) + ce(t-1) \qquad |c| < 1 \tag{3.52}$$

and $\{e(t)\}$ is a zero mean white noise sequence of variance σ_e^2.

The off-line LS estimate of a is

$$\hat{a}(N) = \sum_{t=1}^{N} y(t)y(t-1) \Big/ \sum_{t=1}^{N} y^2(t-1)$$

$$= a + \sum_{t=1}^{N} \xi(t)y(t-1) \Big/ \sum_{t=1}^{N} y^2(t-1)$$

$$\simeq a + \frac{E[\xi(t)y(t-1)]}{E\, y^2(t-1)}$$

for large N, assuming ergodicity.

From (3.52)

$$E[\xi(t-1)\xi(t-i)] = 0 \qquad i = 3,4,5, \ldots$$

Using (3.51) and (3.52)

$$y(t-1) = \xi(t-1) + a\xi(t-2) + a^2\xi(t-3) + \ldots$$

leads to

$$E[\xi(t)y(t-1)] = E[\xi(t)\xi(t-1)] = c\sigma_e^2$$

and

$$E\, y^2(t-1) = (1+a^2+a^4 + \ldots)\, E\, \xi^2(t-1) + 2a(1+a^2+a^4 + \ldots)\, E[\xi(t-1)\xi(t-2)]$$

$$= \frac{1+c^2}{1-a^2}\sigma_e^2 + \frac{2ac}{1-a^2}\sigma_e^2$$

Finally

$$\hat{a}(N) - a \simeq \frac{c(1-a^2)}{1+c^2+2ac} \tag{3.53}$$

This shows that a bias is present unless c is zero. The bias can be demonstrated experimentally as follows. Consider the system

$$y(t) = ay(t-1) + e(t) + ce(t-1)$$

in which $a = 0.7$ and $c = 0.5$. Figure 3.6 shows the evolution of a RLS estimate of $\hat{a}$ in the model

$$y(t) = \hat{a}y(t-1) + \hat{e}(t)$$

Figure 3.6
Parameter
estimate for a
RLS estimator
(described in
Example 3.7)
applied to a
system with
coloured noise.
Note the bias in
the estimate $\hat{a}$.

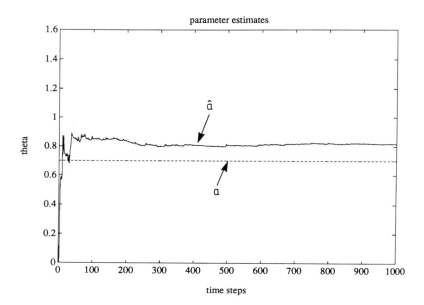

Note that the bias in the estimate $\hat{a}$ is approximately that given by equation (3.53) (in this case $\hat{a}(N)-a = 0.1307$).

$\square$

In order to estimate C in the general case, knowledge of $\hat{e}(t-1)$, ..., $\hat{e}(t-n_c)$ is required. However, $\{\hat{e}(t)\}$ is an *unobservable* error process. A number of estimation procedures exist, however, which replace $\hat{e}(t)$ by an estimate, usually taken to be the prediction error $\epsilon(t)$ or the residual $\eta(t)$.

For example, the Recursive Extended Least Squares estimation algorithm (RELS or RML₁) is as follows: Define

$$\hat{\boldsymbol{\theta}} = [-\hat{a}_1, \ldots, -\hat{a}_{n_a}, \hat{b}_0, \ldots, \hat{b}_{n_b}, \hat{c}_1, \ldots, \hat{c}_{n_c}]$$

$$\boldsymbol{\phi}^T(t) = [y(t-1), \ldots, y(t-n_a), u(t-1), \ldots, u(t-n_b-1), \epsilon(t-1), \ldots, \epsilon(t-n_c)] \quad (3.54)$$

where

$$\epsilon(t) = y(t) - \boldsymbol{\phi}^T(t)\hat{\boldsymbol{\theta}}(t-1) \quad (3.55)$$

is the *prediction error* using output prediction based on information up to the time step $t-1$.

The RELS algorithm is then:

Algorithm: Recursive Extended Least Squares

At time step $t+1$

(i) Form $\boldsymbol{\phi}(t+1)$ using new data $u(t+1)$, $y(t+1)$ and

$$\epsilon(t+1) = y(t+1) - \boldsymbol{\phi}^T(t+1)\hat{\boldsymbol{\theta}}(t)$$

(ii)

$$\mathbf{P}(t+1) = \mathbf{P}(t)\left[\mathbf{I}_m - \frac{\boldsymbol{\phi}(t+1)\boldsymbol{\phi}^T(t+1)\mathbf{P}(t)}{1 + \boldsymbol{\phi}^T(t+1)\mathbf{P}(t)\boldsymbol{\phi}(t+1)}\right]$$

(iii)

$$\hat{\boldsymbol{\theta}}(t+1) = \hat{\boldsymbol{\theta}}(t) + \mathbf{P}(t+1)\boldsymbol{\phi}(t+1)\epsilon(t+1) \tag{3.56}$$

$\square$

An alternative algorithm is Approximate Maximum Likelihood (AML or RML_2) which replaces $\epsilon(t)$ (the prediction error) in $\boldsymbol{\phi}(t)$ (equation (3.54)) by the residual $\eta(t)$ given by

$$\eta(t) = y(t) - \boldsymbol{\psi}^T(t)\hat{\boldsymbol{\theta}}(t) \tag{3.57}$$

where

$$\boldsymbol{\psi}^T(t) = [y(t-1), \ldots, y(t-n_a), u(t-1), \ldots, u(t-n_b-1), \eta(t-1) \ldots \eta(t-n_c)] \tag{3.58}$$

and the recursion proceeds as below:

$$\mathbf{P}(t+1) = \mathbf{P}(t)\left[\mathbf{I}_m - \frac{\boldsymbol{\psi}(t+1)\boldsymbol{\psi}^T(t+1)\mathbf{P}(t)}{1 + \boldsymbol{\psi}^T(t+1)\mathbf{P}(t)\boldsymbol{\psi}(t+1)}\right] \tag{3.59}$$

$$\hat{\boldsymbol{\theta}}(t+1) = \hat{\boldsymbol{\theta}}(t) + \mathbf{P}(t+1)\boldsymbol{\psi}(t+1)\epsilon(t+1) \tag{3.60}$$

The AML/RML_2 algorithm can be expressed as a modification of the RELS/RML_1 algorithm as follows:

Algorithm: Approximate Maximum Likelihood

Modify the RELS algorithm as follows

(i) Replace vector $\boldsymbol{\phi}[(t+1)$ by $\boldsymbol{\psi}(t+1)$ computed according to equation (3.58).

(ii) Replace the covariance and parameter update equations by those given in equations (3.59) and (3.60) respectively.

$\square$

Using the AML algorithm it is possible to improve convergence behaviour. Also it is easy to compute the residual $\eta(t)$ directly from $\epsilon(t)$ using

$$\eta(t) = \epsilon(t) / \left[1 + \boldsymbol{\psi}^T(t)\mathbf{P}(t-1)\boldsymbol{\psi}(t) \right] \tag{3.61}$$

(cf. equation (3.38)). A similar expression is valid for RELS with $\boldsymbol{\psi}(t)$ replaced by $\boldsymbol{\phi}(t)$.

The RML algorithms can be interpreted in terms of data filtering. For AML, equation (3.57) can be expressed in the form

$$\eta(t) = \frac{1}{\hat{C}} [\hat{A}y(t) - \hat{B}u(t-1)] \tag{3.62}$$

where the estimated polynomials are formed from $\hat{\boldsymbol{\theta}}(t)$. For slowly varying estimates, we can write

$$\eta(t) = \hat{A}\left(\frac{y(t)}{\hat{C}}\right) - \hat{B}\left(\frac{u(t-1)}{\hat{C}}\right) \tag{3.63}$$

A similar expression can be derived for $\epsilon(t)$ where the estimated polynomials are formed from $\hat{\boldsymbol{\theta}}(t-1)$.

This suggests the following filtering operation on the data:

$$\left. \begin{array}{l} y_f(t) = \dfrac{y(t)}{\hat{C}} \\[3mm] u_f(t) = \dfrac{u(t)}{\hat{C}} \end{array} \right\} \tag{3.64}$$

where $u_f(t)$, $y_f(t)$ are then used in a conventional least squares algorithm. Clearly, if $\hat{C}$ is an accurate estimate then the system can be cast in the form

$$Ay_f(t) = B u_f(t-1) + \frac{C}{\hat{C}} e(t) \simeq B u_f(t-1) + e(t) \qquad (3.65)$$

and the least squares model (3.49) is valid. This filtering interpretation is widely used to illuminate and justify in engineering terms the recursive maximum likelihood approach. We end the chapter with some simple examples comparing the performances of the two RML algorithms.

Example 3.8

Consider the system

$$y(t) = ay(t-1) + e(t) + ce(t-1) \qquad (3.66)$$

in which the true parameters are $a = 0.7$ and $c = -0.5$.

Figures 3.7(a) and (b) show the behaviour of the parameter estimates $\hat{a}(t)$, $\hat{c}(t)$ respectively using RELS and AML and starting from the same initial conditions.

□

Example 3.9

This example illustrates the behaviour of a RELS or AML algorithm when the noise generating mechanism is nonminimum phase. Consider the system (3.66) in which $a = 0.7$ (as in Example 3.8) and $c = -2$ (i.e. a nonminimum phase C polynomial).

Recalling the discussion of Section 2.3.2. (Example 2.4), there exists a minimum phase C polynomial which is equivalent in spectral density terms to the nonminimum phase C which is used to generate the data. The RELS/AML algorithms estimate this minimum phase parameter. Figure 3.8 shows the parameter estimates for RELS. As expected, the estimate $\hat{c}$ converges to $c' = -\frac{1}{2}$, the inverse of -2.

□

Example 3.10

This final example illustrates the covariance structure of the residual $\eta(t)$ (equation (3.35)) and the prediction error $\epsilon(t)$ (equation (3.29)).

Figures 3.9(a), (b) and (c) compare the autocorrelation functions R of $\eta(t)$ and $\epsilon(t)$ calculated for Example 3.8 at time steps $t = 40, 100, 200$, respectively. Note that the correlations are slightly different at $t = 40$ and still at $t = 100$. By $t = 200$, however, they are indistinguishable.

□

Figure 3.7
Parameter
estimates for (a)
RELS (RML₁)
and (b) AML
(RML₂) applied
to the AR system
described in
Example 3.8.

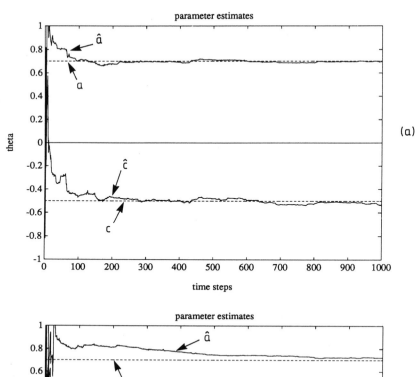

(a)

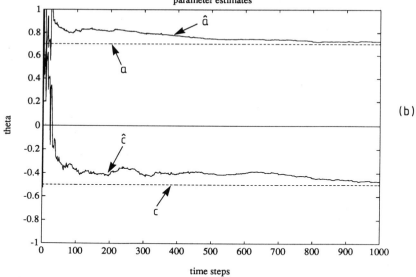

(b)

3.7 SUMMARY

The chapter has developed the basic ideas and algorithms associated with recursive estimation. Specific points covered are

● The basic least squares estimation procedure

Figure 3.8
Parameter
estimate for
RELS (RML$_1$)
applied to a
nonminimum
ARMA system
(Example 3.9).
Note the $\hat{c}$
estimate
converges to the
minimum phase
equivalent value
c'.

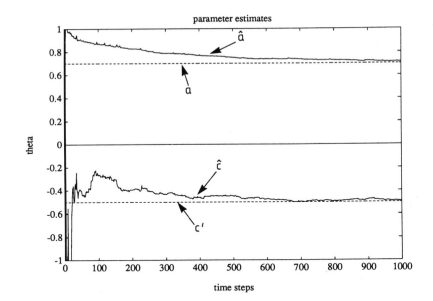

- The development of the matrix inversion recursive least squares (RLS) algorithm
- A review of related gradient-based recursive estimators
- The recursive extended least squares algorithm (RELS) and the approximate maximum likelihood algorithm (AML).

Little has been said about how to use these algorithms in practical situations. This will be covered in Chapter 4. Likewise, different algorithmic versions of recursive least squares exist and these are covered in Chapter 5.

3.8   PROBLEMS

Problem 3.1

Consider a system

$$Ay(t) = Bu(t-1) + e(t)$$

in the usual notation. Find conditions on the degrees of the polynomials F, G in the feedback law

$$Fu(t) = Gy(t)$$

Figure 3.9
Illustrating the
autocorrelation
functions R for
$\eta(t)$ (*a posteriori*
error) and $\epsilon(t)$ (*a
priori* error) from
a RELS(RML$_1$)
estimator after:
(a) 40 time
steps; (b) 100
time steps; and
(c) 200 time
steps.

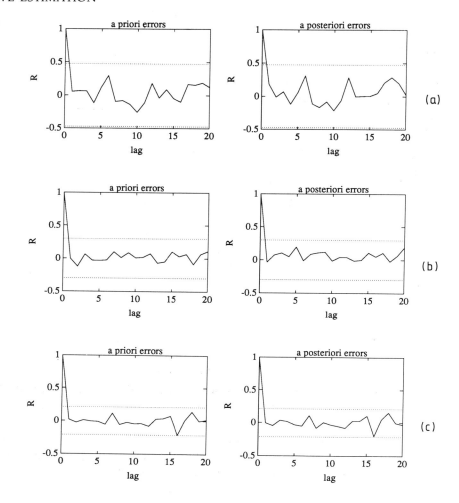

where $F(0) = 1$, such that the coefficients of A and B can be uniquely estimated using least squares.

Problem 3.2

Write out in detail the RLS algorithm appropriate for the iterative estimation of the parameter β in the system model

$$y(t) = \beta u(t) + \hat{e}(t)$$

Problem 3.3

Assuming the model form

$$y(t) = \beta u(t) + \hat{e}(t)$$

calculate the first four iterations of the RLS algorithm using the data set given in the table below.

Table 3.1

t	0	1	2	3	4	5
$y(t)$	2.15	−1.94	−2.05	+1.98	−2.1	+2.1
$u(t)$	+1	−1	−1	+1	−1	+1

Assume $P(0) = 10$, and plot $P(t)$ as a function of t. Repeat with $P(0) = 100$ and $P(0) = 0.1$. Discuss the results.

Problem 3.4

For RLS, show that the residual $\eta(t)$ can be obtained from the prediction error $\epsilon(t)$ using

$$\eta(t) = \epsilon(t) \left[1 - \mathbf{x}^T(t)\mathbf{P}(t)\mathbf{x}^T(t)\right]$$

Problem 3.5

For the system model

$$y(t) = \beta u(t) + \hat{e}(t)$$

and the data in Table 3.1, calculate the first few iterations of the recursive estimator

$$\hat{\beta}(t) = \hat{\beta}(t-1) + \gamma(t)\mathbf{x}(t)\epsilon(t)$$

where

$$\gamma(t) = 1/t$$

Compare convergence rates with those in Problem 3.3.

Problem 3.6

For a system defined by

$$y(t) = \mathbf{x}^T(t)\boldsymbol{\theta} + e(t)$$

where $\{e(t)\}$ is a zero mean white noise sequence and $x(t)$ is deterministic, show that the least squares estimator of θ is unbiased and that the estimate has covariance

$$\text{cov}(\hat{\theta}) = \sigma_e^2 [X^T X]^{-1}$$

where $\text{var}\{e(t)\} = \sigma_e^2$.

Problem 3.7

Assuming the correct model form

$$y(t) = \beta u(t) + \hat{e}(t)$$

where $u(t)$ is a deterministic function of time, show that the LS estimator $\hat{\beta}(N)$ is unbiased with variance

$$\text{var}\hat{\beta}(N) = \left(\sum_{t=1}^{N} u(t)^2 \right)^{-1}$$

What happens if $u(t)$ converges to zero too rapidly as t increases? Simulate the cases:

(i) $u(t) = t^{-\frac{1}{2}}$

(ii) $u(t) = t^{-1}$

using RLS.

Problem 3.8

In Problem 3.7, estimate β recursively using the constant scalar gain algorithm (3.45), (3.47) where $0 < \gamma < 2$. If $u(t)$ is constant and equal to unity, show that the parameter estimator is asymptotically unbiased and derive the steady state variance of the estimation error.

Problem 3.9

The system

$$y(t) = ay(t-1) + d + e(t) \qquad |a| < 1$$

where $e(t)$ is white noise with mean d and variance σ_e^2, is modelled by

$$y(t) = \alpha y(t-1) + \hat{e}(t)$$

and α is estimated using least squares. Show that the nonzero mean disturbance leads to a bias of magnitude

$$d^2 \left(\frac{d^2}{1-a} + \frac{\sigma_e^2}{1+a} \right)^{-1}$$

[*Hint*: Use ergodicity as in Example 3.7.]

Problem 3.10

During each revolution of a single cylinder petrol engine the pressure in the cylinder varies with the crankshaft angle. In addition the pressure will also depend upon the amount of fuel delivered at the tth revolution and the timing of the ignition spark at the tth revolution. Thus, at the tth revolution the pressure is a function of the crankshaft angle α, the point $u_s(t)$ at which the ignition spark is applied and the position $u_f(t)$ of the fuel supply pedal. Suppose that for specific crank angle $\alpha(i)$ the following linear model of the pressure variation applies,

$$y_i(t) = a_i \alpha(i) + b_i u_s(t) + g_i u_f(t) + e_i(t)$$

where a_i, b_i, g_i are unknown parameters and are functions of the crank angle, $\{e_i(t)\}$ is a zero mean white noise sequence and $y_i(t)$ is the pressure at angle $\alpha(i)$ during the tth revolution.

In addition, suppose that the engine is equipped with pressure sensors which measure the pressures at crankshaft angles $\alpha(1)$ and $\alpha(2)$.

You are required to

(a)   formulate a recursive estimator for the unknown parameters associated with the pressure variation at angles $\alpha(1)$ and $\alpha(2)$.

Discuss the identifiability aspects of your solution and

(b)   state how you would modify the estimation scheme to account for dynamics in the fuel supply system; and

(c)   state how you would modify the estimation scheme if the corrupting noise $e_i(t)$ is not white.

3.9   NOTES AND REFERENCES

The recursive least squares algorithm was first proposed by:

Plackett, R.L.
    Some theorems in least squares, *Biometrika*, **37**, 149–57, 1950.

A good readable introduction to recursive least squares is:

Young, P.C.
    Recursive Estimation and Time-Series Analysis, Springer Verlag, 1984.

An excellent introductory work on system identification including a discussion on model order selection is the textbook:

Norton, J.P.
An Introduction to Identification, Academic Press, 1986.

The problem of feedback and identifiability in least squares is discussed further in:

Wellstead, P.E. and Edmunds, J.M.
Least squares identification of closed loop systems, *Int. J. Control*, **21**(4), 689–99, 1975.
Soderstrom, T., Gustavsson, I. and Ljung, L.
Identifiability conditions for linear systems operating in closed loop, *Int. J. Control*, **21**(2), 243–55, 1975.

A unified discussion of RML and RELS type algorithms is in:

Solo, V.
Some aspects of recursive parameter estimation, *Int. J. Control*, **32**, 395–410, 1980.

An interesting unified approach to the construction of recursive estimation algorithms is discussed in:

Graupe, D. and Fogel, E.
A unified sequential identification structure based on convergence considerations, *Automatica*, **12**, 53–9, 1976.

APPENDIX 3.1

The following PASCAL procedure is for the matrix inversion form of RLS. It executes the algorithmic steps given in the text as the 'Matrix Inversion Lemma RLS Algorithm' (Section 3.3).

The object VECT1TONPAR is defined as:

REAL ARRAY [1 .. NPAR].

The object VECTCOVPAR is defined as:

REAL ARRAY [1 .. NCOV].

NCOV is the number of parameters in the upper triangular covariance matrix {e.g. NCOV=$\frac{1}{2}$(NPAR*(NPAR+1))}. NPAR is the number of parameters to be estimated. NC is the number of noise parameters. If NC > 0 then recursive maximum likelihood is used; if NC=0 then normal recursive least squares. COV is the upper triangular covariance matrix stored as a long vector. TH is the vector of unknown parameters. DAT is the vector of observed data. R is the vector of past residuals. YT is the new output observation. The R vector is defined by VECONCMAX, which in turn is defined as:

REAL ARRAY [0 .. NC].

FF is the forgetting factor {for use of this object see Chapter 4; for basic recursive least squares FF=1}.

```
PROCEDURE RECURSIVELEASTSQUARES(VAR DAT,TH : VECT1TONPAR;
                                 VAR COV : VECTCOVPAR;
                                 VAR R : VECONCMAX;
                                 NC,NPAR : INTEGER;
                                 YT,FF : REAL);

VAR
  I,J,K : INTEGER;
  CP: REAL
  COVDAT,GAIN : VECT1TONPAR;

BEGIN
  (* form the current prediction error r{0} *)
  R[0]:=YT;
  FOR I:=1 TO NPAR DO
    R[0]:=R[0] − DAT[I]*TH[I]

  (* update prediction error vector if rml used *)
  IF NC >0 THEN
    FOR I:=NC DOWNTO 1 DO
      R[I]:=R[I−1];

  (* form current covariance data vector covdat{i} *)
  CP:=1

  FOR I:=1 TO NPAR DO
  BEGIN
    COVDAT[I]: = 0;
    FOR K:=1 TO NPAR DO
    BEGIN
      IF K<I THEN J:=(K−1)*(NPAR−1)−(K−1)*(K−2) DIV 2+I
             ELSE J:=(I−1)*(NPAR−1)−(I−1)*(I−2) DIV 2+K
      COVDAT[I]:=COVDAT[I]+COV[J]*DAT[K]
    END;
    CP:=CP+COVDAT[I]*DAT[I]
  END;
  (* form kalman gain vector gain{i}, and update parameter vector th{i} *)
  FOR I:=1 TO NPAR DO
  BEGIN
    GAIN[I]:=COVDAT[I]/CP;
    TH[I]:=TH[I]+GAIN[I]*R[0]
  END;
```

```
(* update covariance matrix stored in vector cov{i} *)
K:=0;
FOR I:=1 TO NPAR DO
    FOR J:=1 TO NPAR DO
        IF NOT(J<I) THEN
        BEGIN
            K:=K+1;
            COV[K]:=COV[K]-GAIN[I]*COVDAT[J]
        END;

(* apply forgetting factor ff to diagonal of covariance matrix *)
IF FF<1 THEN
BEGIN
    K:=1;
    FOR I:=1 TO NPAR DO
    BEGIN
        COV[K]:=COV[K]/FF;
        FOR J:=I TO NPAR DO K:=K+1
    END
END
END; (* of recursiveleastsquares *)
```

4 Using Recursive Estimators

4.1 OUTLINE AND LEARNING OBJECTIVES

The material in Chapter 3 described the recursive form of least squares estimation and its algorithmic basis. The aim of this chapter is to describe how these recursive estimation algorithms are applied in practical situations and the main learning objective is to understand these techniques. The discussion will focus upon the estimation of the parameters for dynamical systems which can be described adequately by the linear discrete time models developed in Chapter 2.

The decision process to be followed in setting up and operating a self-tuning estimator falls roughly into four parts A–D. These are informally summarized in Table 4.1. The ranking of these issues follows the chronological order in which they present themselves to the self-tuning analyst.

Table 4.1 Operating a self-tuning estimator

A:	Initializing the estimator	Section 4.2
B:	Specification of the model	Section 4.3
C:	Operational conditions	Section 4.4
D:	Manipulation of the covariance matrix	Section 4.5

The material presented here refers to the use of recursive estimators as isolated algorithms. When they are used as part of self-tuning control or signal processing algorithms, then the methods and ideas described here will still apply. Some additional 'know-how' will be required, however, depending upon the form of self-tuning algorithm being considered. This extra know-how will be dealt with as the self-tuning methods are described.

4.2 INITIALIZING THE ESTIMATOR

In this section the problem of initializing the recursive estimator is considered. The issues which arise here are the choices of values for the data vector, parameter vector and covariance matrix at the first time step.

4.2.1 Initial values for the data vector $x(t)$

The usual way to fill the data vector with initial values is to commence sampling the information for a few time steps *before* the recursive estimator is started. The sampling process is continued for as many steps as are required to fill the data vector.

In general, for a model of the form

$$Ay(t) = Bu(t-1) + \mathcal{D}(t) + Dv(t) + Ce(t) \tag{4.1}$$

the sampling process must run for n steps, where

$$n = \max \{n_a, n_b+1, n_d, n_d, n_c\} \tag{4.2}$$

The recursive estimator is then started with $x(0)$ filled with valid system data. Figure 4.1 illustrates this procedure.

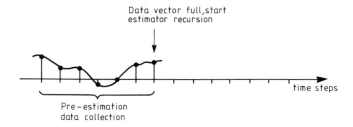

Figure 4.1 Illustrating presampling to load the data vector $x(t)$ with initial data. The number of pre-estimation samples as $\max(n_a, n_b+1, n_d, n_d, n_c)$.

Example 4.1   Loading the data vector

Suppose the system model is given by

$$y(t) = x^T(t)\theta$$

$$\theta^T = [a_1, b_0, b_1, b_2]$$

$$x^T(t) = [y(t-1), u(t-1), u(t-2), u(t-3)]$$

Filling the data vector requires three time steps in order to collect sufficient past values of $u(t)$. At the fourth sample interval, the recursive estimator may be started.

□

In order to familiarize oneself with recursive estimation procedures, it is usual to use a digital simulation in order to generate data for the recursive estimator. In this case it is important to start the digital simulation well before the estimator data collection begins. This ensures that the initial recursions of the estimator are not influenced significantly by the initial conditions of the simulation.

4.2.2   Initial parameter values

Recursive algorithms require an initial estimate $\hat{\mathbf{\theta}}(0)$ of the parameter vector as a starting point for the recursion. A number of possibilities exist here:

(i) Sometimes prior knowledge will exist which enables good initial values to be selected for the estimated coefficients of A, B, C, etc. For example, if a physical model of the system can be constructed, then we can determine a set of discrete time transfer function coefficients using z-transform tables.

(ii) A technique which is useful when no prior knowledge exists is to assume that the system is a single integrator with unit gain. Thus from Table 2.2 the initial parameter estimates would be selected as:

$$a_1 = -1; \qquad b_0 = \tau_\mathrm{s};$$
$$a_i = 0 \ (i \neq 1); \qquad b_i = 0 \ (i \neq 0)$$

The initial values of other parameter estimates (i.e. the noise and disturbance coefficients) are usually set at zero. Once one recursive estimation run of a particular system has been completed, it is standard practice to use the final estimates to initialize subsequent runs. In general, however, the choice of initial parameter estimates is not crucial to convergence behaviour.

4.2.3   Initial values of the covariance matrix

In almost all practical situations in self-tuning we will use a recursive least squares (RLS) or recursive maximum likelihood (RML) algorithm. Therefore, an important object which will require initial setting is the covariance matrix $\mathbf{P}(t)$. The main point here is to note that the size of $\mathbf{P}(t)$ reflects our uncertainty concerning the unknown parameters. If we have no prior knowledge of the system/signal parameters in the model, then a large initial covariance would reflect this. Conversely, if the initial parameters $\hat{\mathbf{\theta}}(0)$ are known to be close to the true values, then a small covariance matrix should be used. A useful way to visualize how the initial covariance matrix $\mathbf{P}(0)$ influences subsequent covariance values is to recall that the covariance matrix at time t can be written as

$$\mathbf{P}(t) = \left[\mathbf{P}^{-1}(0) + \sum_{i=1}^{t} \mathbf{x}(i)\mathbf{x}^{\mathsf{T}}(i) \right]^{-1} \tag{4.3}$$

If $\mathbf{P}(0)$ is chosen large, therefore, its influence upon $\mathbf{P}(t)$ is much less than the information in the data $\{\mathbf{x}(i) \ (i=1, \ldots, t)\}$. Conversely, if $\mathbf{P}(0)$ is small, its influence upon $\mathbf{P}(t)$ is correspondingly greater.

A standard choice for $\mathbf{P}(0)$ is the unit matrix scaled by a positive scalar r, i.e.

$$\mathbf{P}(0) = r\mathbf{I}_{\mathrm{m}} \tag{4.4}$$

Typically, for large $\mathbf{P}(0)$, set r in the region 100 to 1000. For small $\mathbf{P}(0)$, set r in the region 1 to 10.

Example 4.2   Influence of various choices of $\mathbf{P}(0)$

Consider the two-parameter simulated autoregressive (AR) system defined by

$$y(t) = -a_1 y(t-1) - a_2 y(t-2) + e(t)$$

where $a_1 = -1$, $a_2 = 0.3$ and   the   initial   parameter   estimates   are $a_1(0) = a_2(0) = 0$. The input process $u(t)$ is a zero mean, unit amplitude square wave with period 40 and the disturbance $e(t)$ is zero mean and normally distributed with variance 0.1.

Figure 4.2 shows the evolution of the parameter estimates for this process for various initial covariance matrix settings. Note that a large setting for $\mathbf{P}(0)$ (Figure 4.2(a)) causes rapid and large excursions in the parameter estimates, but that (as indicated by equation (4.3)) the influence of $\mathbf{P}(0)$ upon the parameter estimates soon reduces. Compare this with Figure 4.2(b) in which a small initial $\mathbf{P}(0)$ is used. Note that the parameter estimates are correspondingly docile in the first few recursions and change slowly as the influence of $\mathbf{P}(0)$ is gradually worked out of the recursion.

□

In some situations the parameters associated with the A, B, C polynomials have different initial uncertainties associated with them. In such circumstances $\mathbf{P}(0)$ is chosen diagonal but with each diagonal entry reflecting the uncertainty

Figure 4.2
Illustrating the
influence of
initial covariance
matrix sizes for
the system of
Example 4.2: (a)
$\mathbf{P}(0) = 1000\mathbf{I}_2$;
(b) $\mathbf{P}(0) = 0.1\mathbf{I}_2$.

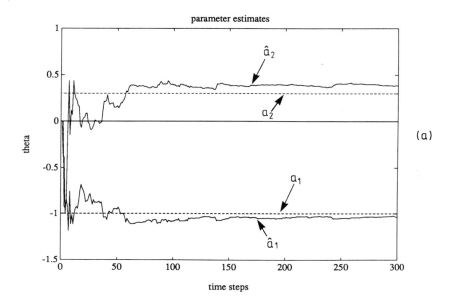

(a)

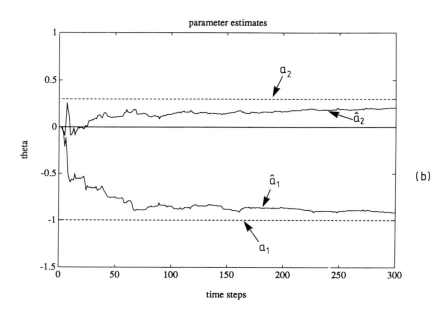

(b)

associated with the corresponding parameter. In this context it is not uncommon to give different initial weights to the coefficient estimates of the *A* polynomial, *B* polynomial and so on.

Example 4.3

Consider the system

$$y(t) = \mathbf{x}^{\mathrm{T}}(t)\boldsymbol{\theta}$$

where

$$\boldsymbol{\theta}^{\mathrm{T}} = (a_1, a_2, b_0, b_1)$$

If the coefficients b_0, b_1 of the *B* polynomial are poorly known (reflecting poor knowledge of the system gain and system zeroes), while prior transient experiments establish the approximate values of the system poles (and hence a_1, a_2), then a typical initial $\mathbf{P}(0)$ in this case would be

$$\mathbf{P}(0) = \mathrm{diag}\ \{r_a, r_a, r_b, r_b\}$$

where r_a is small (around unity, for example) and r_b is large (around 100).

☐

In general it is difficult to be more specific than the terms 'large' and 'small'. Much depends upon experiment, experience and the particular application.

In a practical self-tuning experiment an initial $\mathbf{P}(0)$ should be chosen which is sufficiently large to allow parameter estimates to tune, but not so large that high amplitude parameter variations occur, possibly leading to correspondingly high control signal fluctuations. If the initial parameter estimates are poor but a small $\mathbf{P}(0)$ is chosen to limit parameter fluctuations, then a 'start-up' forgetting factor may be employed to facilitate tuning (see Section 4.5.9).

4.2.4 Initial performance of a recursive estimator

The initial performance of a recursive estimator is also dependent upon the type of system or signal under analysis and the form of excitation which is applied. These issues will be dealt with in Section 4.5 at a general level. However, it is informative to explain some of these issues as they influence the initialization of an estimator.

Although the *general* estimation model is given by

$$Ay(t) = Bu(t-1) + Dv(t) + \mathscr{D}(t) + Ce(t) \tag{4.1bis}$$

the initial behaviour of the estimates associated with this model will vary according to the *specific* form that is used.

Influence of random noise

If the term $Ce(t)$ in equation (4.1) is absent or very small, then the model is deterministic. As a result the corresponding recursive estimator will converge very quickly to the correct parameter values from almost *any* initial settings of $\mathbf{x}(t)$, $\boldsymbol{\theta}(t)$ and $\mathbf{P}(t)$. The convergence time (in multiples of τ_s) should just exceed the number of unknown parameters. This is because, in the absence of noise, the RLS algorithm is essentially solving a set of linear deterministic equations for the unknown parameters.

Example 4.4

Consider the deterministic system

$$(1 + a_1 z^{-1})y(t) = (b_0 + b_1 z^{-1})u(t-1)$$

where $a_1 = -0.9$, $b_0 = 0.5$, $b_2 = 0.8$ and $\mathbf{P}(0) = 200\mathbf{I}_3$ $\boldsymbol{\theta}(0) = 0$. The input $u(t)$ is a square wave of period 20 sample intervals. The parameter evolution for a RLS algorithm is shown in Figure 4.3. Note the rapid convergence of the RLS estimates $\hat{a}_1$, $\hat{b}_0$, $\hat{b}_1$ to their correct values within four time steps. As noted

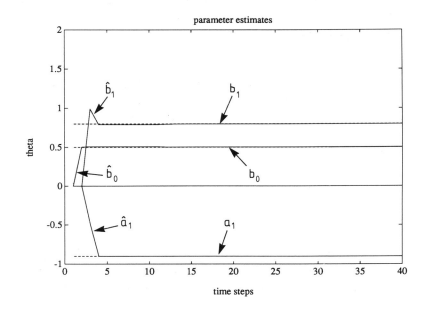

Figure 4.3 Illustrating parameter convergence for the deterministic system of Example 4.4.

previously, the convergence time is determined by the number of parameters (three in this case).

$\square$

Presence of a noise model

If the term $Ce(t)$ in equation (4.1) is present, then the convergence of the parameters will be slower than in Figure 4.3. If $C = 1$, then RLS may still be applied.

Example 4.5

Consider the system

$$(1 + a_1z^{-1})y(t) = (b_0 + b_1z^{-1})\, u(t-1) + e(t)$$

The parameter values and initial conditions are the same as those used in Example 4.4. The input $u(t)$ is a unit amplitude square wave of period 40 sample intervals and $e(t)$ is zero mean white noise with variance σ_c^2.

Figure 4.4 shows the evolution of the parameter estimates for an RLS estimator with two values of noise variance σ_c^2. Note the slower response compared with the deterministic case. In general, as the variance of the noise increases, the convergence rate steadily worsens in comparison to that for the deterministic case. Note also the characteristic 'kick' in the parameter estimates each time a transition in the square wave $u(t)$ occurs. Essentially, the transitions introduce new information into the algorithm. The excitation properties of the square wave are discussed in Section 4.5.3.

$\square$

In the coloured noise case (i.e. $C \neq 1$), the RELS or AML algorithm is appropriate. Note that the convergence rate of the C coefficients for such estimators will be noticeably slower than for the other parameters.

Example 4.6

Consider the system

$$(1 + a_1z^{-1})y(t) = (1 + c_1z^{-1} + c_2z^{-2})e(t)$$

The parameter values used are $a_1 = 0.9$, $c_1 = 0.5$, $c_2 = 0.8$. The initial covariance matrix is $\mathbf{P}(0) = 200\,\mathbf{I}_3$ and the initial parameter estimates are $\hat{\boldsymbol{\theta}}(0) = 0$. The driving noise $e(t)$ is a zero mean white noise signal with variance $\sigma_c^2 = 0.1$.

Figure 4.4
Illustrating the
comparative
convergence
rates of an RLS
algorithm
applied to the
system of
Example 4.5. (a)
$\sigma_e^2 = 0.1$ and (b)
$\sigma_e^2 = 1$.

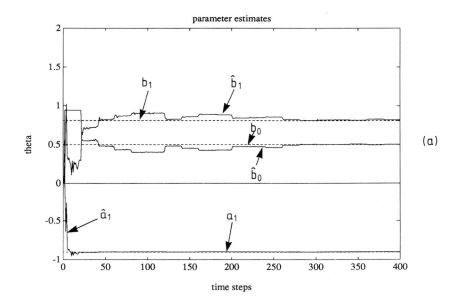

(a)

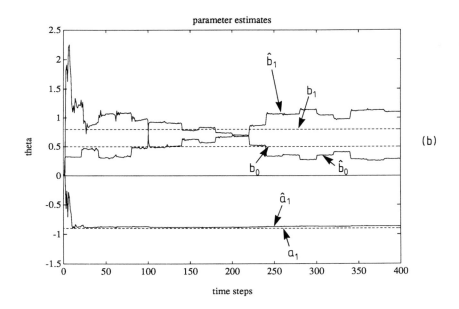

(b)

Figure 4.5
Illustrating the
comparative
convergence
rates for the
RELS algorithm
applied to the
system of
Example 4.6.

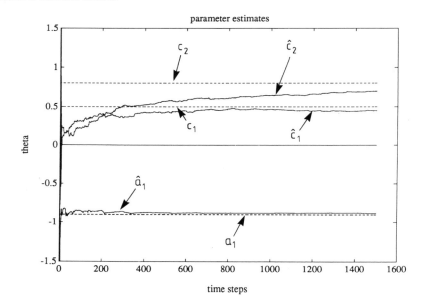

Figure 4.5 shows the evolution of the parameter estimates when the RELS algorithm is applied to the system. Note the slow convergence of the c_1, c_2 estimates when compared with RLS-based estimates.

□

4.3   SPECIFICATION OF THE ESTIMATOR

In this section it is assumed that the reader now has a basic familiarity with recursive estimation algorithms. As noted previously, this preliminary experience will have been obtained using digitally simulated difference equation models. We can therefore move on to consider the issues which arise when the estimator is coupled to the real world. Specifically, the questions addressed concern the estimator model selection task under the ideal assumption that the real world is linear. (Section 4.4 considers practical issues when the linear assumption is not unconditionally valid.)

4.3.1   Sample interval selection

There are no hard and fast rules for selecting the sample interval in recursive estimation for self-tuning. There are, however, some useful guidelines which relate the sample interval to the response rate of the system to be identified. For example, Figure 4.6 illustrates rules of thumb for selecting the sample

Figure 4.6
Sample interval
selection:
illustrating
reasonable
sample intervals
in (a) the time
domain and (b)
the frequency
domain.

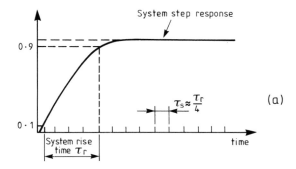

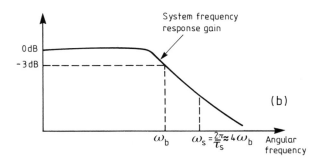

interval τ_s. The assumption is that we have an approximate idea of the system step response. It is then reasonable to use a discrete time model which will (in the time domain) give at least four points on the initial rise of the system transient response (Figure 4.6(a)). Viewed in the frequency domain, if the system bandwidth is known to be ω_b, then a reasonable way to capture the information in this frequency band is to use a sample frequency ω_s which is four times greater (Figure 4.6(b)). It is stressed that the above rules are an initial guide. During recursive estimation and self-tuning the sample interval can be changed on-line to achieve the best subjective results. In addition, other self-tuning systems engineers will have different sample interval selection rules from the ones given here. This diversity of viewpoints underlines the subjective nature of sample rate selection. The rules given here are meant only to get you into the right range of values.

Certain symptoms will appear in the estimated model which indicate that the wrong sample interval has been selected:

(i) If the discrete time model obtained from recursive estimation has all its poles clustered tightly around the $z = 1$ point in the z-plane, then you are sampling too rapidly (Figure 4.7(a)).

Figure 4.7
Showing typical
pole/zero
patterns for (a)
fast, (b) slow
and (c)
acceptable
sampling rates.

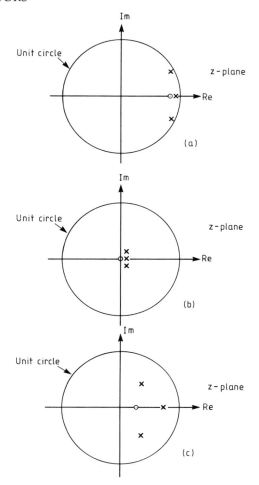

(ii) If the estimated model poles are clustered tightly around the origin of
the z-plane, then you are sampling too slowly (Figure 4.7(b)).

The ideal aim is for a set of estimated model parameters which correspond
to a reasonable spread of pole positions in the z-plane (e.g. Figure 4.7(c)).
   The above remarks are generally true for recursive estimation in self-tuning
systems. Some additional considerations apply, however, in certain self-tuning
applications:

Bandwidth limitations

In self-tuning control, the objective is often to change the response rate or
bandwidth of the overall system by a feedback controller based on a self-

tuning design. In this case the sample interval selection process must take account of the *desired* closed-loop control system response rate as well as the actual (open loop) system response rate. In particular, if the closed loop response is required to be faster than the open loop response, then the sample interval should be decreased accordingly. The point to remember here is that the sample interval of a digital control loop sets a limit on the attainable speed of response. This is illustrated in Figure 4.8 where it is possible to see that the zero order hold (ZOH) introduces a low pass filtering action into the control loop. The first zero of this filter is at $1/\tau_s$, and hence it can be seen that the sample interval τ_s strongly influences the attainable closed loop bandwidth (see also Section 4.4.2).

External constraints

Some self-tuning systems are such that the sample interval is fixed or constrained by external considerations. In many self-tuning prediction examples

Figure 4.8
Illustrating the bandwidth limiting properties of the ZOH: (a) as a feedback loop element; (b) its frequency response.

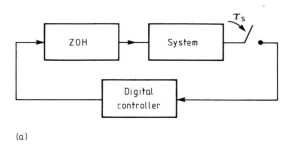

(a)

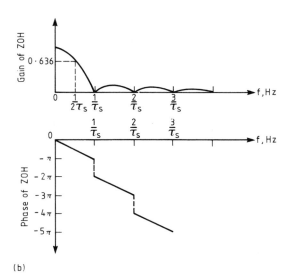

(b)

the sample interval is fixed by the way in which statistical data are collected. For example, Problem 10.7 in Chapter 10 refers to a prediction problem in which the data concern the prices of company shares which are traditionally collected at three-monthly intervals. In the same spirit, certain control applications have the sample rate fixed by external considerations. For example, in the control of engines the sample rate is always geared to the speed of rotation of the engine crank or the firing time of each cylinder. The automotive application of self-tuning extremum control in Chapter 13 illustrates this point. Other applications which involve externally constrained sample rates occur when the self-tuning system must be integrated into an existing control and signal processing system. For example, the control and instrumentation systems for process plant are often operated at sample rates which are fixed by the instrument manufacturers in consultation with process designers. Usually, however, in such systems the existing sample interval will have been wisely chosen by experienced engineers and will be quite suitable for self-tuning.

Certain cases arise in which the required bandwidth of a self-tuning system is such that the system model exhibits the pole/zero clustering effect illustrated in Figure 4.7. In such cases the estimator and associated self-tuning synthesis rules may be poorly conditioned numerically. In such cases the delta model discussed in the final section of Chapter 2 should be considered. In addition, the special synthesis algorithms of Appendix 7.1 should be used.

Despite the constraints and precautions mentioned above, practical experience indicates that self-tuning controllers in particular can give satisfactory performance over a wide range of sample intervals. In some cases this range extends far beyond that conventionally recommended by sampled data theory.

4.3.2 Model type

The discussion in Chapters 2 and 3 referred to the possibility of a model which included the system, drift, measurable disturbances and noise. As a result, the general system model form (equation (4.1)) is rather too general. The first decision, therefore, is to decide which of the components of equation (4.1) are present in a particular practical situation. This can only be done by a consideration of the application in hand. The questions which arise include:

(i) Is the application a control or signal processing one? If it is a control problem, the control input term must be included.

(ii) Is the system output subject to noise? If it is, then it is reasonable to include a representation of the random noise component in the model.

(iii) Is the system output subject to drift? Often the only drift component is associated with a constant term in the process output. This is represented by a drift term $\mathscr{D}(t) = d_0$ in the model.

(iv) Are there components of output disturbance associated with a measurable signal apart from the main input signal $u(t)$? In control system applications (to be considered in Part 2) such a measurable signal would be associated with a signal available for feedforward control. In signal processing applications (to be considered in Part 3) such a measurable signal occurs in interference cancelling filters and some prediction applications.

Assuming a control application, for example, a prototype model with which to begin recursive estimation is

$$Ay(t) = Bu(t-1) + Ce(t) \qquad (4.5)$$

A term d_0 would be added if a non-zero constant component of output $y(t)$ exists when $u(t) = 0$.

4.3.3   Choice of recursive estimator

There are two decisions to be made in selecting a recursive estimator. The first relates to the *algorithm*, the second to the *computational method of implementation*. These two issues are dealt with separately below.

The recursive estimators described in Chapter 3 offer three algorithm types: Recursive Least Squares (RLS), Recursive Extended Least Squares (RELS) or Approximate Maximum Likelihood (AML). If the system has no random noise or the noise is thought to be white then RLS should be used. If the system noise is coloured, then RELS or AML may be used. In practice both algorithms work well, but both are slower to converge than RLS. Note that in some self-tuning algorithms it will be possible to use RLS *even though the noise is coloured*. These special forms are called implicit algorithms and will be given special mention later at appropriate points.

Chapter 5 gives some computational alternatives for the implementation of RLS, RELS and AML. There are thus three computational forms for each algorithm and the prospective user needs to know which one is best. The following guidance is offered in this respect. The Matrix Inversion Lemma (MIL) implementation was used in Chapter 3 to introduce RLS. It is the easiest to implement and is therefore recommended for those readers who are just getting started and need to become familiar with RLS and self-tuning algorithms. In the same spirit, the authors have used the MIL implementation in numerous self-tuning applications with good results. The MIL implementation, however, can run into problems when short word length computations are being used or when a recursive estimator is left to run for very long periods without management. In such circumstances, the U–D factorization implementation (see Chapter 5) with its superior numerical properties is recommended. Best professional practice dictates that any self-tuning system intended for a permanently installed application should use the U–D factorization method.

The 'fast' version of RLS (see Chapter 5) is of use when the number of parameters is very large. Specifically, fast algorithms are very efficient when high order MA or AR models are estimated. This rarely happens in control applications but some signal processing self-tuners often use this form of model. In particular, the interference cancelling filters discussed in Chapter 11 can use high order AR models as an approximate representation of signal sources or systems which are inverse unstable.

4.3.4 Model order

The model order decision concerns the selection of the integers n_a, n_b, n_c, etc. The primary information used in making these decisions should be the physical consideration of the system to be identified. The information in Chapter 2 is helpful in this respect since it tells us that the degree n_a will be the same as that of the underlying continuous time system. Likewise, the degree n_b of B will be less than or equal to n_a.

The degree of C, the noise model, is less easy to determine from physical considerations. The following remarks are appropriate:

(i) If the system is deterministic or there are very low levels of noise, then assume $n_c = 0$.

(ii) If the random noise level on the output is significant then try to determine its source, since this may give an indication of its structure. If this is not possible then a visual inspection of the output data can prove helpful as can inspection of the output autocorrelation function.

Figure 4.9 shows some 'typical' noise processes which may prove helpful in a visual inspection. If the noise appears coloured (Figure 4.9(b)), then begin by choosing $n_c = 1$. It may be necessary to use $n_c = 2$ but this is the maximum value which you should need in practice. In all the authors' experience the degree of the C polynomial has only rarely been greater than two. If the noise process tends to drift around (Figure 4.9(c)), then you should check for a drift component which can be correctly modelled by a $\mathcal{D}(t)$ polynomial. If this does not provide a reasonable solution given the practical setting, then try an integrated noise model (see Section 2.4). In general, the choice of noise model order is not crucial and a value of $n_c = 1$ or 2 is usually sufficient.

In many self-tuning situations it is possible to choose model orders which are significantly less than the technically correct values. This is because it is only necessary to capture the dominant dynamical features of a system or

Figure 4.9
Typical noise
signals
encountered in
recursive
estimation. (a)
Probably white
noise; (b)
probably
coloured noise;
and (c) probably
an integrated
noise process.

(a)

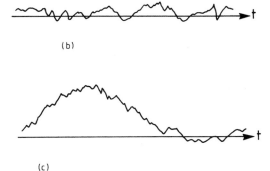

(b)

(c)

signal generating mechanism in any control or signal processing application. For this reason it is good practice to consider whether *all* the dynamics of a system are significant for achieving a *specific* experimental goal. Simplified models of lower order may be adequate.

4.3.5   System delay

The delay through a system in discrete time can be calculated from a knowledge of the continuous time transport delay and the sample interval (cf. the discussion in Section 2.2.1). If the transport delay is τ_d then the corresponding discrete time delay k is given by

$$k = \text{int}\left\{\frac{\tau_d}{\tau_s}\right\} + 1 \tag{4.6}$$

It is important to note that some forms of self-tuning algorithms do not require knowledge of the time delay while others need accurate knowledge of k in order to achieve their design objective.

It is also necessary to recall that, if a partial time delay is present (i.e. a delay which is not an integer multiple of the sample interval), then an extra zero in the B polynomial will be present (see Chapter 2). In a self-tuning application it is also important to account properly for computation delays in the self-tuning algorithms when calculating the effective time delay through a system.

4.4 OPERATIONAL CONDITIONS

In this section we extend beyond the assumptions of the previous discussion and suppose that the system and signal with which we are dealing are real processes which usually implies the presence of many practical complexities. Mainly this involves nonlinear behaviour and inadequate excitation signals. The material which follows shows how these phenomena are dealt with within the assumption of a linear system model which is required for our recursive estimation procedure.

4.4.1 System linearity and signal limiting

The main assumption in all the discussions throughout this text is that the signal sources and systems considered are linear dynamical processes. In reality, this is only *conditionally* true. The best we can hope for is that the systems and signals are approximately linear in the operating range which we consider important in our self-tuning experiments. This reality has two major implications in setting up and running a recursive estimator for self-tuning.

The first concerns the permitted levels of test/dither signals and has been hinted at above. The second concerns the operating point of the system and signal sources. We deal with these issues in turn.

Signal levels

Many systems and signal sources have nonlinear characteristics associated with actuation and sensing (Figure 4.10).

Figure 4.11 shows a possible input nonlinearity which illustrates some typical problems. Specifically, if the test signal $u(t)$ contains components which exceed the saturation level of $\pm U_s$, then the signal which is transmitted to the system dynamics will be subject to a nonlinear distortion known as *saturation*. This is dealt with in practice by obtaining an estimate of U_s and limiting the amplitude of the test signal. Test signals composed from the square wave, PRBS and uniformly distributed noise described in Section 4.5.3 allow this to be done easily.

Signal limiting must be applied if another form of $u(t)$ is used or occurs naturally. If a feedback controller is present, for example, the $u(t)$ signal may

Figure 4.10
Illustrating the
possible
presence of
input (actuation)
nonlinearity and
output (sensor)
nonlinearity.

Figure 4.10
Illustrating the
possible
presence of
input (actuation)
nonlinearity and
output (sensor)
nonlinearity.

Figure 4.11
Showing a
typical actuator
or input
nonlinearity
with saturation
limits $\pm U_s$ and
dead-zone limits
of $\pm U_d$.

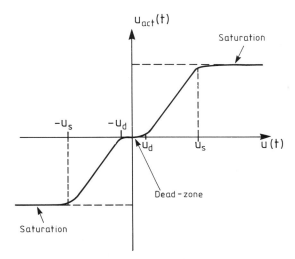

be the output of a digital controller. In such cases, the controller output is limited at U_s within the control computer before being output to the actuator. A good computer control practice, which would certainly be used in the commissioning of a self-tuning controller, would be to count the number of times the control law output had been limited by software. A controller which is frequently producing outputs which would saturate an actuator or system input may be malfunctioning or be inappropriate. In this context, a software count of the average number of saturating control signals is a useful statistic in the assessment of controller performance. Typically, if the control law saturates during steady state then the loop gain is too high and/or the controller bandwidth is too high and is feeding high frequency disturbances back through the control law.

A further typical input nonlinearity is the *dead-zone* $\pm U_d$ shown in Figure 4.11. This represents a signal level below which no actuation signal $u_{act}(t)$ is sent to the system. The signal $u(t)$ must be designed to ensure that it has a minimum amount of its amplitude content in this region. Square waves and PRBS are binary signals and hence avoid this dead-zone problem completely. Uniform and gaussian distributions are subject to nonlinear distortion due to

Figure 4.12
Showing a
dead-zone
compensator
used when
significant input
dead-zones $\pm U_d$
exist and are
known.

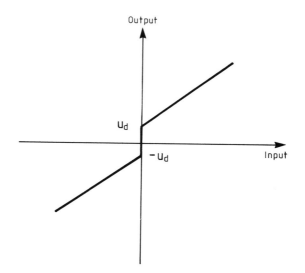

system dead-zones. If $u(t)$ has low amplitude components which are within the dead-zone, then a dead-zone compensator may be placed between the signal $u(t)$ and the system input. The compensator characteristic is shown in Figure 4.12. If U_d is known exactly then the compensator exactly eliminates the dead-zone effect. However, dead-zone compensation should be used with care if the dead-zone is not accurately known. If $u(t)$ is generated by a feedback control law then a dead-zone compensator may be used to eliminate large dead-zones. Be careful, however. In some control schemes a dead-zone is deliberately built into the control law! This is to prevent continuous low amplitude actuation signals being unnecessarily applied to actuators which would otherwise wear out quickly.

Another input nonlinearity concerns the maximum achievable rate of change of the input actuator. This is called the *slew-rate limit*. With small test or dither signals the slew-rate limit will not usually be encountered. However, for large and rapidly changing inputs slew-rate limiting of the actuator is common. This usually occurs in high performance servomechanisms when a large setpoint change occurs. Indeed, most high performance servoamplifiers of electronic and hydraulic type will incorporate slew-rate limits. These limits will be part of the technical specification and therefore known. It is normal in digital controllers to incorporate a software slew-rate limit on setpoint changes to avoid driving actuators too hard (Figure 4.13 illustrates this).

Sensor or output nonlinearities are usually less severe than input nonlinearities. If they are known to exist, then nonlinear compensation can sometimes be applied to remove their effect.

Figure 4.13
Software slew-
rate limiting of
setpoint changes
in a digital
controller. (a)
Setpoint rate
limiter; (b) effect
of rate limiting
on square wave
setpoint.

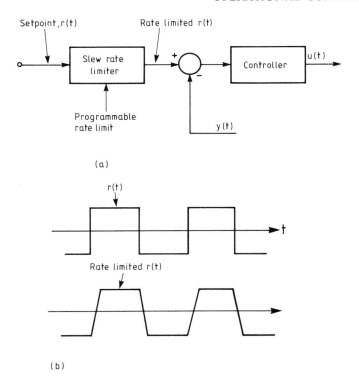

(a)

(b)

Operating levels

A system may be highly nonlinear over its full operating range. In many control situations, however, the system normally functions at a specific operating level. When considered in the vicinity of this level, the standard assumption is that, for small signal excursions, the system is approximately linear. In applying recursive estimation we usually attempt to estimate the coefficients of an approximate linear representation. This approximate linear model will subsequently be used for self-tuning control or signal processing around the normal operating point. This procedure has some implications as follows.

(i) The signal $u(t)$ should not be of such a large amplitude that the system is driven significantly away from the normal operating conditions. The input $u(t)$ or dither amplitude should ideally be selected to be sufficiently large so that corresponding changes of about 10% of the normal operating value of $y(t)$ are visible. If this cannot be achieved without saturating the actuator, then something is wrong. Probably the test or dither signal has a bandwidth which is too large.

(ii) If the operating point is changed during recursive estimation or self-tuning then the equivalent locally linearized model will also change. In an experimental situation where the analyst has complete control, the correct procedure is to maintain a constant operating point throughout an experiment. In a self-tuning application, however, the recursive estimator should be configured to adapt the estimated coefficients during an operating point change. In this scheme of things a change in operating point of a nonlinear system is treated as a change in the parameter values of a linear system. Techniques for allowing a recursive estimator to track changes in parameter values are given in Section 4.5.

4.4.2 Signal filtering, test signal bandwidth and further sample rate considerations

In some recursive estimation and self-tuning applications, the output variable $y(t)$ is subject to significant wide-band noise which is associated with the process of measuring $y(t)$ (Figure 4.14). This form of noise is called *sensor noise* and should be removed or minimized since it has no relationship with the underlying process (unlike the process noise) and can (if left untreated) cause problems in a self-tuning system. If the spectrum of the sensor noise is known, it can be treated using the filtering techniques of Chapter 10. In many situations, however, a simple low pass filter will be sufficient. For recursive estimation and self-tuning signal processing the measured output $y(t)$ should be filtered with a low pass filter which has a bandwidth which is flat over the bandwidth of the system/signal to be estimated and then decays rapidly. This will ensure that the sensor noise components outside the frequency range of interest are removed. The sensor noise filter should be situated *before* the output sampling process to avoid aliasing high frequency noise components. In situations where the test signal contains components outside the bandwidth

Figure 4.14 Illustrating sensor noise.

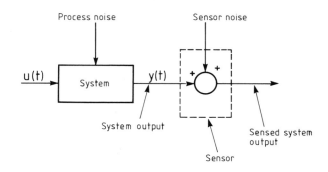

required, then it too may be filtered. The filter, however, should be identical to the filter applied to the output measurement.

The bandwidth of the test signal or dither signal has been discussed previously. It does, however, require further comment and also coordination with the sample rate selection process. In the foregoing paragraphs the sample rate and dither signal bandwidth were linked to the bandwidth of the system to be identified. Certain other considerations apply, however, when the recursive estimator is used in a self-tuning control application where there are *unmodelled system dynamics.*

Consider first the case of self-tuning control where the estimation model is a correct representation of the underlying system (i.e. the model degrees and structure correspond to those of the physical system). In this situation, as noted in Section 4.3.1, the sample rate selection and dither signal bandwidth should be selected not only with regard to the open loop system bandwidth but also the desired closed loop bandwidth. In particular, if the self-tuning controller is intended to increase the closed-loop system bandwidth, then the sample rate should be selected such that it is about four times the desired closed loop bandwidth. If the aim is to maintain or even reduce the closed loop bandwidth then the sample rate selection is determined by the open loop system bandwidth.

Different considerations apply if the recursive estimator is based upon a model which excludes certain dynamical features of the system. This may occur if the system includes dynamics which are of second order interest or cause a problem if they are excited. One example is the position control of an object through a compliant connection. The connection will typically add a high frequency oscillatory mode in the open loop response. In self-tuning control we may choose to ignore this in order to obtain a reduced order model which will consequently require the estimation of fewer parameters. If this procedure is adopted, then care should be taken to use a test/dither signal which does not excite these unmodelled dynamics. Likewise, in increasing the closed loop bandwidth, care should be taken to avoid extending the closed loop bandwidth in a way which will excite these unwanted dynamical modes.

4.4.3 Quantization

A further problem in recursive estimation and self-tuning concerns the quantization process which occurs when sampling a real signal with an analogue/digital (A/D) converter and representing it in the control/estimation computer by an integer. With modern A/D converters, which typically use 12 bit conversion, the quantization effect is quite small. Nonetheless, care must be taken to ensure that the A/D converters are operating over their full dynamic range. Specifically, if the amplitude of variations in $u(t)$ and $y(t)$ are small, then the A/D conversion process is probably only using a few of the least significant bits of the converter. As a result quantization errors will be relatively large.

Quantization errors can be represented for analysis purposes as a further source of sensor noise on the measurement process. In general, however, a proper scaling of measured variables and range selection of A/D converters should be used to avoid quantization problems. Note that the U–D factorization algorithm of Chapter 5 will not solve quantization problems. The U–D factorization is only used to reduce the build-up of round-off error within the recursive estimator.

4.5 COVARIANCE MANAGEMENT

4.5.1 Basic aims of covariance management

If we assume that the chosen recursive estimator is an RLS or RML type algorithm, then the recursion can be cast in the form:

$$\hat{\boldsymbol{\theta}}\,(t+1) = \hat{\boldsymbol{\theta}}(t) + \mathbf{P}(t+1)\mathbf{x}(t+1)\epsilon(t+1) \tag{4.7}$$

$$\mathbf{P}(t+1) = \mathbf{P}(t)\left[\mathbf{I}_m - \frac{\mathbf{x}(t+1)\mathbf{x}^T(t+1)\mathbf{P}(t)}{1+\mathbf{x}^T(t+1)\mathbf{P}(t)\mathbf{x}(t+1)}\right] \tag{4.8}$$

We know from our discussions in Chapter 3 that the matrix $\mathbf{P}(t)$ becomes ill-conditioned if the system modes are not sufficiently excited. This can lead to loss of parameter identifiability and algorithmic instability. Our primary task, therefore, is to make sure that operational conditions are such that the covariance matrix remains in good shape, with eigenvalues that are all nonzero but not too large. We first consider, therefore, some operational issues which influence the evolution and conditioning of the covariance matrix: feedback and test signal structure.

4.5.2 Presence of feedback

It may well be that the system we wish to recursively estimate is part of a feedback system (Figure 4.15). This will certainly be true if the self-tuning system is set up for control purposes. It may also be true in a signal processing context and in the self-tuning systems described in the 'Special Topics' of Part 4. In Chapter 3 some of the conditions were studied under which a feedback system can be identified. As demonstrated there the basic problem is that the feedback signal can cause ambiguity in the relationship between the input

Figure 4.15
Operational
condition
involving
feedback.

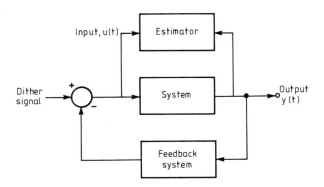

and output signals. The simplest and most effective way round this problem is to add an independent signal into the loop as indicated in Figure 4.15. This extra *dither* signal should satisfy the requirement for persistent excitation (see Section 4.5.3) and should not have large amplitude components which drive the system into nonlinear operation (see Section 4.4.1).

Example 4.7

Consider the system

$$(1 + a_1 z^{-1})y(t) = (z^{-1}b_0 + z^{-2}b_1)u(t) + e(t)$$

which is subject to the proportional controller

$$u(t) = g(r(t) - y(t))$$

where $r(t)$ is a reference or setpoint signal which is normally zero. Figure 4.16 shows the system parameter estimates for the case (a) $r(t) = 0$ and (b) $r(t) = $ square wave dither signal of period 40 time steps and unit amplitude. Note that the parameters are unidentifiable when $r(t) = 0$ and that the dither signal restores identifiability. The system parameters in this example are $a_1 = -0.7$, $b_0 = 0.3$, $b_1 = 0.7$ and the initial conditions are $\mathbf{P}(0) = 200\mathbf{I}_3$ and $\hat{\boldsymbol{\theta}}(0) = 0$. The feedback coefficient g is 0.3.

□

This discussion concerned the case in which feedback caused incorrect parameter estimation. In the next section we consider the open loop case and examine suitable choices of $u(t)$.

Figure 4.16
Illustrating the
use of a dither
signal added to
the reference
signal in order
to ensure
identifiability of
system
parameters. (a)
$u(t) = -0.3y(t)$;
(b) where $r(t)$
is a square wave
dither signal of
unit amplitude
and period 40
time steps.

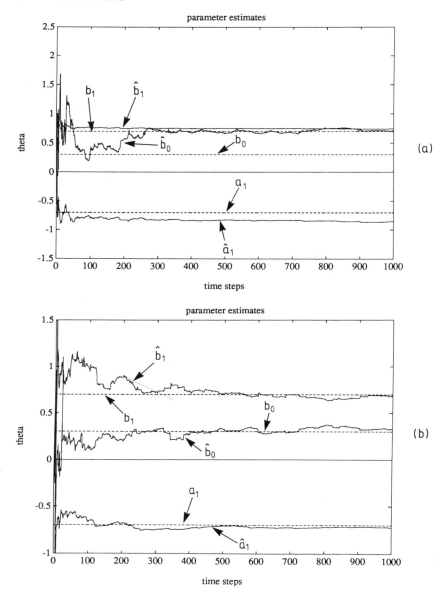

4.5.3   Test signals and excitation levels

In recursive estimation there is often an assumption that the input signal $u(t)$ is 'sufficiently exciting', so that a unique set of parameter estimates will result from the estimation process. Moreover, if the choice of model structure is correct, these estimates will then correspond to the physically correct parameters for the system.

The above requirement on $u(t)$ is met if it is *persistently exciting*. In broad terms, a signal is persistently exciting if it excites all the modes of the system. This is in principle a simple notion, but the concept becomes subject to different detailed definitions and interpretations in a self-tuning context (see Section 4.8 Notes and references). In practical terms the following forms of signal are suitable for recursive estimation. If the system is under closed loop control the signals noted in Table 4.2 are suitable for use as the dither signal added to the system in the manner discussed in the last section.

Table 4.2 Commonly used input or dither signals for recursive estimation

Signal	Comments
Square wave	See Remark 1
Pseudo-random binary noise	See Remark 2
Uniformly distributed noise	See Remark 3
Gaussian distributed noise	See Remark 4

Remark 1

Square wave signals are easy to generate and have a strictly limited amplitude range. This is useful since it is generally necessary to limit the amplitude excursions of a test signal to avoid taking a system outside its linear operating range (see Section 4.4.1). The frequency of the square wave needs to be selected such that the system dynamics are adequately excited. A reasonable rule of thumb is that the square wave period T_{sw} should be approximately six times the dominant time constant of the system. Alternatively, the square wave frequency should be approximately 0.16 of the system bandwidth. These are rough guides aimed at ensuring that most of the square wave power (associated with the first three harmonic components) is inside the system bandwidth. Figure 4.17 gives a visual guide to the acceptability of a square period based upon the appearance of the system output signal.

Remark 2

Pseudo-Random Binary Sequences (PRBS) are popular signals for the identification of system dynamics. Consequently, programmes to implement them are commonly available in books and software libraries (see Section 4.8 Notes and references).

PRBS share an advantage of square waves in that they have only two amplitude levels. It is thus relatively easy to select a PRBS amplitude which

Figure 4.17
Illustrating the
selection of a
square wave test
signal period by
inspection of the
corresponding
system output
shape. Square
wave period: (a)
acceptable; (b)
too large; and (c)
too small.

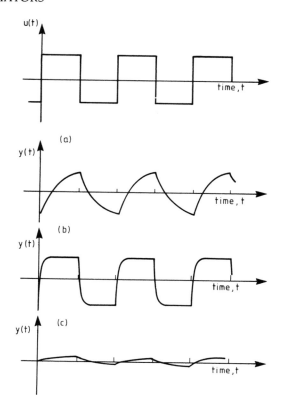

does not exceed the bands of linearity of the system under test. The advantage of a PRBS over a square wave is that it has a richer spectrum. The disadvantage is that PRBS are more difficult to implement.

The basic decisions in PRBS test signal design concern the so called clock frequency and the sequence length. Detailed information on these choices is given in the references (see Section 4.8) but a reasonable rule of thumb is to select the clock frequency to match the bandwidth of the system to be estimated. The question of test signal bandwidth is further discussed in Section 4.4.2. The sequence length is not of great practical importance in recursive estimation applications. An indication of the relevance of this parameter, however, may be gained by noting that the sequence length gives the number of harmonic frequencies in the PRBS spectrum range from zero to the clock frequency. A choice of sequence length of 63 or greater is reasonable.

Remark 3

Uniformly distributed noise is widely used because most software libraries have a random number programme which generates uniformly distributed real numbers. These programmes can be easily modified to generate uniformly distributed noise with any required maximum amplitude range. The limited maximum amplitude is advantageous for the same reasons as given above for PRBS signals. However, the amplitude range of the signal is uniformly distributed and the presence of very low amplitudes can give distortion if the system has dead-zone nonlinearities (see Section 4.4.1 and the remarks on quantization in Section 4.4.3). Most random number programmes give test signals which are approximately white noise sequences. In practice, these sequences should be digitally filtered to produce a test signal which has the desired bandwidth.

Remark 4

Gaussian distributed noise is used by many as a test signal or dither signal. It is to be avoided in general, however, because the gaussian distribution involves high amplitude outliers which can excite any nonlinear modes of a system. Also the concentration of signal energy at low amplitudes will cause nonlinear distortion if dead-zones are present in the system. Despite its popularity, gaussian distributed noise is not a recommended dither or test signal in practical situations. Gaussian noise computer programmes usually produce white noise sequences. If it is to be used, the sequence should be filtered to produce the test signal bandwidth required in the recursive estimation.

To summarize, the test or dither signal should have an amplitude level and distribution which does not violate the linear operating range of the system or drive the system into undesirable operating regions. When setting up a test signal for recursive estimation, it is most important that the signal $u(t)$ is constant between sample intervals. In practice this means transitions in the test or dither signal should be synchronized with the sampling device. This will occur naturally if the test/dither signal is generated in the computer used for recursive estimation.

If feedback is present and the feedback is not a digital controller, then $u(t)$ will not be constant between sample intervals. In this case a ZOH may be introduced as shown in Figure 4.18.

4.5.4 Tracking parameter changes

The recursive estimation procedures discussed so far have been aimed at the determination of a *constant* parameter vector. In many cases, however, the estimator will be required to track changes in a set of system/signal parameters. These changes may be due to nonlinear phenomena which influence local linear models after an operating condition change (see the discussion in Section 4.4). Alternatively, the parameter changes in a system or signal may

Figure 4.18
Showing the use
of a ZOH in a
feedback system
to ensure that
the input $u(t)$ is
constant
between
sampling
intervals.

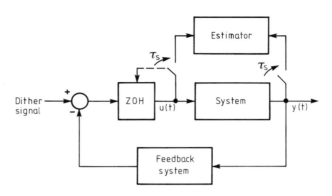

be due to variations which occur over time due to external factors. These changes may be associated with known phenomena, in which case the time at which parameters change will be known. Often, however, there will be no prior knowledge of when and what changes will occur, so that the recursive estimator will be required to adapt its estimate without this prior information. Indeed, many applications monitor the variations in a set of recursive parameter estimates in order to diagnose the onset of fundamental changes (faults, degradation) in the underlying mechanism which generates the data.

In a self-tuning control or signal processing environment the recursive estimator generates a model which can be used for implementing in real-time the corresponding controller/signal processor. When a system or signal source changes, it is therefore important that the recursive estimator rapidly adapts its estimates in order that the corresponding controller/signal processor adapts its coefficient settings and maintains its design role.

In this connection it is important to note that the essential difference between a self-tuning system and an adaptive system is in the recursive estimator. A *self-tuning* system need not include a provision for parameter variation, whereas an *adaptive* system must make such provisions. How this is done is discussed later.

Note that, provided that there is sufficient excitation, then the recursion (4.8) ensures that the elements of the $\mathbf{P}(t)$ matrix decrease in size as the number of time steps increases. Thus as the estimates become more accurate they require less adjustment. In this context $\mathbf{P}(t)$ can be interpreted as acting as an adjustment gain in equation (4.7). Within a recursion which allows adaption, the manipulation of the size of $\mathbf{P}(t)$ is the basic mechanism for controlling the adaptive capabilities of the estimator. The basic mechanism, therefore, for parameter tracking is to control the size of the adjustment gain $\mathbf{P}(t)$. In the following we consider two approaches for doing this. The first (Section 4.5.5) is the most direct in that it explicitly modulates the covariance matrix. The second (Section 4.5.6) arises from a consideration of how data are weighted in the cost function minimized by RLS.

4.5.5 Random walk

A symmetric positive definite matrix $\mathbf{R}$ can be added to $\mathbf{P}(t)$ to prevent it becoming too small. Usually, $\mathbf{R}$ is added at a particular time step to allow the estimator to respond to a sensed change in the system. For example, if at time step t a change in the system parameters is detected, the covariance matrix update, for that iteration only, becomes

$$\mathbf{P}(t) = \mathbf{P}(t-1)\left[\mathbf{I}_m - \frac{\mathbf{x}(t)\mathbf{x}^{\mathrm{T}}(t)\mathbf{P}(t-1)}{1 + \mathbf{x}^{\mathrm{T}}(t)\mathbf{P}(t-1)\mathbf{x}(t)}\right] + \mathbf{R} \tag{4.9}$$

If the parameters are known to be continually drifting then (4.9) can be used for each time t. In this case, it turns out that (4.9) can lay claim to optimality if the parameter variations are described by the random walk model

$$\boldsymbol{\theta}(t+1) = \boldsymbol{\theta}(t) + \mathbf{w}(t)$$

where $\mathbf{w}(t)$ is zero mean white noise and $\mathbf{R}$ is interpreted as its covariance matrix for each t. The justification for this lies in the realm of optimal state estimation (Kalman filtering) and is not pursued here (see Problem 4.4). The interpretation, however, is useful in that it links the diagonal elements of $\mathbf{R}$ to the variances of individual parameters.

It is usual to make $\mathbf{R}$ a diagonal matrix for simplicity. If only certain parameters change, then $\mathbf{R}$ can be selected to have zeroes on the diagonal at all positions except those which correspond to a changing parameter. For example, if it is known that only the first and third parameters in the parameter vector have changed in a system with four parameters then $\mathbf{R}$ would take the form

$$\mathbf{R} = \operatorname{diag}(r_1, 0, r_3, 0)$$

The values of r_1, r_3 should reflect the anticipated magnitude of the parameter changes. For example, if parameter θ_1 changes a little, $r_1 = 0.1$ might be tried; if parameter θ_3 changes by 100%, $r_3 = 0.5$ might be tried. There is, however, considerable freedom in selecting the sizes of the diagonal elements of $\mathbf{R}$. Just remember that they must be positive because $\mathbf{P}(t)$ is a covariance matrix. Also note that $\mathbf{R}$ will increase the step length of adjustment and a large value will cause a large perturbation in the corresponding parameter estimates.

Example 4.8   Influence of a random walk on parameter estimates

Consider a system defined by

$$y(t) = -a_1 y(t-1) + b_0 u(t-1) + b_1 u(t-2) + e(t)$$

where

$$b_0 = 1, \qquad b_1 = +0.5$$

and

$$a_1 = 0.7 \qquad 0 < t < 200$$
$$= 0.35 \qquad t \geqslant 200$$

Figure 4.19 shows the parameter estimates for this system using RLS. In Figure 4.19(a) no random walk is applied. In Figures 4.19(b) and (c) a random walk is applied at the time step $t = 200$ when parameter a_1 changes. The random walk used is given by

$$\mathbf{R} = r_1 \mathbf{I}_3 \qquad\qquad (4.10)$$

with $r_1 = 10$ for Figure 4.19(b) and $r_1 = 0.1$ for Figure 4.19(c).

Figure 4.19
Illustrating the
influence of a
random walk on
the RLS
algorithm, as
described in
Example 4.8. In
(a) no random
walk is applied;

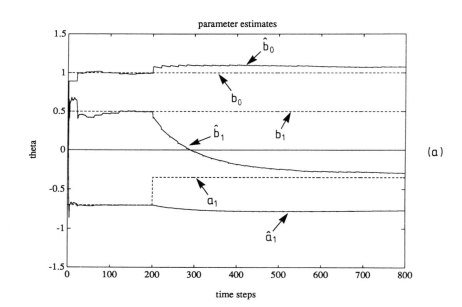

(a)

Figure 4.19
(cont.)
(b) $\mathbf{R} = 10\mathbf{I}_3$; (c)
$\mathbf{R} = 0.1\mathbf{I}_3$.

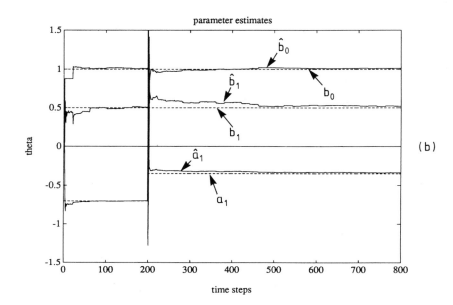

(b)

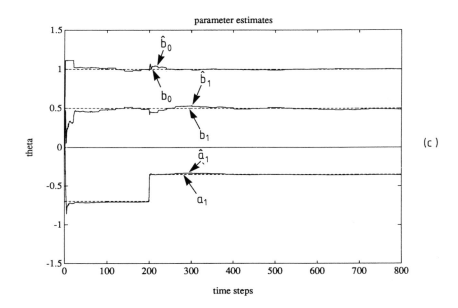

(c)

Note that in this case we have used the prior knowledge that a_1 is the parameter which is subject to change and that the change will occur at the time step $t = 200$.

The general effect of parameter changes on conventional RLS is, as indicated in Figure 4.19(a), to worsen estimation performance. The estimates will recover but only after a long time interval. The addition of a large random walk (Figure 4.19(b)) causes large excursions in the estimates at the transition point. The smaller random walk (Figure 4.19(c)) allows a rapid yet smooth transition to the new a_1 parameter value.

□

4.5.6 Forgetting factors

The second and most popular way of getting a recursive estimator to adapt is the forgetting factor technique. A forgetting factor λ is a number between 0 and 1 which is used to progressively reduce the emphasis placed on past information. The idea of a forgetting factor can best be understood by considering the way in which information is weighted in the least squares cost function. Specifically, the normal least squares approach is to minimize the cost function

$$J_t = \sum_{i=1}^{t} \hat{e}^2(i) \tag{4.11}$$

at each time t. This choice implies that all values of $\hat{e}(i)$ from $i = 1$ to t carry an equal weighting.

The forgetting factor approach applies a differential weighting to the data by use of the modified cost function

$$\bar{J}_t = \sum_{i=1}^{t} \lambda^{t-i} \hat{e}^2(i) \tag{4.12a}$$

The discounting effect of λ upon past errors can be seen by writing (4.12a) in the form

$$\bar{J}_t = \lambda \bar{J}_{t-1} + \hat{e}^2(t) \tag{4.12b}$$

The forgetting mechanism, therefore, uses the influence of λ to progressively reduce the importance given to old data. The way in which this weighting varies with time is illustrated for various values of λ in Figure 4.20. Values commonly used in practice lie between 0.95 and 1. It is clear that we can

Figure 4.20
Variation of data
weighting in the
least squares
cost function
with constant
forgetting
factor λ.

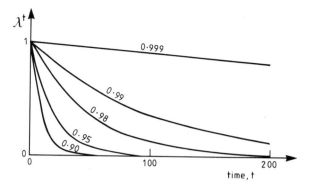

interpret λ as a forgetting 'time constant' and associate it with a *data memory* of, roughly, $(1-\lambda)^{-1}$.

By rederiving the least squares equation (Section 3.3) but using the modified cost function of equation (4.12a), it is possible to show that the modified least squares estimate becomes (cf. equation (3.18))

$$\hat{\boldsymbol{\theta}}(t) = [\mathbf{X}^T(t)\boldsymbol{\Lambda}(t)\mathbf{X}(t)]^{-1}\,\mathbf{X}^T(t)\boldsymbol{\Lambda}(t)\mathbf{y}(t) \qquad (4.13a)$$

where

$$\boldsymbol{\Lambda}(t) = \mathrm{diag}(\lambda^{t-1}, \lambda^{t-2}, ..., \lambda^2, \lambda, 1) \qquad (4.13b)$$

The key practical reason for applying the particular form of weighting used in equation (4.12) is that the corresponding recursive formulation is particularly simple. Specifically, step (iii) in the RLS algorithm becomes:

$$\mathbf{P}^{-1}(t) = \lambda\mathbf{P}^{-1}(t-1) + \mathbf{x}(t)\mathbf{x}^T(t) \qquad (4.14)$$

i.e.

$$\mathbf{P}(t) = \lambda^{-1}\mathbf{P}(t-1)\left\{\mathbf{I}_m - \frac{\mathbf{x}(t)\mathbf{x}^T(t)\mathbf{P}(t-1)}{\lambda + \mathbf{x}^T(t)\mathbf{P}(t-1)\mathbf{x}(t)}\right\} \qquad (4.15)$$

The fact that it is implemented so easily in recursive form makes the forgetting factor technique very popular. It has the disadvantage, however, that it is indiscriminate in that *all* elements of $\mathbf{P}(t)$ are scaled by the same amount. In order to target the effect upon selected parameters of a model, some users scale only the diagonal elements of $\mathbf{P}(t)$. This allows them to use a different forgetting factor for each parameter, achieving a similar effect to the use of the random walk matrix **R**.

In some situations it is of practical importance to use different forgetting factors for different parameters. In particular, if a disturbance model $\mathcal{D}(t)$ is present, then the constant term d_0 (and possibly the other coefficients of $\mathcal{D}(t)$) will require a different forgetting factor from other parameters.

Example 4.9 *Use of forgetting factors*

Consider a system represented by

$$y(t) = -a_1y(t-1) - a_2y(t-2) + b_0u(t-1) + e(t)$$

Figure 4.21 shows the influence on RLS of various choices of forgetting factor. In this example, the b_0 coefficient changes at $t = 200$ and the a_1, a_2 coefficients are constant:

$$a_1 = -1$$

$$a_2 = 0.25$$

$$b_0 = 1 \qquad t < 200; \qquad b_0 = 2 \qquad t \geqslant 200$$

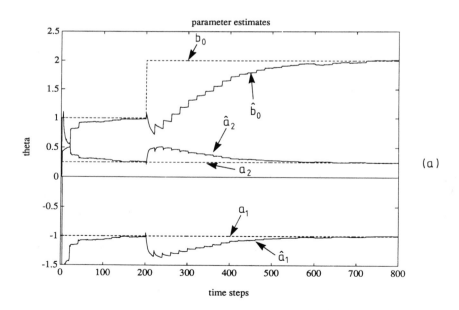

Figure 4.21
Illustrating the influence on RLS of different forgetting factor choices (Example 4.9). Note that the true value of the b_0 parameter changes at $t = 200$ and that in (a) $\lambda = 0.99$,

Figure 4.21
(cont.)
(b) $\lambda = 0.98$ and
(c) $\lambda = 0.97$.

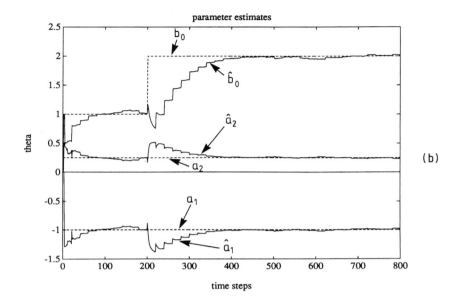

(b)

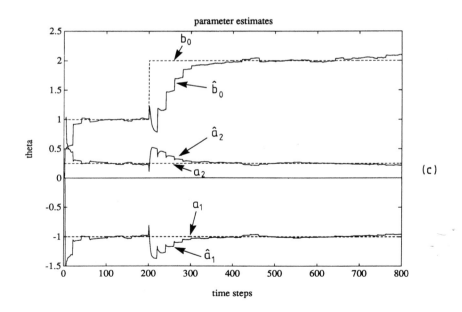

(c)

Note that the effect of reducing λ is to increase the sensitivity of the estimates to the change in b_0. An associated penalty, however, is that the variability of the estimates is also increased. This latter point reflects the fact that as λ is decreased the effective amount of information in the estimator is also decreased. This in turn increases the variability of the estimates, as can be seen from (4.14) or (4.15).

□

The use of either a random walk or forgetting factor in RLS creates conflicting tendencies in the formation of the covariance matrix which must be carefully balanced. Examining equation (4.14), we see that new information introduced via the data vector drives $\mathbf{P}(t)$ towards zero, while the use of λ with a value less than unity has precisely the opposite effect.

If the data vector $\mathbf{X}(t)$ brings little or no new information into the estimator then the bracketed term in equation (4.15) will not act to decrease $\mathbf{P}(t)$. In this case $\mathbf{P}(t) \simeq \mathbf{P}(t-1)/\lambda$. The forgetting factor will therefore have the effect of increasing the size of $\mathbf{P}(t)$ from one recursion to the next. A problem which can arise from this is known as 'covariance blow-up' or *estimator wind-up*. Specifically, if no new information enters the estimator over a long period, then the continual division by λ can cause the elements of $\mathbf{P}(t)$ to become very large, leading to numerical overflow.

Example 4.10   *Illustrating estimator wind-up*

Consider the system used in Example 4.9 in which the input signal $u(t)$ is a unit amplitude square wave of period 40 time steps for the first 400 steps and 0.1 amplitude thereafter. Figure 4.22 shows the recursive estimation of the process parameters. Note that the covariance trace decreases at each transition of the square wave but otherwise increases during periods when no new information enters the data vector. Thus, when the input excitation reduces in amplitude the covariance matrix (as indicated by its trace) increases and the parameter estimates become correspondingly erratic until the algorithm fails.

□

In general, forgetting factors must be used with due attention to the level and type of excitation applied to the system. In particular, a dither signal should be used where possible to provide a persistently exciting signal. Even so, care must be taken to balance the amplitude of the dither against the size of forgetting factor.

Following these general remarks, we now focus on specific mechanisms for preventing wind-up, such as constant trace algorithms, directional forgetting and variable forgetting.

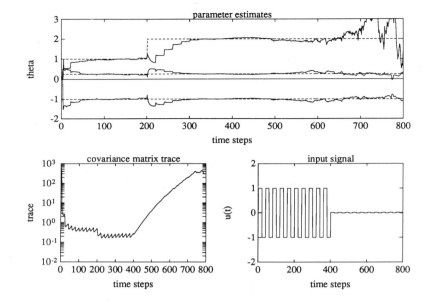

Figure 4.22 Illustrating covariance wind-up (Example 4.10). Note that the covariance matrix entries (as measured by the trace of $\mathbf{P}(t)$) increase rapidly as the amplitude of the input $u(t)$ is decreased at time step 400.

4.5.7   Constant trace algorithms

The first method involves replacing $\mathbf{R}$ in (4.9) by $\mathbf{R}(t)$ chosen so that the trace of $\mathbf{P}(t)$ is held constant. This leads to the constraint

$$\text{tr}[\mathbf{R}(t)] = \frac{\mathbf{x}^{\text{T}}(t)\mathbf{P}^2(t-1)\mathbf{x}(t)}{1 + \mathbf{x}^{\text{T}}(t)\mathbf{P}(t-1)\mathbf{x}(t)} \tag{4.16}$$

which is satisfied by setting

$$\mathbf{R}(t) = m^{-1}\text{tr}[\mathbf{R}(t)]\mathbf{I}_m \tag{4.17}$$

If more detailed information is available concerning possible parameter variation, then the trace value can be distributed more appropriately amongst the diagonal elements of $\mathbf{R}(t)$. The constant trace algorithm is incorporated into the recursive estimator by (i) calculating $\text{tr}[\mathbf{R}(t)]$ using equation (4.16); (ii) determining the corresponding random walk matrix $\mathbf{R}(t)$ from equation (4.17); and (iii) adding $\mathbf{R}(t)$ to the $\mathbf{P}(t)$ at each time step as in a normal random walk procedure.

This approach eliminates blow-up and gives the estimation algorithm a tracking capability. Note, however, that lack of new information can lead to $\mathbf{P}(t)$ becoming almost singular and $\mathbf{P}(t)\mathbf{x}(t+1)$ tending to zero. This is not ruled out by demanding that $\mathbf{P}(t)$ has a positive trace. In this case $\mathbf{R}(t)$ tends to zero so that $\mathbf{P}(t) \simeq \mathbf{P}(t-1)$.

Figure 4.23
Illustrating a
constant trace
algorithm using
a random walk.
The system is
that described in
Example 4.10.
Note that the
covariance trace
is held constant
despite the
reduction in
amplitude of $u(t)$
at $t = 400$.

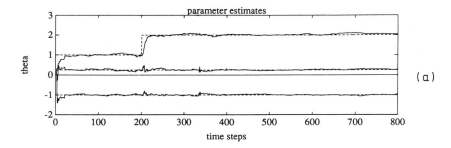

(a)

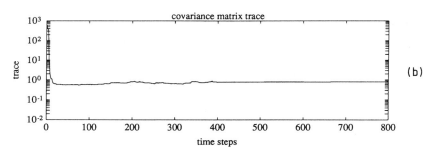

(b)

To avoid this phenomenon an excitation signal should be input to the system or $\mathbf{R}(t)$ should be bounded below.

Returning to Example 4.10 and employing the random walk (4.17) at time step $t = 20$, the simulation results show (Figure 4.23) that the covariance trace is indeed held constant (Figure 4.23(b)) and the parameter estimate $\hat{b}_0$ tracks the change which occurs at $t = 200$. Note also that the reduction in excitation level at time step 400 does not influence the covariance object. In particular, the estimation remains satisfactory and wind-up is avoided.

Other algorithms exist that attempt to bound $\mathbf{P}(t)$ in similar ways (see Section 4.8 Notes and references).

4.5.8   Directional forgetting

A second technique, termed directional forgetting, avoids covariance wind-up by applying a variable forgetting factor only in those directions in parameter space in which new information becomes available.

Using directional forgetting the covariance matrix is updated as follows (cf. (4.15)):

$$\mathbf{P}^{-1}(t) = \mathbf{P}^{-1}(t-1) + r(t-1)\mathbf{x}(t)\mathbf{x}^{\mathsf{T}}(t) \qquad (4.18)$$

leading to

$$P(t) = P(t-1) \left[I_m - \frac{x(t)x^T(t)P(t-1)}{r^{-1}(t-1) + x^T(t)P(t-1)x(t)} \right]$$ (4.19)

Note that equation (4.18) controls the *inflow* of information associated with the *single* direction (parallel to the vector $x(t)$) associated with the rank one matrix $x(t)x^T(t)$, whereas the conventional forgetting factor approach controls the *outflow* of information in *all* directions. The directional forgetting factor is therefore selectively sensitive to incoming information.

A possible choice for $r(t)$ is

$$r(t) = \lambda' - \frac{(1-\lambda')}{x^T(t+1)P(t)x(t+1)}$$ (4.20)

The scalar λ' is similar to a fixed forgetting factor. If λ' is unity the normal infinite memory recursion is obtained. As λ' is decreased care must be taken since too small a value will cause $P(t)$ to lose its positive definite property. Figure 4.24 shows the effect of using directional forgetting in Example 4.10. In this figure λ' was set at 0.8. Note that, after the excitation level decreases in amplitude at time step 400, the covariance trace rises but stabilizes at a higher level. This behaviour should be compared with the continuously

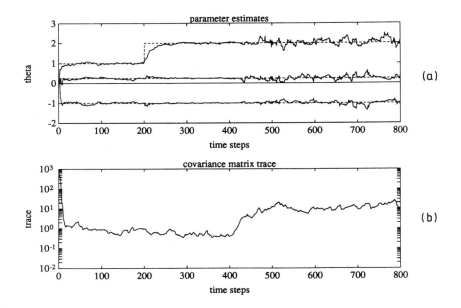

Figure 4.24 Illustrating the effect of directional forgetting. The system used is that described in Example 4.10. Note that the covariance trace does not now 'blow-up' as with the standard constant forgetting factor. In this simulation $\lambda' = 0.8$.

increasing covariance trace when using a conventional forgetting factor (Figure 4.22).

4.5.9   Other variable forgetting factors

Changes in system parameters may occur at unknown times, separated by periods over which the parameters remain steady. In such cases, it is useful to be able to automatically vary the forgetting factor as in the last two sections. When parameters are constant the forgetting factor should be equal to unity, but when a change is sensed it should automatically decrease. There are two situations when a lower forgetting factor is desirable.

Start-up forgetting factor

At start-up the estimator may have little or no knowledge of the system. Here it is useful to have a forgetting factor which is initially small and increases at a prespecified rate towards unity. A variable forgetting factor which does this is given by

$$\lambda_1(t) = \gamma + (1-\gamma)\,[1 - \exp(-t/\tau_f)] \tag{4.21}$$

where γ is the initial value (at $t = 0$) of the forgetting factor and τ_f is the time constant of the forgetting factor determining how fast $\lambda_1(t)$ changes. This form of forgetting factor can also be used when the identifier is reset, for whatever reason.

The expression (4.21) can be cast in the simple recursive form

$$\lambda_1(t) = \alpha\,\lambda_1(t-1) + (1-\alpha)$$
$$\alpha = \exp(-1/\tau_f), \qquad \lambda_1(0) = \gamma \tag{4.22}$$

Adaptive or variable forgetting factors

Secondly, during an identification run it is useful if the forgetting factor can be decreased when a change in the system is sensed. In the absence of prior knowledge of change times, one way of detecting change is through the prediction error $\epsilon(t)$. If this error grows, it may mean that the identified model is incorrect and needs adjustment. At such times we decrease the forgetting factor and allow the model to adapt. A variable forgetting factor which allows this is

$$\lambda_2(t) = \frac{\tau_f}{\tau_f - 1} \left[1 - \frac{\epsilon^2(t)}{\tau_f s_f(t)} \right] \tag{4.23}$$

where τ_f again determines the rate of adaption and $s_f(t)$ is a weighted average of the past values of $\epsilon^2(t)$. If $s_f(t)$ is calculated according to

$$s_f(t) = \frac{\tau_f - 1}{\tau_f} s_f(t-1) + \frac{\epsilon^2(t)}{\tau_f} \tag{4.24}$$

then equation (4.23) can be simplified to give

$$\lambda_2(t) = \frac{s_f(t-1)}{s_f(t)} \tag{4.25}$$

Finally, it is possible to combine the start-up/reset forgetting factor $\lambda_1(t)$ with the adaptive forgetting factor $\lambda_2(t)$ to give the variable forgetting factor $\lambda_v(t)$, thus:

$$\lambda_v(t) = \lambda_1(t)\lambda_2(t) \tag{4.26}$$

4.5.10   Variable excitation and estimator dead-zones

The possibility of tracking parameter changes hinges crucially on the level of excitation balanced against the rate at which information is discounted from the recursive algorithm. The techniques discussed above have tackled this balance by adjusting the mechanism for discounting data. In some cases more direct methods may be applicable. For example, if the practical situation allows, it is often desirable to alter the amplitude of the excitation or dither. Thus, if an increase in $\mathbf{P}(t)$ beyond a certain magnitude is encountered, then by increasing the excitation level $\mathbf{P}(t)$ can be reduced. Correspondingly, if $\mathbf{P}(t)$ becomes too small to allow parameter adaption then, by decreasing the excitation level, $\mathbf{P}(t)$ will grow if a forgetting factor is in operation.

A further procedure which is widely used employs an estimator dead-zone. In this technique the estimator is switched off when the excitation level is very low. This prevents wind-up due to poor excitation. However, during these periods the estimator is of course not ready or able to adapt to parameter changes. To switch the estimator on again some indicator is required. Typically, the prediction error is monitored and, when this increases beyond some threshold, the estimator is reactivated, the assumption being that the prediction error increase indicates a change in the underlying system.

4.6 SUMMARY

In this chapter we have considered the various measures required to start and run a recursive estimator in a self-tuning environment. The issues covered include:

- Starting up a recursive estimator
- Initial design choices required to set up the estimation model and data collection procedure
- The experimental conditions which may be encountered and the various signal conditioning procedures which can be used
- The modifications required for a recursive estimator to track parameter changes

The guidance provided here is generally useful in estimation. Additional considerations apply when the recursive estimator is used in a self-tuning system. These extra considerations are discussed at appropriate points in the relevant self-tuning chapters.

4.7 PROBLEMS

Problem 4.1

Given the cost function

$$\bar{J}_t = \sum_{i=1}^{t} \lambda^{t-i}\, \hat{e}^2(t)$$

derive the corresponding recursive least squares estimator showing how the forgetting factor λ influences the covariance matrix recursion.

Problem 4.2

A discrete time system is described by the model

$$Ay(t)\, Bu(t-1) + \hat{e}(t)$$

in the usual notation.

The standard RLS algorithm for such a model is based upon the cost function

$$J_t = \sum_{i=1}^{t} \hat{e}^2(i)$$

Consider the alternative cost function

$$J_t(N) = \sum_{i=t-N}^{t} \hat{e}^2(i)$$

where N is a fixed 'window' length.
  Derive

(a) an expression for the modified least squares estimator for the coefficients of A, B;

(b) the modified RLS algorithm.

Problem 4.3

Consider the standard recursive least squares iteration

$$\hat{\boldsymbol{\theta}}(t+1) = \hat{\boldsymbol{\theta}}(t) + \mathbf{P}(t+1)\mathbf{x}(t+1)\epsilon(t+1)$$

where

$$\epsilon(t+1) = y(t+1) - \mathbf{x}^T(t+1)\hat{\boldsymbol{\theta}}(t)$$

(a) Show that the evolution of the vector of parameter estimation errors

$\tilde{\boldsymbol{\theta}}(t) = \boldsymbol{\theta} - \hat{\boldsymbol{\theta}}(t)$ is governed by the state equation

$$\tilde{\boldsymbol{\theta}}(t+1) = [\mathbf{I}_m - \mathbf{P}(t+1)\mathbf{x}(t+1)\mathbf{x}^T(t+1)]\tilde{\boldsymbol{\theta}}(t)$$

(b) If the recursion for $\hat{\boldsymbol{\theta}}(t+1)$ is modified as follows:

$$\hat{\boldsymbol{\theta}}(t+1) = \hat{\boldsymbol{\theta}}(t) + \mathbf{P}(t+1)\mathbf{x}(t+1)\epsilon(t+1) + \mathbf{R}\mathbf{x}(t+1)\mathbf{x}^T(t+1)\tilde{\boldsymbol{\theta}}(t)$$

show how the state equation for $\tilde{\boldsymbol{\theta}}(t+1)$ is changed by this modification.

Problem 4.4

The least squares model

$$y(t) = a_1 y(t-1) + \ldots + a_n y(t-n) + b_0 u(t-1) + \ldots + b_{n-1}(t-n) + e(t)$$

can be written in the 'observation equation' form

$$y(t) = \mathbf{c}^T(t)\mathbf{x}(t) + e(t)$$

where

$$\mathbf{x}(t) = [a_1, \ldots, a_n, b_0, \ldots, b_{n-1}]^\mathsf{T}$$

is the state whose time invariance can be expressed by the 'process equation'

$$\mathbf{x}(t+1) = \mathbf{x}(t) \tag{*}$$

Write down $\mathbf{c}(t)$ and derive the Kalman filter for state estimation corresponding to these state space equations. Show that the Kalman filter is equivalent to the RLS parameter estimation algorithm.

Parameter drift can be modelled by replacing (*) by

$$\mathbf{x}(t+1) = \mathbf{x}(t) + \mathbf{w}(t)$$

where $\{\mathbf{w}(t)\}$ is a vector white noise sequence independent of $\{e(t)\}$. Show that the Kalman filter equations now correspond to the random walk modification of RLS discussed in Section 4.5.5.

Problem 4.5

For the system

$$y(t) = \mathbf{x}^\mathsf{T}(t)\boldsymbol{\theta} + e(t)$$

the parameter n-vector $\boldsymbol{\theta}$ is estimated using off-line LS with constant forgetting factor λ. Show that, if the sequence $\{\mathbf{x}(t)\}$ is deterministic, then the estimator is unbiased with covariance

$$\operatorname{cov} \hat{\boldsymbol{\theta}}(N) = \sigma_e^2\, \mathbf{A}^{-1}(N)\, \mathbf{B}(N)\, \mathbf{A}^{-1}(N)$$

where N is the number of data points and

$$\mathbf{A}(N) = \sum_{t=1}^{N} \lambda^{N-t}\, \mathbf{x}(t)\, \mathbf{x}^\mathsf{T}(t)$$

$$\mathbf{B}(N) = \sum_{t=1}^{N} \lambda^{2(N-t)}\, \mathbf{x}(t)\, \mathbf{x}^\mathsf{T}(t)\,.$$

If $\mathbf{x}(t)$ is of period m and the matrix

$$\mathbf{P}^{-1} = \sum_{t=1}^{m} \lambda^{m-t}\, \mathbf{x}(t)\, \mathbf{x}^\mathsf{T}(t)$$

is invertible, show that $m \geqslant n$ and that, under gaussian assumptions on the estimator, the estimate $\hat{\theta}_i(N)$ lies asymptotically in the interval $\theta_i \pm 2\sigma_i$ with 95% probability, where

$$\sigma_i^2 = \left(\frac{1 - \lambda^m}{1 + \lambda^m} \right) \sigma_e^2\, P_{ii}$$

Calculate the confidence intervals for the process

$$y(t) = b_1 u(t-1) + b_2 u(t-2) + e(t)$$

where $u(t) = 1$ (t odd), $= 0$ (t even), and check your results by Monte Carlo simulation, plotting $\hat{b}_1(500)$ against $\hat{b}_2(500)$ on a phase plane diagram for each RLS run, using your chosen values of b_1, b_2, σ_e, λ.

Problem 4.6

Derive the constant trace algorithm that leads to the time varying random walk satisfying equation (4.17). Find the variable forgetting factor $\lambda(t)$ that has the same effect.

Problem 4.7

The system $y(t) = bu(t) + e(t)$, in the usual notation, contains the unknown parameter b, which is estimated using least squares estimation with a constant forgetting factor λ. If $u(t)$ is a deterministic function of time, derive the parameter estimator $\hat{\beta}(N)$ and calculate var $\hat{\beta}(N)$. Show that, for $\lambda < 1$, a constant input signal does not lead to parameter identifiability.

Problem 4.8

In Problem 4.7, if b is replaced by a time-varying signal $b(t)$ and $u(t) \equiv 1$, calculate $E[\hat{\beta}(N) - b(N)]^2$ as a measure of tracking efficiency in the case:

$$b(t) = b_0 + w(t)$$

where $w(t)$ is a zero mean white noise signal, uncorrelated with $\{e(t)\}$.

Discuss the influence of λ on tracking efficiency.

4.8 NOTES AND REFERENCES

The notion of persistent excitation is discussed in detail in:

Soderstrom, T. and Stoica, P.
  System Identification, Prentice Hall, 1989.
These authors supply full technical details of persistent excitation and discuss the variations in the definition for various applications of recursive estimation.

The generation of PRBS based test signals is discussed in detail in:

Peterson, W.W.
  Error Correcting Codes, MIT Press, 1961.

The use of PRBS in system identification is covered in:

Godfrey, K.R.
  Correlation methods, *Automatica*, **16**, 527–34, 1980.

The representation and analysis of quantization error in A/D conversion processes is dealt with in many books on digital filtering and signal processing. See for example:

Hamming, R.W.
  Numerical Methods for Scientists and Engineers, McGraw Hill, 1962.

Variable forgetting factors which depend upon the estimator error were first suggested in:

Fortescue, T.R., Kershenbaum, L.S. and Ydstie, B.E.
  Implementation of self-tuning regulators with variable forgetting factors, *Automatica*, **17**, 1981.

See also:
Sanoff, S.P. and Wellstead, P.E.
  Comments on 'implementation of self-tuning regulators with variable forgetting factors', *Automatica*, **19**(3), 345–6, 1983.

The directional forgetting discussion used here is based on:

Kulhavy, R.
  Restricted exponential forgetting in real-time identification, *Automatica*, **23**(5), 589–600, 1987.

See also:
Hagglund, T.
  New estimation techniques for adaptive control, Report TFRT-1025, Lund Institute of Technology, 1983.

The management of self-tuning estimators is discussed in the survey articles mentioned in Section 1.8. See also the techniques used in:

Sanoff, S.P.
  A self-tuning instrument based on a personal computer, Control Systems Centre Report 620, UMIST, 1984.
Wittenmark, B. and Astrom, K.J.
  Practical issues in the implementation of self-tuning control, *Automatica*, **20**(5), 595–605, 1984.
Isermann, R. and Lachmann, K-H.
  Parameter adaptive control with configuration aids and supervision functions, *Automatica*, **21**(6), 625–38, 1985.
Salgado, T., Goodwin, G. and Middleton, R.
  Modified least squares algorithm incorporating experimental resetting and forgetting, *Int. J. Control*, **47**, 477–91, 1988.

For a Bayesian approach to initializing a recursive estimator see:

Karny, M.
  Quantification of prior knowledge about global characteristics of linear normal model, *Kybernetika*, **20**(5), 375–85, 1984.

Many modifications to the standard RLS algorithm are surveyed in:

Shah, S.L.
  RLS based estimation schemes for self-tuning control, in *Implementation of Self-Tuning Controllers*, ed. K. Warwick, Peter Peregrinus, 1988.

5 Computational Alternatives for Recursive Estimation

5.1 OUTLINE AND LEARNING OBJECTIVES

The recursive least squares (RLS) algorithm presented in Chapter 3 is based upon the Matrix Inversion Lemma and represents the basic formulation of the least squares recursion. Situations arise, however, when the basic form of the recursion is inadequate due to either (i) restricted computational precision or (ii) restricted computational time.

When increased computational precision is required, use can be made of a class of matrix factorization algorithms. These algorithms have approximately the same computational requirements (in terms of the number of operations) as the conventional matrix inversion RLS. However, by operating upon factors of the covariance matrix to recursively update the parameter estimates, they are less liable to round-up error accumulation within the algorithm.

Some situations exist, particularly in digital signal processing, in which the matrix inversion and factorization methods for implementing RLS are too slow computationally. In particular, the number of multiplications required for one cycle of the matrix inversion RLS is proportional to the square of the number of unknown parameters m. Thus for large m (as required in some signal processors) the computation time can increase enormously. To overcome this problem a class of 'fast' algorithms has been developed which require of the order of m multiplications per recursion.

Computationally robust algorithms based on factorization are discussed in Section 5.2. The computationally fast algorithms are discussed in Section 5.3.

The key learning objective of this chapter is to understand the algorithmic variations on matrix inversion RLS. In addition, the reader should gain an insight into the areas of application of these algorithms and their respective advantages and disadvantages.

5.2   NUMERICALLY ROBUST RECURSION

The matrix inversion lemma version of RLS can be numerically ill-conditioned. In particular, rounding errors can accumulate and cause errors to occur in both the estimated parameters and the covariance matrix. Apart from the loss of accuracy, the calculated covariance matrix may become negative semi-definite with consequent divergence of the parameter estimates. This may happen in computers with limited computational accuracy where the round-up error at each recursion is large. In a similar vein, numerical problems can occur during estimator runs involving many tens of thousands of recursions in which normally negligible round-up errors are allowed to accumulate. It should be stressed that in most circumstances the standard recursion is completely adequate. Despite this, most users of self-tuning algorithms and estimators will employ the numerically robust versions of RLS as a matter of good engineering practice. For this reason a Pascal listing of one of the most popular algorithms is included in an appendix to this chapter.

The key idea in numerically robust RLS algorithms is to replace the recursive calculation of a square matrix by a recursive calculation of a factor of the square matrix. For example, the covariance matrix $\mathbf{P}(t)$ is symmetric and can be written in the factored form

$$\mathbf{P}(t) = \mathbf{S}(t)\mathbf{S}^{\mathrm{T}}(t) \tag{5.1}$$

where $\mathbf{S}(t)$ is a triangular matrix and corresponds to the 'square root' of the $\mathbf{P}(t)$ matrix. A least squares algorithm can be constructed which computes and updates $\mathbf{S}(t)$ in each recursion. Since we are effectively operating on the square root of $\mathbf{P}(t)$ the procedure is roughly equivalent to computing $\mathbf{P}(t)$ in double precision. In addition, forming the product in equation (5.1) ensures that $\mathbf{P}(t)$ remains positive definite.

There are numerous factorization algorithms for computing the RLS recursion. A widely used version in self-tuning control is the Bierman U–D factorization algorithm. This method operates upon a factored form of the covariance matrix $\mathbf{P}(t)$. The Bierman algorithm emerged from Kalman filtering work and the need to create numerically stable state estimators in aerospace applications. A complementary but essentially equivalent line of development has occurred in numerical analysis where the object $\mathbf{R}(t) = \mathbf{P}^{-1}(t)$ is updated in factored form via a modified Givens rotation.

The Bierman algorithm is described in Section 5.2.1 and the modified Givens rotation is outlined in Section 5.2.2.

5.2.1 Bierman U–D factorization

The factorization of $\mathbf{P}(t)$ in terms of $\mathbf{S}(t)$ has been extensively used under the heading of 'square root filtering'. Such algorithms require the square root of m real positive numbers to be computed at each recursion. The neat feature of the Bierman U–D algorithm is that the need for computing square roots is avoided, thus giving a significant additional computational economy.

The main idea in the U–D algorithm is to assume that $\mathbf{P}(t)$ can be factored as

$$\mathbf{P}(t) = \mathbf{U}(t)\mathbf{D}(t)\mathbf{U}^{\mathrm{T}}(t) \tag{5.2}$$

where $\mathbf{U}(t)$ is an upper triangular matrix of the form

$$\mathbf{U}(t) = \begin{bmatrix} 1, & u_{12}(t), & \ldots, & u_{1m}(t) \\ 0, & 1, & u_{23}(t), \ldots, & u_{2m}(t) \\ & 0 & & \\ \cdot & \cdot & \cdot & \\ \cdot & \cdot & \cdot & \\ \cdot & \cdot & \cdot & \\ & & & u_{m-1,m}(t) \\ 0, & 0 & \ldots 0, & 1 \end{bmatrix} \tag{5.3}$$

and $\mathbf{D}(t)$ is a diagonal matrix given by

$$\mathbf{D}(t) = \begin{bmatrix} d_1(t) & 0 & \ldots & 0 \\ 0 & d_2(t) & & \cdot \\ \cdot & & \cdot & \cdot \\ \cdot & & \cdot & \cdot \\ \cdot & & & 0 \\ 0 & \ldots & 0, & d_m(t) \end{bmatrix} \tag{5.4}$$

The essence of the U–D factorization is to recursively compute $\mathbf{D}(t)$ and $\mathbf{U}(t)$ from $\mathbf{D}(t-1)$ and $\mathbf{U}(t-1)$. The Kalman gain $\mathbf{L}(t)$ is then computed directly and the parameter estimates $\hat{\boldsymbol{\theta}}(t-1)$ are updated in the normal way. The algorithm is as follows:

Algorithm: U–D Factorization

If the data vector is denoted by the vector $\mathbf{x}(t)$ with m entries corresponding to m unknown parameters, the U–D update at time t is given by

Step 1 Compute the m vectors $\mathbf{f}$ and $\mathbf{g}$ as

$$\mathbf{f} = \mathbf{U}^{\mathrm{T}}(t-1)\cdot\mathbf{x}(t) \tag{5.5a}$$

$$\mathbf{g} = \mathbf{D}(t-1)\cdot\mathbf{f} \tag{5.5b}$$

and set $\beta_0 = \lambda$, the forgetting factor $\tag{5.5c}$

Step 2 for $j = 1$ to m repeat (i) and (ii).
  (i) Compute

$$\beta_j = \beta_{j-1} + f_j g_j \tag{5.5d}$$

$$d_j(t) = \beta_{j-1} d_j(t-1)/\beta_j \lambda \tag{5.5e}$$

$$v_j = g_j \tag{5.5f}$$

$$\mu_j = -f_j/\beta_{j-1} \tag{5.5g}$$

(ii) For $i=1$ to $j-1$ ($j>1$) compute

$$u_{ij}(t) = u_{ij}(t-1) + v_i \mu_j \tag{5.5h}$$

$$v_i = v_i + u_{ij}(t-1)v_j \tag{5.5i}$$

Step 3 Compute

$$\bar{\mathbf{L}}(t) = [v_1, \ldots, v_m]^{\mathrm{T}} \tag{5.5j}$$

$$\mathbf{L}(t) = \bar{\mathbf{L}}(t)/\beta_m \tag{5.5k}$$

Step 4 Update the parameter estimates

$$\hat{\boldsymbol{\theta}}(t) = \hat{\boldsymbol{\theta}}(t-1) + \mathbf{L}(t)\epsilon(t)$$

where $\mathbf{L}(t)$ is the Kalman gain vector $\mathbf{P}(t)\mathbf{x}(t)$ and as usual

$$\epsilon(t) = y(t) - \mathbf{x}^{\mathrm{T}}(t)\hat{\boldsymbol{\theta}}(t-1)$$

Step 5 $t \rightarrow t+1$ and loop back to Step 1.   □

A proof that the above algorithm does in fact compute $\mathbf{D}(t)$, $\mathbf{U}(t)$ from $\mathbf{D}(t-1)$, $\mathbf{U}(t-1)$ is given in Appendix 5.1. In addition, a Pascal procedure for the U–D factorization is given in Appendix 5.2.

5.2.2 Modified Givens rotation

The main idea in this approach is to operate upon the matrix $\mathbf{R}(t) = \mathbf{P}^{-1}(t)$ and reduce it to an upper triangular form. The parameter estimates are then obtained by a process of back substitution. This is a different philosophy from the RLS material presented so far and is therefore worth considering in some detail. In particular, consider equation (3.6), rearranged thus

$$\hat{\mathbf{e}} = \mathbf{y} - \mathbf{X}\hat{\boldsymbol{\theta}} \qquad (3.6)\text{bis}$$

We recall that the sum of squares of the entries in the error vector $\hat{\mathbf{e}}$ is minimized by

$$\mathbf{X}^{\mathrm{T}}\mathbf{X}\hat{\boldsymbol{\theta}} = \mathbf{X}^{\mathrm{T}}\mathbf{y} \qquad (5.6)$$

Now suppose that we write $\mathbf{X}$ in the form

$$\mathbf{X} = \mathbf{QS} \qquad (5.7)$$

where $\mathbf{Q}$ is an $N \times m$ matrix with orthonormal columns and $\mathbf{S}$ is an upper triangular $m \times m$ matrix.

Then equation (5.6) can be written as

$$\mathbf{S}^{\mathrm{T}}\mathbf{Q}^{\mathrm{T}}\mathbf{QS}\,\hat{\boldsymbol{\theta}} = \mathbf{S}^{\mathrm{T}}\mathbf{Q}^{\mathrm{T}}\mathbf{y} \qquad (5.8)$$

or

$$\mathbf{S}^{\mathrm{T}}\mathbf{S}\hat{\boldsymbol{\theta}} = \mathbf{S}^{\mathrm{T}}\mathbf{Q}^{\mathrm{T}}\mathbf{y}$$
$$\mathbf{S}\hat{\boldsymbol{\theta}} = \bar{\mathbf{y}} \qquad (5.9)$$

where we have rewritten $\mathbf{Q}^{\mathrm{T}}\mathbf{y}$ as the vector $\bar{\mathbf{y}}$.

The triangular system of equations given by equation (5.9) can be rapidly solved for $\hat{\boldsymbol{\theta}}$ by a process of back substitution. Indeed, it is a popular alternative to the RLS solution of the least squares problem which is particularly used in numerical analysis and statistical computation. The objects $\mathbf{S}$ and $\bar{\mathbf{y}}$ are formed from the data and equation (5.9) solved for $\hat{\boldsymbol{\theta}}$.

The method can also be put into a recursive form using a Givens rotation transformation. The Givens rotation operating on two row vectors takes the form:

$$\text{Original vectors} \begin{cases} 0,. . . .,0, r_i, r_{i+1},. . .,r_k,. . . \\ 0,. . . .,0, p_i, p_{i+1},. . .,p_k,. . . \end{cases}$$

$$\text{Transformed vectors} \begin{cases} 0,. . . .,0, r'_i, r'_{i+1},. . .,r'_k,. . . \\ 0,. . . .,0, 0, p'_{i+1},. . .,p'_k,. . . \end{cases}$$

where

$$\begin{rcases} r'_k = cr_k + sp_k \\ p'_k = -sr_k + cp_k \end{rcases} \tag{5.10}$$

and

$$\begin{rcases} r'_i = (r_i^2 + p_i^2)^{1/2} \\ c = r_i/r'_i \\ s = p_i/r'_i \end{rcases} \tag{5.11}$$

In a recursive formulation it is assumed that at the tth recursion, the upper triangular matrix $S(t)$ and $\bar{y}(t)$ exist, together with the sums of squares of errors $V(t)$. The new data vector $x(t+1)$ and the new output observation $y(t+1)$ are also given. These can be arranged in a table as follows.

$$\begin{matrix} \text{Existing data} \begin{cases} \\ \\ \end{cases} \\ \text{New data} \rightarrow \end{matrix} \begin{bmatrix} S(t) & y(t) \\ 0 & V(t) \\ x^T(t+1) & y(t+1) \end{bmatrix} \tag{5.12}$$

The row containing new data can be rotated with the first row of $S(t)$ in order to zero the first element of the new data vector $x^T(t+1)$. Rotation with the second row of $S(t)$ will zero the second element of $x^T(t+1)$ and so on until all the entries in $x^T(t+1)$ have been transformed to zero. One further rotation with the row containing $V(t)$ will incorporate the remaining information in the new data into the cost function. The table now takes the form

$$\begin{bmatrix} S(t+1) & \bar{y}(t+1) \\ 0 & V(t+1) \\ 0 & 0 \end{bmatrix} \tag{5.13}$$

Thus when a new data vector $x^T(t+2)$ and observation $y(t+2)$ become available, they can be inserted into the vacant last row and the update procedure continued.

The Givens rotation procedure described above is a square root algorithm executed on the matrix $\mathbf{R}(t)$. The square root calculations in the Givens rotation occur in equations (5.11). These can be avoided in the same manner as employed in the Bierman algorithm. Specifically, instead of considering the $\mathbf{S}(t)$ upper triangular matrix, we consider a diagonal matrix $\mathbf{D}(t)$ and an upper triangular matrix $\mathbf{U}(t)$ given by

$$\mathbf{S}(t) = \mathbf{D}^{1/2}(t)\mathbf{U}(t) \tag{5.14}$$

A modified Givens rotation can be used for the rows of the product $\mathbf{D}^{1/2}(t)\mathbf{U}(t)$ with a new row vector, such that the modified rotation does not require a square root operation. In particular, consider the two rows

$$0,\ldots,0, (d)^{1/2}, \quad (d)^{1/2} r_{i+1},\ldots,(d)^{1/2}r_k.\ldots$$

$$0,\ldots,0, (\delta)^{1/2}p_i, \quad (\delta)^{1/2} p_{i+1},\ldots,(\delta)^{1/2}p_k.\ldots$$

The rotated rows are

$$0,\ldots.0, (d')^{1/2}, (d')^{1/2} r'_{i+1},\ldots,(d')^{1/2} r'_k.\ldots$$

$$0,\ldots,0, \quad 0, (\delta')^{1/2} p'_{i+1},\ldots,(\delta')^{1/2} p'_k.\ldots$$

where

$$\left.\begin{array}{l} d' = d + \delta p_i^2 \\[4pt] \delta' = d\delta/(d+\delta p_i^2) = d\delta/d' \\[4pt] \bar{c} = d/d' \\[4pt] \bar{s} = \delta p_i/d' \end{array}\right\} \tag{5.15}$$

and

$$\left.\begin{array}{l} p'_k = p_k - p_i r_k \\[4pt] r'_k = \bar{c}r_k + \bar{s}p_k \end{array}\right\} \tag{5.16}$$

Thus the modified rotation eliminates an element of the second vector without recourse to a square root operation. This modified rotation can be applied to the table of data (5.12) to provide an equivalent procedure to the Bierman factorization algorithm. The key difference between this recursive modified Givens rotation and the Bierman algorithm is that the Givens algorithm operates upon the $\mathbf{R}(t)$ matrix of correlations while the Bierman algorithm operates upon $\mathbf{P}(t) = \mathbf{R}^{-1}(t)$. The Givens rotation requires an additional back substitution procedure to solve for $\hat{\boldsymbol{\theta}}$ from the equation (5.9). However, this is negligible computationally and the information contained in the $\mathbf{S}(t)$, $\bar{\mathbf{y}}(t)$ vectors is of intrinsic value to statisticians.

5.3 FAST LEAST SQUARES RECURSION

The Bierman algorithm and the Givens rotation take approximately the same number of computational operations as an efficiently programmed RLS algorithm using the matrix inversion lemma. They do not speed up the computations significantly. However, situations arise in which many parameters are to be estimated and the computationally fast algorithms are required. This class of RLS recursions exploits the fact that some of the information required at the tth recursion is in fact already available in the $(t-1)$th covariance matrix. This leads to a class of 'fast' algorithms which rely on a property known as 'shift-invariance' within the covariance matrix.

The shift invariance property can be readily seen by considering a third order autoregressive model of the data $y(t)$:

$$y(t) = a_1 y(t-1) + a_2 y(t-2) + a_3 y(t-3) + e(t) \tag{5.17}$$

The relevant matrix used to estimate the coefficients (a_1, a_2, a_3) using data in the time span $t \in [0, N-1]$ is

$$\mathbf{R}(N-1) = \begin{bmatrix} \sum_{t=0}^{N-1} y(t)y(t), & \sum_{t-1}^{N-1} y(t)y(t-1) & \vdots & \sum_{t-2}^{N-1} y(t)y(t-2) \\ \sum_{t=1}^{N-1} y(t-1)y(t) & \sum_{t-1}^{N-1} y(t-1)y(t-1) & \vdots & \sum_{t-2}^{N-1} y(t-1)y(t-2) \\ \sum_{t=2}^{N-1} y(t-2)y(t) & \sum_{t=2}^{N-1} y(t-2)y(t-1) & \vdots & \sum_{t-2}^{N-1} y(t-2)y(t-2) \end{bmatrix} \tag{5.18}$$

At the next time step the data available are in the interval $t \in [0, N]$ and the $\mathbf{R}(N)$ matrix is

$$\mathbf{R}(N) = \mathbf{X}^{\mathrm{T}}(N)\mathbf{X}(N) = \begin{bmatrix} \sum_{t=0}^{N} y(t)y(t) & \vdots & \sum_{t-1}^{N} y(t)y(t-1) & \sum_{t-2}^{N} y(t)y(t-2) \\ \sum_{t=1}^{N} y(t-1)y(t) & \vdots & \sum_{t=1}^{N} y(t-1)y(t-1) & \sum_{t-2}^{N} (t-1)y(t-2) \\ \sum_{t=2}^{N} y(t-2)y(t) & \vdots & \sum_{t=2}^{N} y(t-2)y(t-1) & \sum_{t-2}^{N} y(t-2)y(t-2) \end{bmatrix} \tag{5.19}$$

Note that the top left submatrix partitioned off in $\mathbf{R}(N-1)$ is the same as the bottom right submatrix partitioned off in $\mathbf{R}(N)$. Hence, most of the information at time step $N-1$ is preserved (apart from a shift down and along) in the covariance matrix at step N. This is the 'shift invariance' property. Fast RLS algorithms aim to exploit this invariance by saving the preserved information from the previous iteration and introducing an updating which contributes the new information required to create the first row and column in (5.19).

In order to be efficient the shift invariance must be valid over a large area of the $\mathbf{R}(N)$ matrix. This in turn implies that fast algorithms are suited to high order AR or MA models and are widely used in signal processing areas where high order AR models are needed. In ARMA estimation the shift invariance property is not valid over the entire $\mathbf{R}(N)$ matrix. A 'block shift invariance' property is used for the submatrices of $\mathbf{R}(N)$ involving $u(t)$, $y(t)$ and their covariances. The net result is that, while the number of multiplications may be reduced, the indexing and data manipulation overhead increases. The result is that fast algorithms are only of marginal benefit unless very high order AR or MA terms are involved and in such situations research effort is focussed upon creating the required computational speed by parallel processing concepts applied to factorization algorithms.

The basic recursion for an AR system is of use in adaptive signal processors and for this reason is discussed below. The main idea is to retain a model of the AR system in both forward time:

$$y(t) = \mathbf{x}^T(t)\hat{\boldsymbol{\theta}}\,(t-1) + \epsilon(t) \qquad (5.20)$$

and also in backward time

$$y(t-m) = \mathbf{x}^T(t+1)\hat{\boldsymbol{\Phi}}(t-1) + \pi(t) \qquad (5.21)$$

where $\hat{\boldsymbol{\Phi}}(t-1)$ are the backward time parameter estimates given $\mathbf{x}(t+1)$ and $\pi(t)$ is the backward time innovation object.

The fast recursion for $\hat{\boldsymbol{\theta}}(t)$ then proceeds on the basis of using $\hat{\boldsymbol{\theta}}(t-1)$, $\hat{\boldsymbol{\Phi}}(t-1)$ jointly to update the Kalman gain $\mathbf{L}(t)$.

The fast algorithm for model (5.20) is:

Algorithm

At the tth time step:

(i)   Form the forward and backward innovations

$$\epsilon(t) = y(t) - \mathbf{x}^T(t)\hat{\boldsymbol{\theta}}(t-1)$$

$$\pi(t) = y(t-m) - \mathbf{x}^T(t+1)\hat{\boldsymbol{\Phi}}(t-1)$$

(ii)   Update $\hat{\boldsymbol{\theta}}(t)$ and $E(t)$

$$\hat{\boldsymbol{\theta}}(t) = \hat{\boldsymbol{\theta}}(t-1) + \mathbf{L}(t)\epsilon(t)$$

$$E(t) = E(t-1) + \eta(t)\epsilon(t)$$

(iii) Form the extended Kalman gain vector $[\mathbf{l}^*, \, l^{**}]^\mathrm{T}$

$$\begin{bmatrix} \mathbf{l}^* \\ l^{**} \end{bmatrix} = \begin{bmatrix} l' \\ l'' \end{bmatrix} = \begin{bmatrix} E^{-1}(t)\eta(t) \\ \mathbf{L}(t) - \hat{\boldsymbol{\theta}}(t)l' \end{bmatrix}$$

(iv)   Form the Kalman gain $\mathbf{L}(t+1)$ and update $\hat{\boldsymbol{\Phi}}(t-1)$

$$\mathbf{L}(t+1) = \frac{\mathbf{l}^* + \hat{\boldsymbol{\Phi}}(t-1)l^{**}}{1 - \pi(t)l^{**}}$$

$$\hat{\boldsymbol{\Phi}}(t) = \hat{\boldsymbol{\Phi}}(t-1) + \mathbf{L}(t+1)\pi(t)$$

5.4   SUMMARY

In this chapter we have reviewed the main computational alternatives for computing the least squares parameter estimates in a recursive manner. The main points are:

- Factorization methods are used to improve numerical accuracy. They are roughly equivalent to computing Matrix Inversion RLS in double precision and take about the same amount of computational effort
- Two flavours of factorization methods exist. One (the Bierman type algorithm) recursively computes a factored form of the $\mathbf{P}(t)$ matrix. The other (the Modified Givens type algorithm) recursively computes an upper triangularized form of the $\mathbf{R}(t) = \mathbf{P}^{-1}(t)$ matrix
- The Bierman type of U–D algorithm is widely used in self-tuning control as the standard estimator algorithm
- When very large numbers of parameters are to be estimated in an AR model, then fast algorithms are used. This situation occurs in some adaptive signal processing applications
- Fast algorithms suffer from round-up error problems and require special features to numerically stabilize them
- In situations where computational speed is crucial, a current trend is toward the use of parallel computing methods to provide very fast parallel implementations of the factorization algorithms (either Bierman or Modified Givens)

5.5 NOTES AND REFERENCES

The factorization methods used in least squares estimation are exhaustively discussed in:

Bierman, G.J.
Factorization Methods for Discrete Sequential Estimation, Academic Press, 1977.

As noted in the text the U–D factorization algorithm is aimed at reducing round-up error in the computation process. For a fundamental examination of these processes see:

Wilkinson, J.H.
Rounding Errors in Algebraic Processes, HMSO, 1963.

The modified Givens rotation was introduced in

Gentleman, W.M.
Least squares computation by Givens transformation without square roots, *J. Institute of Mathematics and its Applications*, **12**, 329–36, 1973.

and the explanation of the modified Givens rotation in this text was adapted from this paper. The generic area of algorithms for solving least squares problems in the style of Section 5.2.2 is covered in the highly recommended book:

Lawson, C.L. and Hanson, R.J.
Solving Least Squares Problems, Prentice-Hall, 1974.

The use of parallelism in order to speed up recursive computations is discussed in the book:

Kung, S.Y., Whitehouse, H.J. and Kailath, T.
VLSI and Modern Signal Processing, Prentice Hall, 1985.

Fast recursions are based upon the asymptotic algorithm of Levinson for solving the Wiener filter equation. See:

Levinson, N.
The Wiener RMS error criterion in filter design and prediction, *J. Math. Phys.* **25**, 261, 1947.

The development of fast recursions based on shift invariance is due to Kailath and his co-workers. The discussion given here is taken from:

Robins, A.J. and Wellstead, P.E.
Recursive system identification using fast algorithms, *Int. J. Control*, **33**(3), 455–80, 1981.

and this paper gives a comprehensive account of how fast algorithms are derived and used for ARMA system identification. For more depth and interesting relationships with model validation, see:

Robins, A.J.
Data Decomposition in Structural Identification, PhD Thesis, Control Systems Centre, UMIST, 1980.

Fast recursions can have numerical stability problems which are associated with the Kalman gain update. For a discussion of these problems and their solution see:

Botto, J.L. and Moustakides, G.V.
Stabilizing the fast Kalman algorithms, *IEEE Trans.* **ASSP 37**(9), 1342–8, 1989.

APPENDIX 5.1

Proof of the U–D recursion for L(t)

The proof given here follows that of Bierman (Chapter V, Section 3, Theorem V.3.1) and Ljung and Soderstrom (Chapter 6, Section 6.2.2).

The normal recursion for $\mathbf{P}(t)$ is

$$\mathbf{P}(t) = \lambda^{-1}[\mathbf{P}(t-1) - \beta^{-1}\{\mathbf{P}(t-1)\mathbf{x}(t)\mathbf{x}^{\mathrm{T}}(t)\mathbf{P}(t-1)\}] \tag{A5.1}$$

where

$$\beta = \lambda + \mathbf{x}^{\mathrm{T}}(t)\mathbf{P}(t-1)\mathbf{x}(t) \tag{A5.2}$$

Note that $\mathbf{x}(t)$ is the column m vector of data and λ is the forgetting factor. Define the m vectors

$$\mathbf{f} = \mathbf{U}^{\mathrm{T}}(t-1)\mathbf{x}(t) \tag{A5.3}$$

$$\mathbf{g} = \mathbf{D}(t-1)\mathbf{f} \tag{A5.4}$$

and assume that $\mathbf{P}(t)$ can be factored as

$$\mathbf{P}(t) = \mathbf{U}(t)\mathbf{D}(t)\mathbf{U}^{\mathrm{T}}(t) \tag{A5.5}$$

Using (A5.3), (A5.4), (A5.5) in equations (A5.1) and (A5.2) it can be shown that

$$\mathbf{P}(t) = \mathbf{U}(t)\mathbf{D}(t)\mathbf{U}^{\mathrm{T}}(t) = \mathbf{U}(t-1)[\mathbf{D}(t-1) - \beta^{-1}(\mathbf{g}\mathbf{g}^{\mathrm{T}})]\mathbf{U}(t-1)\lambda^{-1} \tag{A5.6}$$

and

$$\beta = \lambda + \mathbf{f}^{\mathrm{T}}\mathbf{g} \tag{A5.7}$$

Now, if the bracketed terms on the right-hand side of equation (A5.6) can be written in the factored form

$$\bar{\mathbf{U}} \, \bar{\mathbf{D}} \, \bar{\mathbf{U}}^{\mathrm{T}} = \mathbf{D}(t-1) - \beta^{-1}(\mathbf{g} \, \mathbf{g}^{\mathrm{T}}) \tag{A5.8}$$

then the values of $\mathbf{U}(t)$ and $\mathbf{D}(t)$ can be found at time t from

$$\mathbf{U}(t) = \mathbf{U}(t-1)\bar{\mathbf{U}}; \qquad \mathbf{D}(t) = \bar{\mathbf{D}}\lambda^{-1} \tag{A5.9}$$

and the Kalman gain $\mathbf{L}(t)$ determined from

$$\mathbf{L}(t) = \mathbf{P}(t-1)\mathbf{x}(t)\beta^{-1} = \mathbf{U}(t-1)\mathbf{D}(t-1)\mathbf{U}(t-1)\mathbf{x}(t)\beta^{-1}$$

$$\mathbf{L}(t) = \mathbf{U}(t-1)\mathbf{g}\beta^{-1} \tag{A5.10}$$

The main part of the U–D factorization proof is in establishing the factorization (A5.8) such that we can determine $\mathbf{U}(t)$, $\mathbf{D}(t)$ from (A5.9).

We first denote $\bar{\mathbf{U}}$, $\bar{\mathbf{D}}$ and $\mathbf{D}(t-1)$ as follows:

$$\bar{\mathbf{U}} = [\bar{\mathbf{u}}_1, \ \bar{\mathbf{u}}_2, \ldots, \bar{\mathbf{u}}_m] = \begin{bmatrix} 1, & \bar{u}_{12}, & \bar{u}_{13}, & & \cdots & & \bar{u}_{1m} \\ 0, & 1, & \bar{u}_{23}, & & & & \\ & \cdot & \cdot & 1 & & & \cdot \\ & \cdot & \cdot & \cdot \cdot & & & \cdot \\ & \cdot & \cdot & \cdot & \cdot & & \cdot \\ & & & \cdot & & \cdot & \\ & & & & & & \bar{u}_{m-1m} \\ 0 & 0 & 0 & & \cdots & & 0 \quad 1 \end{bmatrix}$$

$\bar{\mathbf{D}} = \mathrm{diag} \, \{\bar{d}_j\}$, $\mathbf{D}(t-1) = \mathrm{diag} \, \{d_j\}$.

The factored form (A5.8) can now be written as a summation over the m column vectors $\bar{\mathbf{u}}_i$ which make up $\bar{\mathbf{U}}$. Thus:

$$\sum_{i=1}^{m} \bar{d}_i \bar{\mathbf{u}}_i \bar{\mathbf{u}}_i^{\mathrm{T}} = \sum_{i=1}^{m} d_i \mathbf{e}_i \mathbf{e}_i^{\mathrm{T}} - \beta_m^{-1} \mathbf{g} \mathbf{g}^{\mathrm{T}} \tag{A5.11}$$

where $\mathbf{e}_i$ is the ith unit vector, and β_m is introduced as the value of β defined by

$$\beta_m = \beta = \lambda + \sum_{i=1}^{m} f_i g_i \tag{A5.12}$$

Also, the vector $\mathbf{v}_m$ is defined as

$$\mathbf{v}_m = \mathbf{g} \tag{A5.13}$$

The aim of the proof from here on is to select $\bar{d}_i$ and $\bar{\mathbf{u}}_i$ to equate the ith row and column of the left- and right-hand sides of (A5.11). The method works backwards starting with the final (mth) term in the summation. Define the matrix $\mathbf{M}_m$ as

$$\mathbf{M}_m = \bar{d}_m \bar{\mathbf{u}}_m \bar{\mathbf{u}}_m^T - d_m \mathbf{e}_m \mathbf{e}_m^T + \beta_m^{-1} \mathbf{v}_m \mathbf{v}_m^T \qquad (A5.14)$$

By inspection, the mth row and column of $\mathbf{M}_m$ are zeroed by selecting

$$\bar{d}_m = d_m = \beta_m^{-1} v_{mm}^2 \qquad (A5.15)$$

$$\bar{u}_{mm} = 1 \qquad (A5.16)$$

$$\bar{d}_m u_{im} u_{mm} = -\frac{1}{\beta_m} v_{mm} v_{mi} \qquad (A5.17)$$

Using (A5.16) in (A5.17) gives

$$u_{im} = \frac{v_{mm} v_{mi}}{\beta_m \bar{d}_m} \qquad (A5.18)$$

for $\{i = 1, \ldots, m-1\}$.

Now, note that with the substitutions (A5.15), (A5.16) and (A5.18) the matrix $\mathbf{M}_m$ has the last column and row zero, so that its effective dimension is $(m-1) \times (m-1)$. Accordingly, M_m can be rewritten using (A5.15) and (A5.18) as

$$\mathbf{M}_m = \left[\frac{v_{mm}^2}{\bar{d}_m \beta_m^2} + \frac{1}{\beta_m} \right] \mathbf{v}_{m-1} \mathbf{v}_{m-1}^T \qquad (A5.19)$$

where the vector $\mathbf{v}_{m-1}$ is defined as

$$\mathbf{v}_{m-1} = \begin{bmatrix} v_{m1} \\ \cdot \\ \cdot \\ \cdot \\ v_{m,m-1} \\ \cdots \\ 0 \end{bmatrix}$$

Let the object β_k be defined by

$$\beta_k = \lambda + \sum_{i=1}^{k} f_i g_i \qquad (A5.20)$$

with

$$\beta_m = \beta$$

From (A5.13) and (A5.4)

$$g_m = v_{mm}; \qquad f_m = \frac{g_m}{v_{mm}} \qquad \text{and} \qquad \beta_m - \beta_{m-1} = g_m f_m$$

and it is then possible to show that

$$\frac{v_{mm}^2}{\bar{d}_m \beta_m^2} + \frac{1}{\beta_m} = \frac{1}{\beta_{m-1}} \tag{A5.21}$$

such that from (A5.19) we have

$$\mathbf{M}_m = \beta_{m-1}^{-1} \mathbf{v}_{m-1} \mathbf{v}_{m-1}^{\mathrm{T}} \tag{A5.22}$$

Now equation (A5.11) can be rewritten as

$$\sum_{i=1}^{m-1} \bar{d}_i \bar{\mathbf{u}}_i \bar{\mathbf{u}}_i^{\mathrm{T}} = \sum_{i=1}^{m-1} d_i \mathbf{e}_i \mathbf{e}_i^{\mathrm{T}} - \mathbf{M}_m \tag{A5.23}$$

$$\sum_{i=1}^{m-1} \bar{d}_i \bar{\mathbf{u}}_i \bar{\mathbf{u}}_i^{\mathrm{T}} = \sum_{i=1}^{m-1} d_i \mathbf{e}_i \mathbf{e}_i^{\mathrm{T}} - \beta_m^{-1} \mathbf{v}_{m-1} \mathbf{v}_{m-1}^{\mathrm{T}} \tag{A5.24}$$

Equation (A5.24) is the same form as (A5.11) (recall the definition of $\mathbf{v}_m$ (A5.13)) so that we can repeat the procedure to eliminate the $(m-1)$th row and column and evaluate the corresponding values of d_{m-1}, $u_{i,m-1}$ ($i=1, \ldots, m-2$) and so on for $m-2$, $m-3$, etc.

The values of $\mathbf{D}(t)$ and $\mathbf{U}(t)$ are then computed from (A5.9) and the $\mathbf{L}(t)$ object from (A5.10).

APPENDIX 5.2

Pascal programme for U–D factorization

This programme was written by George Wagner for Turbo-Pascal with credit to Ljung and Soderstrom via their book *Theory and Practice of Recursive Identification*, MIT Press, 1983.

You are encouraged to use this listing as a guide to constructing your own programme rather than copying it wholesale. An old programmer's saying gives the warning 'Copied code rarely works'.

```
Procedure Bierman-update    (Var U_uppertriangle    : array2d;
                             Var D_diagonal          : array1d;
                             Var beta_estimated       : array1d;
                                 little_lambda        : array1t;
                                 psi_gradient         : array1d;
                             Var L_Kalmangain        : array1d;
                                 d                   : integer;
                                 debug               : boolean;
                             Var keyed               : char;
                                 Term                : Charset   );
```

{This procedure updates the covariance matrix implicitly, and the Kalman gain vector explicitly}.

```
Var
                             f_UDfact       : array1d;
                             g_UDfact       : array1d;
                             mu_fbeta       : array1d;
                             i,j,k          : integer;
                             U-old          : real;

Begin
  For i:= d downto 2 do      { create f and g vectors }
    Begin
      f_UDfact[i] := psi_gradient[i];
      For j:= 1 to i−1 do
        f_UDfact[i] := f_UDfact[i]  +  U_uppertriangle[j,i]
                                   * psi_gradient[j]
      g_UDfact[i] := D_diagonal[i] * f_UDfact[i];
    End;

  f_UDfact[1] := psi_gradient[1];
  g_UDfact[1] := D_diagonal[1] * psi_gradient[1];

{ Now do the updates for j = 1 only }

  beta-estimated[1] := little_lambda[1] + f_UDfact[1] *
                        g_UDfact[1];
  D_diagonal[1]     := D_diagonal[1] / beta_estimated[1];
  mu_fbeta[1]       := −f_UDfact[1] / beta_estimated[1];
```

{ and now for the rest of the elements }

```
For j:= 2 to d do
  Begin
  beta_estimated[j] := beta_estimated[j−1] +
                          f_UDfact[j] * g_UDfact[j];

  D_diagonal[j]       := D_diagonal[j] * beta_estimated
                          [j−1] /
                          (beta_estimated[j] * little_
                          lambda[1];

  mu_fbeta[j]         := −f_UDfact[j] / beta_estimated[j−1];
  For i:= 1 to j−1 do
    Begin
      U_old                   := U_uppertriangle[i,j];
      U_uppertriangle[i,j] := U_old + g_UDfact[i] *
                              mu_fbeta[j]
      g_UDfact[i]          := g_UDfact[i]
                              + U_old * g_UDfact[j];

    End { i loop };
  End {j loop };
```

{ and finally create the L vector }

```
For i:= 1 to d do
  L_Kalmangain[i]   := g_UDfact[i] / beta_estimated[d]

If debug then

  Begin
    ClrScr;
    GotoXY (30,1);
    Write ('Estimation..');
    GotoXY (1,3);
    Write (' In procedure Bierman_Update');

    For i:= 1 to d do
      Begin
        GotoXY (1,i+4);
        Write ('D ',D_diagonal[i]:9,
         ' f ',f_UDfact[i]:9,
         ' g ',g_UDfact[i]:9,
         ' mu ', mu_fbeta[i]:9,
         ' U ', U_old:9,
```

```
                        ' beta ', beta_estimated [i]:9);
                  End;

            GotoXY (1,24);
            Write (' Hit ⟨CR⟩, ⟨SPC⟩, ⟨CTRL-N⟩, or ⟨CTRL-E⟩',
                    , to finish viewing.');
            Repeat
              Read (kbd,keyed);
            Until keyed in Term;
            ClrScr;
        End;
```

6 Convergence Analysis for Recursive Algorithms

6.1 OUTLINE AND LEARNING OBJECTIVES

In previous chapters we have discussed recursive algorithms for parameter estimation. An assumption in that discussion has been that the algorithms will exhibit stable behaviour and the parameters converge to produce useful estimates. For self-tuning systems, however, the experimental conditions may be extremely complex. In self-tuning control, for example, the system is driven by an input signal dependent on parameter estimates that are, in turn, functions of the input/output data. This situation may be further complicated by the implementation of necessary management techniques (Chapter 4). The operational conditions may therefore be such that the boundedness of crucial signals or the convergence of parameter estimates cannot be *a priori* guaranteed.

It is against this background that the material contained in this chapter should be viewed. In particular, it is recognized that recursive estimators are nonlinear, time-varying dynamic systems which may or may not be stable and whose stability properties are difficult to analyze in general. This chapter is concerned with the convergence properties of recursive algorithms in the immediate vicinity of a convergence point. The results so obtained give us information concerning the local convergence of the algorithm. The approach adopted is the Ordinary Differential Equation (ODE) method. In this approach, as discussed here, stability is assumed and a set of ODEs is formed which tells us the set of possible convergence points as well as how the recursive algorithm behaves in the vicinity of each convergence point.

The learning objective is to understand how to use the ODE approach as a means of checking analytically and/or graphically the basic viability of a recursive estimator when used alone or as part of a self-tuning system.

6.2 CONVERGENCE ISSUES IN SELF-TUNING

In this section we discuss the issues which give the convergence analysis of recursive algorithms in a self-tuning system its special flavour. In addition we also introduce some of the types of convergence definitions which exist. This is important in self-tuning because it is possible for a self-tuning system to function to the desired design specification even though the recursively estimated parameters are not at their 'physically correct' values.

6.2.1 Systems and models in recursive estimation

A key distinction which must be made in recursive estimation is the distinction between the object which generates the data and the object which we use in the estimator to model the data generating mechanism. The data generating mechanism may be a dynamical system which we wish to control or a signal source which we wish to process in some way. The data generating mechanism is termed the *system*. The representation of the system used in a recursive estimator is termed the *model*. In the discussions so far (and in Chapter 3 especially) the assumption has been usually made that the system and model have the same form or structure. In particular, it has been assumed that the system can be represented by a linear difference equation of some kind and that the estimation model is identical in form but with unknown coefficient values.

In fact, however, while in most situations we will endeavour to make model and system correspond structurally, there are occasions when the model and system will not have the same form, either through error or design.

Example 6.1   Model parametrization incorrect

The system is a random signal source described by the ARMA process

$$y(t) = \frac{1 + c_1 z^{-1}}{1 + a_1 z^{-1}} e(t) \tag{6.1}$$

The assumed model form is

$$y(t) = \frac{1 + \hat{c}_1 z^{-1}}{1 + \hat{a}_1 z^{-1} + \hat{a}_2 z^{-2}} \hat{e}(t) \tag{6.2}$$

The model (equation (6.2)) is an overparametrized representation of the system. The RELS estimates will converge on the true values so that the estimated coefficient $\hat{a}_2$ will tend to zero.   □

Example 6.2   Model structure incorrect

The system is the ARMAX process given by

$$y(t) = \frac{b_0 z^{-1}}{1+a_1 z^{-1}} u(t) + \frac{1+c_1 z^{-1}}{1+a_1 z^{-1}} e(t) \qquad (6.3)$$

The assumed model is the ARMAX process given by

$$y(t) = \frac{\hat{b}_0 z^{-1}}{1+\hat{a}_1 z^{-1}} u(t) + \hat{e}(t) \qquad (6.4)$$

The model (equation (6.4)) is in an incorrect form to represent the system (equation (6.3)) adequately if the experimental conditions are open loop. The estimates $\hat{b}_0$, $\hat{a}_1$ will be biased estimates of b_0, a_1.   □

The two examples of incorrect models are worth noting in the following sense. The first example (6.1) will still yield correct (i.e. unbiased) parameter estimates. The second will yield incorrect (i.e. biased) parameter estimates. In an open loop system identification experiment the model of Example 6.2 would be undesirable and would give an incorrect representation of the underlying system. The model, however, can still be of use if the input used to excite the system takes a specific closed loop form. The implicit self-tuners of Chapters 8 and 10 are examples of situations where this is true. The topic of data structure is discussed next.

6.2.2   Data generation in recursive estimation

As noted previously, the data associated with a *system* may be generated in a number of ways. In particular, consider a control situation in which a system may be open loop (Figure 6.1(a)), under fixed coefficient feedback control (Figure 6.1(b)) or under self-tuning adaptive feedback control (Figure 6.1(c)). In each case the interdependence of data and parameters is rather different (Figures 6.1a', b', c', respectively). In an open loop case the parameter estimates $\hat{\theta}$ are functions of the data (Figure 6.1(a)) in the classical manner studied in system identification. When a fixed coefficient controller is present, then a two-way dependence is set-up between the input $u(t)$ and output $y(t)$. This can cause problems as indicated in Chapter 3, but is nonetheless fairly straightfoward to analyze. The third situation (when a self-tuning controller is present) causes a further two-way interrelation between parameter estimates $\hat{\theta}$ and the input signal $u(t)$. The following example illustrates this interdependence.

Example 6.3

A simple self-tuning (adaptive) regulator takes the following form:

Figure 6.1 Data interrelations for some estimation situations.

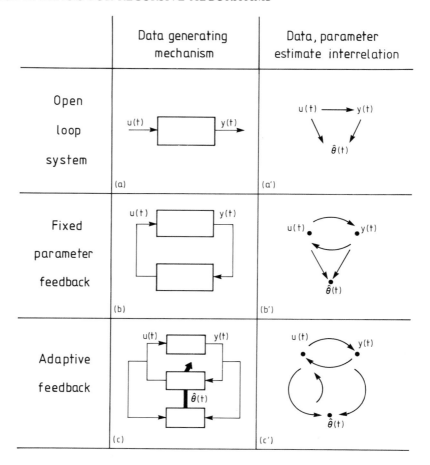

	Data generating mechanism	Data, parameter estimate interrelation
Open loop system	(a)	(a')
Fixed parameter feedback	(b)	(b')
Adaptive feedback	(c)	(c')

System: $y(t) = bu(t-1) + e(t)$ (6.5)

Model: $y(t) = \beta u(t-1)$ (6.6)

The model parameter β is estimated using RLS and used in the control law

$$u(t) = (z_0/\beta)y(t) \qquad (6.7)$$

The control law (equation (6.7)) and hence $u(t)$ is then specifically a function of the model parameter estimate $\hat{\beta}(t)$. The closed loop system obtained by combining (6.5) and (6.7) then becomes

$$y(t) = [bz_0/\hat{\beta}(t-1)]y(t-1) + e(t) \qquad (6.8)$$

This is a simple form of (adaptive) pole assignment control which is discussed in depth in Chapter 7.   □

At first glance, any analysis of the above algorithm may seem intractable given that the system, recursive estimator, and controller together form a time-varying, nonlinear stochastic system. The key, however, to understanding the nature of adaptive control is to recognize that the central feature of each algorithm is the estimation step. Problems of analysis are those associated with the behaviour of an estimation algorithm in which the input signal is generated from the data in a rather complex way. For example, equation (6.7) can be expressed as

$$u(t) = f[\hat{\theta}(t), y(t)] \tag{6.9}$$

where $\hat{\theta}(t)$ depends in a nonlinear way on past data through a recursive estimator.

Before the 1970s, most discussion of estimation algorithms was restricted to cases in which the input was open loop or generated via a constant linear feedback law. The essential difference between these cases and the adaptive situation is in the flow of information as depicted in Figure 6.1. Clearly, what is required is a method of analysis that deals with a wide class of estimation algorithms without specifying too narrowly the input generating mechanism.

6.2.3   Types of convergence of recursive algorithms

We can ask a number of important questions concerning the behaviour of any recursive algorithm. The first concerns stability:

Stability property

(S)   Is the overall system globally (or locally) stable? We would like the input and output signals to be bounded (in some sense) under a wide range of starting-up conditions.

We will not be dealing with the stability problem in this book. Rather we will focus upon the convergence behaviour which can be expected from a recursive algorithm under the assumption that it does in fact converge.

Convergence properties

(C1)   The *system* behaviour is correct (at least asymptotically).

(C2)   The *model* parameter estimates converge.

Figure 6.2 The hierarchy of convergence conditions.

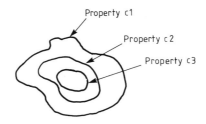

Property c1

Property c2

Property c3

(C3)   The estimates converge to 'correct' values.

These convergence properties can be arranged in a hierarchy (Figure 6.2) as follows:

(i) Property (C1) corresponds to *system identifiability* and is essential for achieving the correct control objective. It is important to note that property (C1) does *not* imply that property (C2) is true (or *vice versa*).

(ii) Property (C3) corresponds to *parameter identifiability* (or *consistency*). Note that property C1 does not imply that property C3 is true.

(iii) Property (C3) is only meaningful if the model is structurally consistent with the controlled system. Then, Property (C3) implies that Properties (C1) and (C2) are valid.

It will be useful to consider examples of Properties (C1), (C2) and (C3). Consider the following example of a self-tuning controller in which Property (C1) is met while (C2) and (C3) are not.

Example 6.4

A basic form of self-tuner is the minimum variance regulator which is considered in detail in Chapter 8. Here we consider a simple form in which a self-tuning algorithm exists for the following *system* and *model* description:

$$\text{System:} \qquad y(t) = y(t-1) - u(t-1) + e(t)$$

$$\text{Model:} \qquad y(t) = \hat{a}y(t-1) + \hat{b}u(t-1) + \hat{e}(t)$$

The minimum variance controller for the *system* is

$$u(t) = y(t)$$

and gives the closed loop equation

$$y(t) = e(t)$$

The minimum variance controller for the estimated *model* is

$$u(t) = - \frac{\hat{a}(t)}{\hat{b}(t)} y(t)$$

yielding *system* behaviour

$$y(t) = \left[1 + \frac{\hat{a}(t)}{\hat{b}(t)} \right] y(t-1) + e(t) \qquad (6.10)$$

Hence Property (C1) holds if

$$\lim_{t \to \infty} \left[1 + \frac{\hat{a}(t)}{\hat{b}(t)} \right] = 0 \qquad (6.11)$$

at a fast enough rate, but this does not imply either (C2) or (C3) since $\hat{a}(t)$, $\hat{b}(t)$ can take any values which satisfy (6.11).

As an illustration of the behaviour which might be expected in this situation, the reader's attention is drawn to Figure 6.3. Note that from Figure 6.3(a) system behaviour (as measured by the controller gain) converges to a steady value. The parameter estimates, however, move freely, constrained only by equation (6.11).

In the remainder of this chapter we will introduce and use a powerful analytical tool to examine the behaviour of a class of recursive estimation algorithms. In many cases we will be able to establish whether Properties (C1)–(C3) hold. Note, however, that our assumption is always that convergence takes place.

6.3   THE ORDINARY DIFFERENTIAL EQUATION (ODE) APPROACH — BASIC IDEAS

In this section we consider some basic concepts in the so-called ODE approach to the analysis of recursive algorithms. It is noted that the basic assumption is that the recursive algorithm is stable. The ODE method will give us a means for studying the recursive algorithm when it is close to a convergence point. Again we remind the reader that, while we will be focusing the discussion upon recursive algorithms, the entire class of self-tuning controllers and signal

Figure 6.3
Evolution of
parameter
estimates and
controller gain
in Example 6.4.

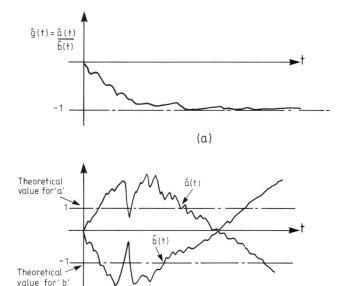

processors are included in the discussion. As pointed out in Section 6.2.2, the function of the controller or signal processor is to modify the way in which the recursive algorithm behaves via its influence upon the data.

6.3.1 Asymptotic behaviour of recursions

This method of analysing discrete-time recursions associates a *deterministic* ODE with each algorithm and derives the required properties of an algorithm by examining the solutions of the ODE.

There are two basic steps in the ODE approach. One concerns a time compression so that the asymptotic behaviour of the algorithm can be examined on a reasonable time scale. The second major step is the averaging out of stochastic and time-varying elements of recursive algorithms, leaving what is effectively the average behaviour of the parameter estimates.

The first of these steps, the time scale compression, can be explained by considering the deterministic scalar difference equation

$$x(t+1) = x(t) + \gamma(t+1) f(x(t)) \tag{6.12}$$

where the positive scalar gain sequence $\{\gamma(t)\}$ satisfies

$$\lim_{t \to \infty} \gamma(t) = 0 \tag{6.13}$$

If $f(\cdot)$ is a 'smooth' function and $\{x(t)\}$ is bounded (an implicit *stability assumption*), then (6.13) implies that $x(t)$ and $f(x(t))$ change slowly for large enough t. This assumption means that we can approximate the equation

$$x(t+s) = x(t) + \sum_{k=t+1}^{t+s} \gamma(k)\, f(x(k-1))$$

by

$$x(t+s) \simeq x(t) + \left[\sum_{k=t+1}^{t+s} \gamma(k) \right] f(x(t)) \tag{6.14}$$

The idea here is that, if $f(x(t))$ is changing slowly, we can assume it is constant over the interval t to $t+s$, provided s is not too large. The next step is to change to a new compressed time scale. To do this we introduce a new continuous 'time' variable, τ, defined by

$$\tau = \sum_{k=1}^{t} \gamma(k) \tag{6.15}$$

Note that the compressed time τ matches the behaviour of the real-time t under certain conditions. That is, $\tau=0$ when $t=0$ and τ tends monotonically to infinity with t, provided

$$\sum_{k=1}^{\infty} \gamma(k) = \infty \tag{6.16}$$

This is a requirement that the gain sequence does not converge to zero too rapidly and is a standard requirement in recursive estimators when the true parameters are known to be constant. To complete the transformation to the new time scale τ we write

$$X(\tau) = x(t) \tag{6.17}$$

$$\tau + \Delta\tau = \sum_{k=1}^{t+s} \gamma(k) \tag{6.18}$$

Equation (6.18) allows us to see how the transformation in time variables acts as a compression of the time scale. As shown in Figure 6.4, the gain sequence decreases rapidly with time. Thus the increment $\Delta\tau$ in the compressed time

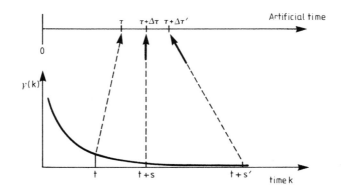

Figure 6.4 Illustrating the time scale compression which takes place in the transformation from t to compressed time τ.

decreases at a similar rate, such that as real-time increases, the increment in real-time required to make a significant change in compressed time τ is increasingly large. The effect is to ensure that we can inspect asymptotic values of k on a reasonable scale in the new time variable τ.

Having rescaled the time variable the *difference* equation (6.14) in the original time scale can be recast in the form

$$X(\tau + \Delta\tau) = X(\tau) + \Delta\tau f(X(\tau))$$

which in turn can be written as a *differential* equation in the τ time scale, i.e.

$$\frac{d}{d\tau} X(\tau) = f(X(\tau)) \tag{6.19}$$

in the limit as s (and hence $\Delta\tau$) go to zero.

The above arguments lead us to assert that the ODE (6.19) in some way represents the behaviour of the original difference equation (6.12). In particular, because of the time compression, it is suggested that the *asymptotic* behaviour of $x(t)$ in (6.12) is close to the trajectories of the ODE (6.19).

Example 6.5

Consider a recursion to solve

$$x^2 = 1 \tag{6.20}$$

Propose the recursive solution

$$\hat{x}(t+1) = \hat{x}(t) + \gamma(t+1) \left[\hat{x}^2(t) - 1\right] \tag{6.21}$$

The associated ODE is (from equation (6.17))

$$\frac{d}{d\tau} X(\tau) = f(X(\tau)) = X^2(\tau) - 1 \tag{6.22}$$

and, assuming that $\hat{x}(0)$ is zero, the solution of (6.22) is

$$X(\tau) = -\tanh \tau \tag{6.23}$$

and hence $X(\infty) = -1$, which is one of the two solutions of (6.20).

Figure 6.5(a) shows the recursion (6.21) graphically when $\gamma(t)$ is chosen to be $(t+1)^{-1}$. The estimate decreases monotonically from zero to -1, passing

Figure 6.5
Illustrating the recursive solution to Example 6.5. (a) The evolution of the recursion; (b) the solution to the equivalent ODE.

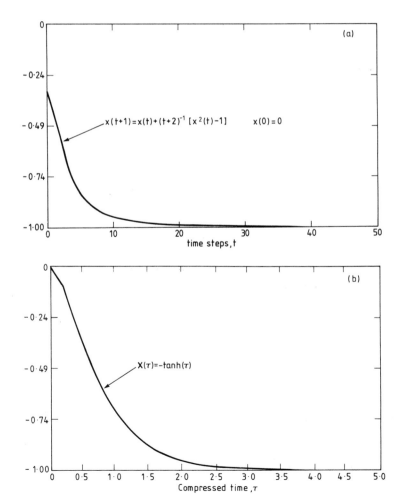

the value -0.99 when $t = 14$. This can be compared with the ODE trajectory (6.23) depicted in Figure 6.5(b) where $X(2.647) = -0.99$. The reason for the more rapid convergence of the ODE trajectory is the relationship between τ and t.

In this particular case, because of the choice of gain sequence $\gamma(t)$, the two time scales are logarithmically related. Thus

$$\tau = \sum_{i=1}^{t} \gamma(i) = \sum_{i=1}^{t} (i+1)^{-1} \simeq \ln t \qquad (6.24) \ \square$$

As noted previously this time scale compression is one important advantage of investigating the convergence behaviour of a recursion through simulation of the ODE rather than running the recursion itself. The reader is reminded that it is analogous to the logarithmic scaling of frequency in transfer function studies.

6.3.2   Local convergence

In a stable recursive algorithm the parameters will usually converge to some fixed point. In many cases the recursion is such that a unique convergence point is attained whatever the initial conditions. In this case the recursion is said to be *globally convergent*. However, many of the recursions of interest to us in estimation and self-tuning will have multiple solutions and in such cases they may only be *locally convergent*. The distinction betwen global and local convergence is important, since the ODE is mainly a tool for examining local convergence. The difference between local and global convergence may be clarified by reference to Figure 6.6. The figure draws an analogy between the movement of a ball on a surface and the solution to a recursive algorithm. In Figure 6.6(a), the ball representing the trajectory will move eventually to the global solution from any point on the surface. In Figure 6.6(b), the ball locally converges on solution θ_0 for θ less than the boundary value θ_d and on solution θ_1 for θ greater than θ_d.

The important point about the associated ODE is that it can give information concerning the local stability of a recursion. To reinforce this point it is informative to revisit Example 6.5.

Example 6.5   (continued)

It can easily be checked that the general solution of (6.22) has the form

$$X(\tau) = \left[\frac{1-\alpha\exp(2\tau)}{1+\alpha\exp(2\tau)} \right] \qquad (6.25)$$

Figure 6.6
Illustrating the
global and local
convergence
concepts via the
analogy to a ball
rolling on a
curved surface.

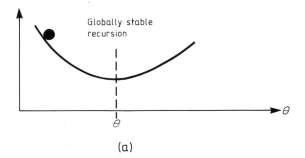

(a)

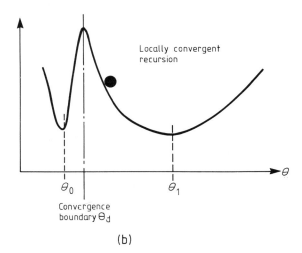

(b)

where α is fixed by initial conditions. Therefore, unless α is zero (i.e. the recursion is started at the solution $x = 1$ and remains there), the asymptotic value will be -1 whatever the starting value. Thus, the recursion will **never** converge to the solution $X = 1$, no matter how close the initial value. The reason can be seen from a closer investigation of the ODE (6.22). Any equilibrium point X^* is given by setting the left-hand side to zero and therefore

$$(X^*)^2 - 1 = 0 \tag{6.26}$$

Clearly both solutions of (6.20) satisfy this stationarity condition. In the neighbourhood of X^*, to the first order in

$$\bar{X}(\tau) = X(\tau) - X^*$$

(6.22) becomes

$$\frac{d}{d\tau} \bar{X}(\tau) = (2X^*) \bar{X}(\tau) \tag{6.27}$$

Therefore, small perturbations from X^* will grow exponentially unless X^* is negative, i.e. *only local convergence to $X^* = -1$ is possible.* □

Returning to the more general ODE (6.19), we can likewise state that, if $X(\tau)$ converges to X^*, then

$$f(X^*) = 0 \tag{6.28}$$

and, linearizing (6.19) around X^*,

$$\frac{d}{dx} f(x)\big|_{x=X^*} < 0 \tag{6.29}$$

Equations (6.28) and (6.29) are necessary conditions for the local convergence of (6.12) to X^*.

It is *most important* to note that these conditions are *not sufficient* and therefore do not guarantee convergence. The significance of (6.28) and (6.29) is as a check on *possible* local convergence points.

Note that, if $f(\cdot)$ is reversed in sign in the recursion (6.21) (equivalent to taking *negative* steps γ), then (6.28) is unchanged, i.e. the same equilibrium points appear. However, the two sets of such points corresponding to convergence and nonconvergence are interchanged. For example, if (6.21) is replaced by

$$\hat{x}(t+1) = \hat{x}(t) + \gamma(t+1) [1 - \hat{x}^2(t)] \tag{6.30}$$

then only the solution $x = 1$ to (6.20) is a possible point of convergence if (6.16) holds.

6.3.3. Some generalizations

The development of the ODE method thus far has been in terms of the simple scalar recursion given in equation (6.12). It is possible to generalize the method in two important ways: first, by making the function $f(x(t))$ in equation (6.12) dependent upon a further variable $y(t)$ and, second, by introducing a stochastic component into the recursion.

If we replace (6.12) by

$$x(t+1) = x(t) + \gamma(t+1) f[x(t), y(t)] \qquad (6.31)$$

where $y(t)$ is the 'state' of some system of interest that is interconnected with the recursion, then the arguments that led to the ODE are still valid provided that $y(t)$ 'settles down', i.e. if $y(t)$ is the state of an asymptotically stable system, so that $y(t)$ converges to some value $\bar{y}$. Then (6.19) is replaced by

$$\frac{dX}{d\tau} = f(X, \bar{y}) \qquad (6.32)$$

If $y(t)$ is in turn driven by $x(t)$, then $\bar{y}$ will be a function of X. This is the usual situation when an adaptive feedback law is coupled to an estimator.

If $f(\cdot)$ includes *stochastic* elements as arguments, then the right-hand side of (6.19) is replaced by the time-averaged expectation

$$\bar{E}(\cdot) = \lim_{N \to \infty} \frac{1}{N} \sum_{t=1}^{N} E(\cdot) \qquad (6.33)$$

where the expectation averages *all* stochastic sources, i.e.

$$f(X) = \bar{E} f(X, \xi(t)) \qquad (6.34)$$

where $\xi(t)$ represents the stochastic effects. If $E(\cdot)$ is independent of t (for example, if the basic stochastic processes are stationary), then $\bar{E}(\cdot)$ can be replaced by $E(\cdot)$ in (6.34).

The importance of introducing the expectation operation into the ODE is that it smooths the influence of random components which may disguise the convergence characteristics of the recursive algorithm. More will be said of this later; for the moment the following example will illustrate the procedure.

Example 6.6

System: $y(t) = b + e(t) \qquad E(e(t)) = 0$

Model: $y(t) = \beta$

Estimator: $\hat{\beta}(t) = \hat{\beta}(t-1) + \gamma(t)[y(t) - \hat{\beta}(t-1)]$

ODE: $\dfrac{d\beta}{d\tau} = f(\beta) = \bar{E}[y(t) - \beta]$

$= E[b + e(t) - \beta] = b - \beta$

For local convergence to β^*

$$f(\beta^*) = 0 \qquad \text{implies that} \qquad \beta^* = b$$

Moreover, the convergence point is stable since from equation (6.29) we have:

$$f'(\beta^*) = -1 < 0$$

Therefore, if convergence takes place,

$$\lim_{t \to \infty} \hat{\beta}(t) = b$$

in some sense. Note that the estimator recursion has a stochastic component whereas the ODE is *deterministic*. Thus there is no ambiguity in the time evolution of the ODE trajectories:

$$\frac{d\beta}{d\tau} = b - \beta$$

On the other hand, the estimates satisfy the stochastic recursion:

$$\hat{\beta}(t) - b = [1 - \gamma(t)] [\hat{\beta}(t-1) - b] + \gamma(t)e(t)$$

The time evolution of the estimate $\hat{\beta}$ is corrupted by noise and it may be difficult to determine convergence behaviour in a particular run.   □

The previous example illustrated the averaging effect which occurs in an ODE when stochastic terms appear in the recursion. The next example is selected to show how the $f(\cdot)$ object can be generalized, as in equation (6.32) to include additional variables.

Example 6.7

Consider the system used in Example 6.6, in which we have added an additional variable $r(t)$ on the right hand side.

System: $y(t) = b + r(t) + e(t)$ \qquad $E[e(t)] = 0$

$\{r(t)\}$ deterministic sequence

$\{e(t)\}$ not necessarily uncorrelated

Now, assuming that the model and estimator are as in Example 6.6, the ODE can be written as

$$\text{ODE:} \qquad \frac{d\beta}{d\tau} = f(\beta) = \bar{E}[y(t)) - \beta]$$

$$= b - \beta + r$$

where

$$\bar{r} = \lim_{N \to \infty} \frac{1}{N} \sum_{t=1}^{N} r(t)$$

For convergence to β^* the limit $\bar{r}$ must exist and be finite. For example, if $r(t)$ consists of a sum of sinusoids, then r exists and is zero. However, if $r(t)$ is a ramp function where $r(t) = \alpha t$, then $\bar{r}$ is infinite.

When $\bar{r}$ exists and is finite then

$$\beta^* = b + \bar{r}$$

and the convergence point is stable since

$$f'(\beta^*) = -1 < 0$$

Thus β^* is a possible point of convergence, but is not equal to b unless $\bar{r}$ is zero. The time evolution of the estimate and the ODE trajectory are shown in Figures 6.7(a) and (b), respectively when $\bar{r}$ is zero. Note that only the asymptotic behaviour displayed in Figures 6.7(a) and (b) are identical. The transient performances are not covered by the ODE framework and here differ considerably. □

6.3.4 Choice of gain sequence

The gain sequence $\{\gamma(t)\}$ used in a recursion does not appear in the ODE but it must be strongly emphasized that the validity of the ODE derivation depends on the properties of the gain sequence.

At a functional level, the gain sequence will (via the transformation 6.15) alter the scaling of the new compressed time variable. However, some formal conditions on $\gamma(t)$ are needed. The conditions

$$\text{(Ca)} \quad \lim_{t \to \infty} \gamma(t) = 0$$

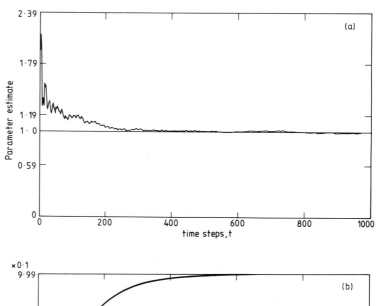

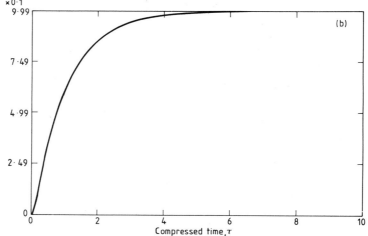

Figure 6.7 Showing the time evolution of the original recursion for the system of Example 6.7. (a) The corresponding ODE; (b) In this example, $r(t) = \sin(0.5t)$ and $b = 1$. $X(\tau) = 1 - \exp(-\tau)$.

(Cb) $\displaystyle\sum_t \gamma(t) = \infty$

(Cc) $\displaystyle\sum_t \gamma^2(t) < \infty$

are always required but may not be sufficient for global convergence. These conditions are satisfied by the sequence

$$\gamma(t) = \frac{1}{t}$$

It is not difficult to show, however, that, if (Ca) holds, then (Cb) *is a necessary and sufficient condition for local convergence* (see Section 6.10 Problems).

The following example illustrates what happens if the gain sequence used in a recursive algorithm breaks one of the above conditions.

Example 6.8

Consider the linear recursion

$$x(t+1) = x(t) + \gamma(t+1)[1 - x(t)] \tag{6.35}$$

If the gain sequence satisfies (Ca) and (Cb) then the only convergence point is $x^* = 1$, satisfying (6.28) and (6.29). The associated ODE is:

$$\frac{d}{d\tau} X(\tau) = 1 - X(\tau) \tag{6.36}$$

which has the solution:

$$X(\tau) = 1 - \exp(-\tau) \tag{6.37}$$

if we select initial condition $X(0) = 0$. The solution (6.37) gives $X(\infty) = 1$ as expected. However, writing $y(t) = 1 - x(t)$, (6.35) can be written as:

$$y(t+1) = [1 - \gamma(t+1)] y(t) \tag{6.38}$$

and if the gain sequence is chosen not to satisfy (Cb), e.g.

$$\gamma(t) = \frac{1}{(t+1)^2} \tag{6.39}$$

then, for arbitrary $x(0)$,

$$y(\infty) = y(0) \prod_{t=2}^{\infty} \left[1 - \frac{1}{t^2}\right] = 0.5 \, y(0) \tag{6.40}$$

i.e.

$$x(\infty) = 0.5[1 + x(0)] \tag{6.41}$$

and the recursion therefore does not converge to the correct value of unity but to a value dependent on the starting point.   □

6.4 THE PHASE PLANE AS A SIMULATION TOOL

Phase plane diagrams are a useful way of presenting information about the convergence behaviour of recursive estimation algorithms when more than a single parameter is being estimated. Each diagram plots one parameter estimate against another, where each estimate can be generated either by the ODEs or from the original discrete-time recursion. The most obvious advantage of a phase plane presentation is that any convergence point appears as a *finite* point on the graph.

Example 6.9

Consider the system

$$y(t) = b_0 u(t-1) + b_1 u(t-2) + e(t) \tag{6.42}$$

where b_0, b_1 are unknown with true values $b_0 = 1$, $b_1 = 2$ and $\{e(t)\}$ is zero mean white noise with variance $\sigma^2 = 0.3$.

Assuming a model of the correct structure, recursive estimates $\hat{\beta}_0(t)$, $\hat{\beta}_1(t)$ of b_0, b_1 are generated by a stochastic gradient algorithm:

$$\begin{bmatrix} \hat{\beta}_0(t) \\ \hat{\beta}_1(t) \end{bmatrix} = \begin{bmatrix} \hat{\beta}_0(t-1) \\ \hat{\beta}_1(t-1) \end{bmatrix} + \gamma(t) \begin{bmatrix} u(t-1) \\ u(t-2) \end{bmatrix}$$

$$\times [y(t) - \hat{\beta}_0(t-1)u(t-1) - \hat{\beta}_1(t-1)u(t-2)] \tag{6.43}$$

Provided that $\{\gamma(t)\}$ satisfies appropriate conditions for convergence (see Section 6.3.4), the associated ODE is

$$\frac{d}{d\tau} \begin{bmatrix} \beta_0(\tau) \\ \beta_1(\tau) \end{bmatrix} = \mathbf{f}(\beta_0, \beta_1) \tag{6.44}$$

where

$$\mathbf{f}(\beta_0, \beta_1) = \bar{E} \left\{ \begin{bmatrix} u(t-1) \\ u(t-2) \end{bmatrix} [y(t) - \beta_0 u(t-1) - \beta_1 u(t-2)] \right\} \tag{6.45}$$

$$= \bar{E} \left\{ \begin{bmatrix} u(t-1) \\ u(t-2) \end{bmatrix} [(b_0 - \beta_1)u(t-1) + (b_0 - \beta_1)u(t-2)] \right\}$$

using (6.42) and assuming that $\{u(t)\}$, $\{e(t)\}$ are not cross correlated.

Assuming that $\{u(t)\}$ is stationary in the sense that finite

$$\rho(i) = \bar{E}[u(t)u(t-i)] \qquad (i=0,1) \tag{6.46}$$

exist and are independent of t, then (6.44) becomes

$$\frac{d\beta_0}{d\tau} = \rho(0)\,(b_0-\beta_0)+\rho(1)\,(b_1-\beta_1) \tag{6.47a}$$

$$\frac{d\beta_1}{d\tau} = \rho(1)\,(b_0-\beta_0)+\rho(0)\,(b_1-\beta_1) \tag{6.47b}$$

These two interlinked ODEs can be simulated to show the 'time' trajectories of the estimates but such diagrams are not always satisfactory in revealing the asymptotic behaviour of the algorithm (corresponding to infinite τ). In this case, however, the ODEs are linear and convergence behaviour can be easily analyzed. The ODEs (6.47) can be cast in the standard state-space form

$$\frac{d}{d\tau}\begin{bmatrix} b_0-\beta_0 \\ b_1-\beta_1 \end{bmatrix} = -\begin{bmatrix} \rho(0) & \rho(1) \\ \rho(1) & \rho(0) \end{bmatrix}\begin{bmatrix} b_0-\beta_0 \\ b_1-\beta_1 \end{bmatrix} \tag{6.48}$$

In (6.48), the system matrix is negative definite provided

$$|\rho(1)| \neq \rho(0) \tag{6.49}$$

(prove it!). Under this condition, the state in (6.48) decays to zero and therefore (6.49) is a 'persistent excitation' condition on the input that ensures consistency, i.e.

$$\lim_{t \to \infty} \hat{\beta}_i(t) = b_i \qquad (i=0,1) \tag{6.50}$$

If (6.49) is not satisfied, then an identifiability problem occurs and (6.50) is not guaranteed.

These phenomena are clearly revealed by phase plane diagrams. The phase plane trajectories for (6.47) are now calculated for two different input choices, only one of which satisfies (6.49). In all cases

$$\gamma(t) = 1/t \tag{6.51}$$

Case (a)   The input is a zero mean white noise signal with unit variance. Then

$$\rho(0) = 1, \qquad \rho(1) = 0 \tag{6.52}$$

so that (6.49) holds and (6.50) results. From (6.47):

$$\frac{d\beta_1}{d\beta_0} = \frac{\beta_1 - b_1}{\beta_0 - b_0} \tag{6.53}$$

so that

$$(\beta_1 - b_1)/(\beta_0 - b_0) = \text{constant} \tag{6.54}$$

Hence the ODE phase plane trajectories are a family of straight lines passing through the point $(b_0, b_1) = (1,2)$ (Figure 6.8). Each straight line corresponds to a particular starting point. The actual trajectories $\beta_0(\tau), \beta_1(\tau)$ are shown in Figure 6.9.

Figure 6.8 ODE phase plane diagram with white noise input (Example 6.9).

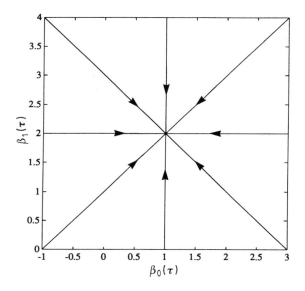

Figure 6.10 is the phase plane picture based on plotting $\hat{\beta}_1(t)$ against $\hat{\beta}_0(t)$ for $t=1,2,\ldots$, and does not give clear convergence information. Figure 6.11, however, presents the same plot for $t=1000,\ldots,3000$ and it is clear that this is approximated well by the ODE trajectories in Figure 6.8 and, if simulated for a longer time, would yield the same convergence point $(1,2)$.

To emphasize the use of phase plane diagrams, let us now assume that a constant forgetting factor λ is used. The gain choice (6.51) can be expressed in the recursive form

Figure 6.9 ODE trajectories corresponding to Figure 6.8.

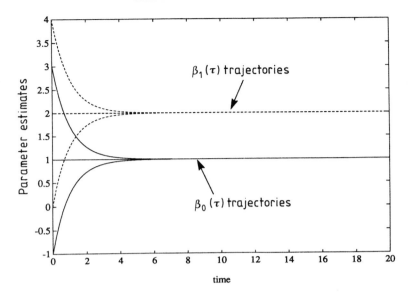

Figure 6.10 Phase plane diagram of parameter estimates starting from $t = 1$ and corresponding to Figure 6.8.

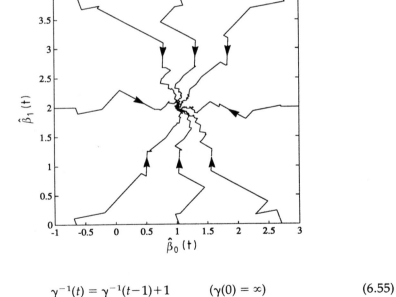

$$\gamma^{-1}(t) = \gamma^{-1}(t-1)+1 \qquad (\gamma(0) = \infty) \tag{6.55}$$

and the rate at which the gain sequence decreases can be slowed down by replacing (6.55) by the recursion

$$\gamma^{-1}(t) = \lambda\gamma^{-1}(t-1)+1 \qquad (\lambda<1) \tag{6.56}$$

Figure 6.11
Phase plane
diagram of
parameter
estimates
($1000 \leqslant t \leqslant 3000$)
and
corresponding to
Figure 6.8.

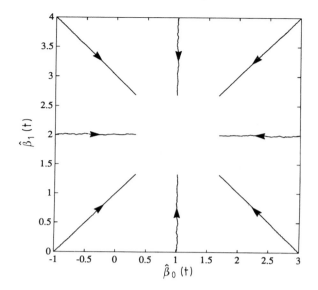

(cf. RLS estimation, Section 4.5.6). Then

$$\lim_{t \to \infty} \gamma(t) = 1-\lambda \neq 0 \tag{6.57}$$

and the arguments used to justify the ODE approximation are no longer valid. A phase plane picture of $\hat{\beta}_0(t)$, $\hat{\beta}_1(t)$, however, is still useful. Figures 6.12 and 6.13 are a replay of Figure 6.10 with $\lambda=0.96$ and 0.90 respectively. We can see that

(i) the estimates converge to a region around the point (1,2); and

(ii) the area of the convergence region increases as λ decreases.

Case (b)   The input is constant and equal to unity.
Then

$$\rho(0) = 1 = \rho(1) \tag{6.58}$$

and therefore (6.49) is violated.
From (6.47)

$$\frac{d\beta_1}{d\beta_0} = 1 \tag{6.59}$$

Figure 6.12
Phase plane
diagram for
forgetting factor
$\lambda = 0.96$
corresponding to
Figure 6.10.

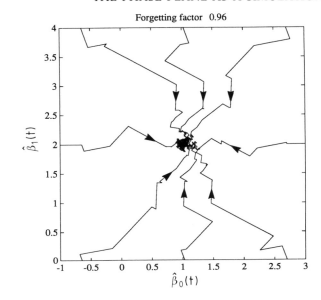

Figure 6.13
Phase plane
diagram for
forgetting factor
$\lambda = 0.90$
corresponding to
Figure 6.10.

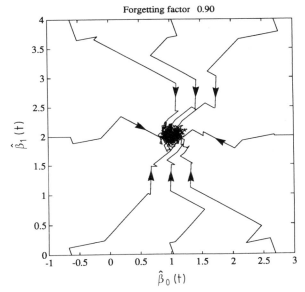

and therefore

$$\beta_1 = \beta_0 + \text{constant} \qquad (6.60)$$

a family of parallel straight lines at 45° to the axes.

Figure 6.14
ODE phase
plane diagram
with constant
input (Example
6.9 case (b)).

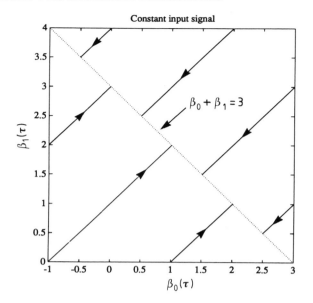

Figure 6.14 is the ODE phase plane diagram and shows this clearly. Note that all the trajectories end on the line

$$\beta_0 + \beta_1 = 3$$

which is the true value of $b_0 + b_1$. Therefore we still have system identifiability although we have lost parameter identifiability (consistency).

Figure 6.15 shows clearly that the parameter estimates converge to values that depend on initial conditions.   □

We may summarize the advantages of using phase plane diagrams as follows:

- Convergence behaviour is clearly shown if phase plane diagrams are based on ODEs.
- Identifiability problems are clearly revealed.
- Phase plane diagrams based on the *original recursion* approximate those generated from the ODEs (for *large* values of time) and give useful information if
  (i) the ODEs do not strictly exist (e.g. presence of constant forgetting factor) or
  (ii) there is difficulty in forming the ODEs (e.g. expectation cannot be carried out analytically due to nonlinear terms).

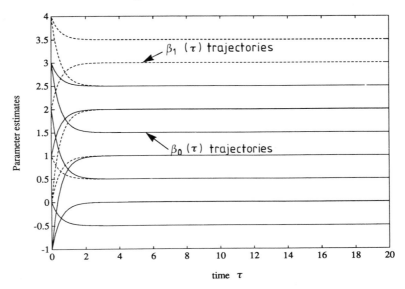

Figure 6.15
ODE trajectories of β_0, β_1 corresponding to Figure 6.14.

6.5 THE ORDINARY DIFFERENTIAL EQUATION APPROACH — GENERAL RESULTS

In this section we take the basic concepts developed in Section 6.3 and show how they relate to general results for ODEs as applied to recursive estimation algorithms.

6.5.1 Stochastic gradient methods

As a preliminary to developing general ODE results for recursive estimators it will be useful to reconsider the recursive algorithms themselves and to cast them in a standard form suitable for ODE analysis. In particular, it is useful to start by considering recursive estimators as hill-climbing or gradient algorithms. In this connection, recall that (Section 3.5), recursive least squares (RLS) can be considered as a hill-climbing algorithm within a stochastic framework. This approach is useful in defining a class of on-line estimation algorithms that can be covered by the available methods of analysis.

Consider the standard problem of fitting a linear-in-parameters model

$$y(t) = \mathbf{x}^{\mathrm{T}}(t)\boldsymbol{\theta} \tag{6.61}$$

to discrete data indexed by $t = 1, 2, \ldots, N$, where $\boldsymbol{\theta}$ is a column m vector of parameters (for example, see Section 3.2). If the data generator (system) is stochastic, then it is reasonable to choose $\boldsymbol{\theta}$ to minimize the variance of the 'equation error'

$$v(t,\boldsymbol{\theta}) = y(t) - \mathbf{x}^T(t)\boldsymbol{\theta} \tag{6.62}$$

i.e.

$$\min_{\boldsymbol{\theta}} \{V(\boldsymbol{\theta}) = \tfrac{1}{2}E[y(t) - x^T(t)\boldsymbol{\theta}]^2\} \tag{6.63}$$

Assuming a calculus minimum exists,

$$-\frac{\mathrm{d}}{\mathrm{d}\boldsymbol{\theta}} V(\boldsymbol{\theta}) = E\,\mathbf{x}(t)\,[y(t) - \mathbf{x}^T(t)\boldsymbol{\theta}] = 0 \tag{6.64}$$

Clearly $\boldsymbol{\theta}$ cannot be calculated from (6.64) unless the probability distributions are known, but it is usually reasonable to approximate by replacing the expectation by a time average, so that

$$\frac{1}{N}\sum_{t=1}^{N} \mathbf{x}(t)y(t) = \frac{1}{N}\sum_{t=1}^{N} \mathbf{x}(t)\mathbf{x}^T(t)\boldsymbol{\theta}$$

i.e. the standard *off-line* least squares solution.

The *direct* solution of (6.64) is a particular case of the following problem studied by Robbins and Monro in 1951 (see Section 6.11 Notes and References). In particular, Robbins and Monro considered the following problem:

Let $\{w(t)\}$ be a sequence of random variables (rv) each with the same distribution. Find the fixed parameter θ such that

$$E[Q(\theta,w(t))] = 0 \tag{6.65}$$

for a given function Q.

Robbins and Monro suggested the following recursive solution:

$$\hat{\theta}(t) = \hat{\theta}(t-1) + \gamma(t)Q(\hat{\theta}(t-1),w(t)) \tag{6.66}$$

where $\{\gamma(t)\}$ is a sequence of positive scalars converging to zero, but not too rapidly. Under certain conditions, the sequence of estimates generated by the recursion (6.66) will converge to a solution of (6.65).

Note that, although $w(t)$ may not be directly observable at t, it is assumed that Q is calculable from data once θ is fixed.

Example 6.10

Find the constant mean $Ew(t)$ from data $w(1)$, $w(2)$. . .,. This is equivalent to finding θ such that

$$E[w(t) - \theta] = 0 \tag{6.67}$$

From (6.66)

$$\hat{\theta}(t) = \hat{\theta}(t-1) + \gamma(t)[w(t) - \hat{\theta}(t-1)] \tag{6.68}$$

Choosing $\gamma(t) = 1/t$, then, for any $\hat{\theta}(0)$,

$$\hat{\theta}(t) = \frac{1}{t} \sum_{k=1}^{t} w(k) \tag{6.69}$$

i.e. the sample mean.   □

Returning to (6.64) and generalizing to the vector $\mathbf{Q}$ case:

$$-\frac{\mathrm{d}}{\mathrm{d}\boldsymbol{\theta}} V(\theta) = E[\mathbf{Q}(\boldsymbol{\theta},\mathbf{w}(t))] = 0$$

where

$$\mathbf{Q}(\boldsymbol{\theta},\mathbf{w}(t)) = \mathbf{x}(t)[y(t) - \mathbf{x}^{\mathrm{T}}(t)\boldsymbol{\theta}]$$

and $\mathbf{w}(t)$ represents the 'randomness' in $y(t)$, $\mathbf{x}(t)$. Then

$$\hat{\boldsymbol{\theta}}(t) = \hat{\boldsymbol{\theta}}(t-1) + \gamma(t)\mathbf{x}(t)[y(t) - \mathbf{x}^{\mathrm{T}}(t)\hat{\boldsymbol{\theta}}(t-1)] \tag{6.70}$$

which is a standard 'stochastic gradient' algorithm and is of the predictor–corrector type.

Note that 'on average' we are moving in the steepest descent direction. This is the stochastic analogue of *steepest descent*. It is well known, however, that steepest descent is inefficient when close to a minimum and that so-called Newton or quasi-Newton methods give better results. We therefore replace (6.70) by

$$\hat{\boldsymbol{\theta}}(t) = \hat{\boldsymbol{\theta}}(t-1) + \gamma(t)\mathbf{R}(t)^{-1}\mathbf{x}(t)[y(t) - \mathbf{x}^{\mathrm{T}}(t)\hat{\boldsymbol{\theta}}(t-1)] \tag{6.71}$$

where $\mathbf{R}(t)$ represents the hessian (second derivative) matrix of the cost function:

$$\frac{d^2}{d\boldsymbol{\theta}^2} V(\boldsymbol{\theta}) = E[\mathbf{x}(t)\mathbf{x}^T(t)] \tag{6.72}$$

We can use a Robbins–Monro scheme to generate $\mathbf{R}(t)$ iteratively by solving for $\mathbf{R}$ from the equation

$$E[\mathbf{x}(t)\mathbf{x}^T(t) - \mathbf{R}] = 0$$

i.e.

$$\mathbf{R}(t) = \mathbf{R}(t-1) + \gamma(t)[\mathbf{x}(t)\mathbf{x}^T(t) - \mathbf{R}(t-1)] \tag{6.73}$$

using the same gain sequence for simplicity.

It turns out that, for the $\gamma(t) = 1/t$, (6.70) and (6.73) yield the standard RLS estimation algorithm. This is easily shown by making the substitution

$$\mathbf{P}(t) = \frac{1}{t}\mathbf{R}(t)^{-1} \tag{6.74}$$

which transforms (6.71) to standard RLS form while (6.73) becomes

$$\mathbf{P}(t)^{-1} = \mathbf{P}(t-1)^{-1} + \mathbf{x}(t)\mathbf{x}^T(t)$$

as required.

Remarks

(i) In (6.71), note that the (symmetric positive definite) matrix $\mathbf{R}(t)$ modifies the descent direction and can assist the rate of convergence.

(ii) The choice of gain sequence $\{\gamma(t)\}$ determines the step length of the algorithm but, unlike the deterministic situation, must converge to zero to progressively suppress the effect of noise on the estimates.

(iii) It can be shown that the choice $\gamma(t) = 1/t$ is optimal in the sense that most information is extracted from the data asymptotically.

(iv) The 'stochastic gradient' approach makes it easier to see why some algorithms work better than others and to suggest modifications and generalizations.

(v) For analysis purposes, the standard form of RLS will be taken as (6.71) and (6.73) with $\gamma(t) = 1/t$, i.e.

$$\hat{\boldsymbol{\theta}}(t) = \hat{\boldsymbol{\theta}}(t-1) + \frac{1}{t} \mathbf{R}(t)^{-1} \mathbf{x}(t)[y(t) - \mathbf{x}^{\mathrm{T}}(t) \hat{\boldsymbol{\theta}}(t-1)] \qquad (6.75\mathrm{a})$$

$$\mathbf{R}(t) = \mathbf{R}(t-1) + \frac{1}{t} [\mathbf{x}(t)\mathbf{x}^{\mathrm{T}}(t) - \mathbf{R}(t-1)] \qquad (6.75\mathrm{b})$$

6.5.2 A class of recursive estimation algorithms

We now turn to a more general class of recursions for which the ODE approach offers a means of analysis and which covers most of the adaptive control algorithms of interest.

Consider the following parameter estimation algorithm:

$$\hat{\boldsymbol{\theta}}(t) = \hat{\boldsymbol{\theta}}(t-1) + \gamma(t)\mathbf{R}(t)^{-1} \mathbf{Q}(\hat{\boldsymbol{\theta}}(t-1),\mathbf{x}(t)) \qquad (6.76\mathrm{a})$$

$$\mathbf{R}(t) = \mathbf{R}(t-1) + \gamma(t) \mathbf{F}(\hat{\boldsymbol{\theta}}(t-1), \mathbf{R}(t-1), \mathbf{x}(t)) \qquad (6.76\mathrm{b})$$

Clearly, the RLS algorithm (6.75) can be cast in this form. The system state $\mathbf{x}(t)$ may be generated via an adaptive feedback law, e.g.

$$u(t) = g_1^{\mathrm{T}}(\hat{\boldsymbol{\theta}}(t))\mathbf{x}(t) + g_2(\hat{\boldsymbol{\theta}}(t))r(t) \qquad (6.77)$$

where $r(t)$ is an external input (reference signal). It is therefore essential, if we wish to carry through the transformation to an ODE, that the system controlled by (6.77), together with (6.76), should give rise to closed loop stability so that $\mathbf{x}(t)$ 'settles down' (Section 6.3.1). Formally, if the closed loop system matrix is denoted by $\mathbf{A}(\hat{\boldsymbol{\theta}}(t))$ at time t, then we must demand that, after some finite time, $\hat{\boldsymbol{\theta}}(t)$ belongs to the set

$$D_{\mathrm{s}} = \{\boldsymbol{\theta}|\mathbf{A}(\boldsymbol{\theta}) \text{ has all eigenvalues inside the unit circle}\}.$$

This is a strong condition involving closed loop stability under the adaptive feedback law.

It is a weakness of the ODE method that, in general, stability can only be assumed and not established within the ODE framework. At this stage, we assume that the stability assumption holds without further analysis.

Finally, note that, in most cases, the state $\mathbf{x}(t)$ may not have to be constructed explicitly but it is used here for analytical purposes.

6.5.3   The general ODE

Given the assumption of stability, let $\boldsymbol{\theta} \in D_s$ and let $\bar{x}(t,\boldsymbol{\theta})$ denote $x(t)$ with $\hat{\boldsymbol{\theta}}(t)$ replaced by $\boldsymbol{\theta}$. This rather clumsy notation is introduced here to emphasize the dependence of certain signals (state, input, output) on the parameter estimate when an adaptive feedback law is in operation. It will be omitted later.

To derive the ODE from (6.76), define

$$\mathbf{f}(\boldsymbol{\theta}) = \bar{E}[\mathbf{Q}(\boldsymbol{\theta},\bar{x}(t,\boldsymbol{\theta}))] \tag{6.78}$$

$$\mathbf{G}(\boldsymbol{\theta},\mathbf{R}) = \bar{E}[\mathbf{F}(\boldsymbol{\theta},\mathbf{R},\bar{x}(t,\boldsymbol{\theta}))] \tag{6.79}$$

(see Section 6.3.1). Then the ODE corresponding to (6.76) is given by:

$$\frac{d\boldsymbol{\theta}}{d\tau} = \mathbf{R}^{-1}\mathbf{f}(\boldsymbol{\theta}) \tag{6.80a}$$

$$\frac{d\mathbf{R}}{d\tau} = \mathbf{G}(\boldsymbol{\theta},\mathbf{R}) \tag{6.80b}$$

where τ is given by equation (6.15) as before, $\boldsymbol{\theta}(\tau) = \hat{\boldsymbol{\theta}}(t)$ and $\mathbf{R}(\tau)$ replaces $\mathbf{R}(t)$.

The right-hand side of (6.80a) is simply the answer to the question: 'What would be the average updating direction for the estimates under a constant feedback law corresponding to a nominal $\boldsymbol{\theta}$ value?'

We now present two results that correspond to those derived in our preliminary discussion of Section 6.3.

Result 1

The trajectories of the ODE (6.80) are the asymptotic paths of the estimates generated by (6.76).

Simulation of the ODE is computationally heavier than simulation of the original recursion. For m parameters (6.80) comprises $m(m+3)/2$ simultaneous nonlinear ODEs (allowing for the symmetry of $\mathbf{R}$). However, as previously noted ODE simulation is advantageous for two reasons in particular:

(i) The ODE time scale is compressed

$$\tau = \sum_{i=1}^{t} \gamma(i) \simeq \ln t \qquad \text{if } \gamma(t) = \frac{1}{t}$$

(ii) The behaviour of the estimates may be disguised by an unusual noise realization in a particular simulation, whereas the ODE is deterministic.

6.5.4   Local convergence

Result 2

If $\hat{\boldsymbol{\theta}}(t) \to \boldsymbol{\theta}^*$ and $\mathbf{R}(t) \to \mathbf{R}^*$ (> 0) as $t \to \infty$ (with a probability greater than zero), then

$$\mathbf{f}(\boldsymbol{\theta}^*) = 0 \qquad \text{and} \qquad \mathbf{G}(\boldsymbol{\theta}^*, \mathbf{R}^*) = 0 \tag{6.81}$$

and the matrix

$$\mathbf{H}(\boldsymbol{\theta}^*) = (\mathbf{R}^*)^{-1} \frac{\mathrm{d}}{\mathrm{d}\boldsymbol{\theta}} \mathbf{f}(\boldsymbol{\theta}) \big|_{\boldsymbol{\theta} = \boldsymbol{\theta}^*} \tag{6.82}$$

must have all its eigenvalues in the left half-plane (including the imaginary axis).

This result states that $\boldsymbol{\theta}^*$ is a locally stable, stationary point of the ODE. Such points are the only possible convergence points of the estimation algorithm. From (6.80), it is clear that convergence implies that the left-hand side is zero at a convergence point and (6.81) follows immediately. Linearizing around $(\boldsymbol{\theta}^*, \mathbf{R}^*)$ leads to

$$\frac{\mathrm{d}\boldsymbol{\theta}}{\mathrm{d}\tau} = \mathbf{H}(\boldsymbol{\theta}^*) \, (\boldsymbol{\theta} - \boldsymbol{\theta}^*) \tag{6.83}$$

and the required stability of $\mathbf{H}(\boldsymbol{\theta}^*)$ follows.

(i) Note that, in order to use the ODE approach, we have to transform the estimation algorithm into the standard form (6.76); and

(ii) calculate the expectations (6.78) and (6.79) to derive $\mathbf{f}(\boldsymbol{\theta})$ and $\mathbf{G}(\boldsymbol{\theta}, \mathbf{R})$.

In order to flesh out these ideas, we now return to the simple self-tuning regulator introduced in Example 6.3. This is probably the most elementary example of the adaptive pole assignment controllers discussed and analyzed in the next chapter.

Example 6.3 (continued)

Consider the *system*

$$y(t) = bu(t-1)+e(t) \tag{6.84}$$

where $\{e(t)\}$ is a zero mean white noise sequence with variance σ_c^2. A *model* of the system is proposed of the form

$$y(t) = \beta u(t-1) \tag{6.85}$$

and the regulator

$$u(t) = \frac{z_0}{\hat{\beta}(t)}\, y(t) \tag{6.86}$$

is synthesized and implemented where $\hat{\beta}(t)$ is the RLS estimate of β in (6.85). Note that, if $\hat{\beta}(t)$ converges to b, then the closed loop behaviour of the system is given by

$$y(t) = \frac{1}{1-z_0 z^{-1}}\, e(t) \tag{6.87}$$

so that the closed loop pole is at $z=z_0$, as required, and gives closed loop stability provided $|z_0|<1$. Our task is to show that, if convergence takes place, then only (6.87) can result.

To complete the description of the algorithm, the RLS estimation step is required in the form (6.75):

$$\hat{\beta}(t) = \hat{\beta}(t-1)+\frac{1}{t}R(t)^{-1}u(t-1)[y(t)-\hat{\beta}(t-1)\,u(t-1)] \tag{6.88a}$$

$$R(t) = R(t-1)+\frac{1}{t}[u^2(t-1)-R(t-1)] \tag{6.88b}$$

To derive the ODEs, fix the estimate $\hat{\beta}$ at some nominal value β everywhere. Then

$$f(\beta) = \bar{E}\, u(t-1)[y(t)-\beta u(t-1)] \tag{6.89}$$

where $y(t)$ is generated by (6.84) under the 'frozen' version of (6.86), i.e.

$$u(t) = (z_0/\beta)y(t) \tag{6.90}$$

so that

$$y(t) = \frac{1}{1-\gamma z^{-1}} e(t) \tag{6.91}$$

where

$$\gamma = bz_0/\beta \tag{6.92}$$

Assuming $|\gamma|<1$ for stability, (6.91) represents a stationary process and therefore $\bar{E}$ can be replaced by E in (6.89). Then

$$f(\beta) = E\left\{ \frac{z_0}{\beta} \frac{z^{-1}}{1-\gamma z^{-1}} e(t) \cdot \frac{1-z_0 z^{-1}}{1-\gamma z^{-1}} e(t) \right\}$$

$$= (\sigma_e^2 z_0/\beta)\mathrm{res}\left\{ \frac{(z-z_0)}{(1-\gamma z)(z-\gamma)} \right\}$$

$$= \frac{\sigma_e^2 z_0}{\beta} \frac{(\gamma-z_0)}{1-\gamma^2} \tag{6.93}$$

where $\mathrm{res}\{\cdot\}$ denotes the sum of the residues at the poles of $\{\cdot\}$ lying within the unit circle.

If $\hat{\beta}(t) \rightarrow \beta^*$, then $f(\beta^*)=0$ and, since $|\gamma|<1$, it follows that $\gamma^*=z_0$, i.e.

$$\beta^* = b \tag{6.94}$$

uniquely, leading to (6.87), the correct input/output behaviour.

To check the eigenvalue condition for local convergence, from (6.88b)

$$G(\beta,R) = Eu^2(t-1)-R \tag{6.95}$$

Therefore

$$R^* = [Eu^2(t-1)]_{\beta=\beta^*} = \frac{\sigma_e^2 z_0^2}{b^2(1-z_0^2)} \tag{6.96}$$

using (6.87), (6.90) and (6.94).

From (6.93)

$$f'(b) = -\frac{z_0}{b}\left[\frac{\mathrm{d}}{\mathrm{d}\gamma} f(\beta) \right]_{\beta=b} = -\frac{\sigma_e^2 z_0^2}{b^2(1-z_0^2)} \tag{6.97}$$

Therefore

$$(R^*)^{-1} f'(\beta^*) = -1 < 0 \qquad (6.98)$$

(Note that, in the case of *single parameter* estimation, R^* is a positive constant, not a matrix. It is therefore sufficient to check that $f'(\beta^*)$ is negative.)   □

Thus we have proved that the only possible convergence point in this adaptive control example corresponds to the desired closed-loop behaviour. This corresponds to system identifiability but not parameter identifiability (consistency). We have *not* proved that the algorithm gives rise to stable behaviour or that convergence will in fact take place at all.

6.6   RECURSIVE LEAST SQUARES REVISITED

In this section we reconsider the recursive least squares (RLS) algorithm in the light of the ODE approach. We examine its convergence properties and how they are affected by experimental conditions. In particular, the influence of the state $\mathbf{x}(t)$ excitation level on convergence is considered, along with such phenomena as parameter estimate biassing.

6.6.1   The recursive least squares algorithm

Consider the RLS estimation of $\boldsymbol{\theta}$ in the model

$$y(t) = \mathbf{x}^\mathrm{T}(t)\boldsymbol{\theta} \qquad (6.61)\text{bis}$$

i.e.

$$\hat{\boldsymbol{\theta}}(t) = \hat{\boldsymbol{\theta}}(t-1) + \frac{1}{t}\,\mathbf{R}(t)^{-1}\,\mathbf{x}(t)[y(t) - \mathbf{x}^\mathrm{T}(t)\hat{\boldsymbol{\theta}}(t-1)] \qquad (6.99\mathrm{a})$$

$$\mathbf{R}(t) = \mathbf{R}(t-1) + \frac{1}{t}\,[\mathbf{x}(t)\mathbf{x}^\mathrm{T}(t) - \mathbf{R}(t-1)] \qquad (6.99\mathrm{b})$$

as in (6.75). The data may be the result of applying a causal, adaptive feedback law.

Note that, at this stage, no restriction (such as linearity) is imposed on the functional form of $\mathbf{x}(t)$ as a function of the input/output data available at time $t-1$, i.e. we are not yet restricted to the standard A,B,C polynomial difference equation (CARMA or ARMAX) model introduced in Chapter 3. In order to form the ODE (6.80), fix $\hat{\boldsymbol{\theta}}$ at some nominal value $\boldsymbol{\theta}$ and perform the operations

$$\mathbf{f}(\boldsymbol{\theta}) = \bar{E}\{\mathbf{x}(t)[y(t) - \mathbf{x}^{\mathrm{T}}(t)\boldsymbol{\theta}]\} \tag{6.100a}$$

$$\mathbf{G}(\boldsymbol{\theta},\mathbf{R}) = \bar{E}[\mathbf{x}(t)\mathbf{x}^{\mathrm{T}}(t) - \mathbf{R}] \tag{6.100b}$$

where $y(t)$, $\mathbf{x}(t)$ are generated by the system and controller, both of which must be specified before the expectations can be carried out. Under an adaptive feedback law (but only then) the quantities $y(t)$, $u(t)$, $\mathbf{x}(t)$ will be functions of $\boldsymbol{\theta}$.

In order to perform the operations in (6.100) assume that the *system* belongs to the class of models (6.61), corresponding to the parameter value $\boldsymbol{\theta}_0$, i.e. the data is generated according to

$$y(t) = \mathbf{x}^{\mathrm{T}}(t)\boldsymbol{\theta}_0 + e(t) \tag{6.101}$$

where $\{e(t)\}$ is assumed to be zero mean white noise with variance σ_e^2. If either (i) $\{e(t)\}$ is an i.i.d. sequence (gaussian, say) or (ii) the system and constant coefficient feedback law are linear in the data, then

$$E[\mathbf{x}(t)e(t)] = 0 \tag{6.102}$$

and substituting (6.101) into (6.100) yields

$$\mathbf{f}(\boldsymbol{\theta}) = \mathbf{G}(\boldsymbol{\theta})\,(\boldsymbol{\theta}_0 - \boldsymbol{\theta}) \tag{6.103a}$$

$$\mathbf{G}(\boldsymbol{\theta},\mathbf{R}) = \mathbf{G}(\boldsymbol{\theta}) - \mathbf{R} \tag{6.103b}$$

where

$$\mathbf{G}(\boldsymbol{\theta}) = \bar{E}[\mathbf{x}(t)\mathbf{x}^{\mathrm{T}}(t)] \tag{6.103c}$$

a symmetric, nonnegative definite matrix of dimension m, the number of estimated parameters.

6.6.2 Local convergence of recursive least squares

If $\hat{\boldsymbol{\theta}}(t) \to \boldsymbol{\theta}^*$ and $\mathbf{R}(t) \to \mathbf{R}^*$ for the recursion (6.99) then Result 2 insists that

$$\mathbf{f}(\boldsymbol{\theta}^*) = \mathbf{G}(\boldsymbol{\theta}^*)\,(\boldsymbol{\theta}_0 - \boldsymbol{\theta}^*) = 0 \tag{6.104a}$$

$$\mathbf{R}^* = \mathbf{G}(\boldsymbol{\theta}^*) \tag{6.104b}$$

If $\mathbf{G}(\boldsymbol{\theta}^*)$ is positive definite (and therefore invertible), (6.104) implies that $\boldsymbol{\theta}^* = \boldsymbol{\theta}_0$, i.e. that we have (strong) consistency. We can therefore interpret this condition, i.e.

$$\bar{E}[\mathbf{x}(t)\mathbf{x}^T(t)] > 0 \tag{6.105}$$

as a generalized 'persistent excitation' condition and this usually requires that inputs to the system vary sufficiently to excite all the modes of the system. If (6.105) holds, then

$$\mathbf{H}(\boldsymbol{\theta}^*) = (\mathbf{R}^*)^{-1} \frac{\mathrm{d}}{\mathrm{d}\boldsymbol{\theta}} \mathbf{f}(\boldsymbol{\theta})\,|_{\boldsymbol{\theta}=\boldsymbol{\theta}_0} = -\,\mathbf{I}_m \tag{6.106}$$

all of whose eigenvalues are at -1 in the left half-plane. Thus, $\boldsymbol{\theta}_0$ is the only possible convergence point under the 'boundedness' and 'persistent excitation' conditions whatever the adaptive feedback law (consistent with these conditions).

Whether (6.105) holds or not, it follows from (6.104) that

$$(\boldsymbol{\theta}_0 - \boldsymbol{\theta}^*)^T \mathbf{G}(\boldsymbol{\theta}^*)\,(\boldsymbol{\theta}_0 - \boldsymbol{\theta}^*) = 0$$

i.e.

$$\bar{E}[\mathbf{x}^T(t)\,(\boldsymbol{\theta}_0 - \boldsymbol{\theta}^*)]^2 = 0$$

so that, with probability one,

$$\mathbf{x}^T(t)\boldsymbol{\theta}_0 = \mathbf{x}^T(t)\boldsymbol{\theta}^* \tag{6.107}$$

This implies that, under the feedback law, the estimated model finally predicts the same output as the system (even if $\boldsymbol{\theta}^* \neq \boldsymbol{\theta}_0$). This ensures system identifiability.

Although system identifiability is usually the main goal of an adaptive control algorithm, lack of sufficient excitation to guarantee a nonsingular $\mathbf{R}^*$ indicates that possible ill-conditioning problems may occur in the recursion (6.99) as time evolves. Given that $\mathbf{R}(t)$ is positive semi-definite by construction, a simple way of rendering this matrix positive definite for all t is to boost its diagonal elements by replacing the recursion (6.99b) by

$$\mathbf{R}(t) = \mathbf{R}(t-1) + \frac{1}{t}[\mathbf{x}(t)\mathbf{x}^T(t) + \delta \mathbf{I}_m - \mathbf{R}(t-1)] \tag{6.108}$$

where δ is a small positive scalar. This is known as the *Levenberg–Marquardt regularization*. Although conceptually simple, there is an implementational disadvantage in that $\mathbf{R}(t)^{-1}$ now involves the inversion of a full rank (rather than rank one) update of $\mathbf{R}(t-1)$. This problem, however, can be circumvented (see Section 6.10 Problems).

Finally, note that (6.108) leads to the replacement of (6.104b) by

$$\mathbf{R}^* = \mathbf{G}(\boldsymbol{\theta}^*) + \delta \mathbf{I}_m \qquad (6.109)$$

which is invertible, so that the eigenvalues of $\mathbf{H}(\boldsymbol{\theta}^*)$ can be calculated (see Section 7.7).

6.6.3   Persistent excitation

The nature of the data in the vector $\mathbf{x}(t)$ is of major importance in determining the behaviour of a RLS recursion. In general we require that the data $\mathbf{x}(t)$ is persistently exciting. The following example illustrates how the nature of the data vector under various experimental conditions influence final convergence behaviour. This discussion should be compared with the heuristic analysis of identifiability in Chapter 3.

Example 6.11

$$\begin{aligned} \text{System:} \qquad & y(t) = a_0 y(t-1) + b_0 u(t-1) + e(t) \\ \text{Model :} \qquad & y(t) = a y(t-1) + b u(t-1) \end{aligned}$$

i.e.

$$\mathbf{x}(t) = [y(t-1), u(t-1)]^{\mathrm{T}}, \qquad \boldsymbol{\theta} = [a,b]^{\mathrm{T}}$$

so that

$$\mathbf{G}(\boldsymbol{\theta}) = \begin{bmatrix} E[y^2(t-1)] & E[u(t-1)y(t-1)] \\ E[u(t-1)y(t-1)] & E[u^2(t-1)] \end{bmatrix}$$

assuming that experimental conditions are such that the relevant processes are stationary. Consider the following experimental conditions:

(i) $u(t) = 0$

$$\mathbf{G}(\boldsymbol{\theta}) = \begin{bmatrix} 1 & 0 \\ 0 & 0 \end{bmatrix} E[y^2(t-1)] \qquad |a_0| < 1$$

We cannot identify b, but $a^* = a_0$.

(ii) $u(t) = 1$

$$\mathbf{G}(\theta) = \begin{bmatrix} E[y^2(t-1)] & fE[y(t-1)] \\ E[y(t-1)] & f^2 \end{bmatrix}, \qquad |a_0| < 1$$

Therefore det $\mathbf{G}$ = var $y(t) = \dfrac{\sigma_e^2}{1-a_0^2} \neq 0$

This implies that $a^* = a_0, \; b^* = b_0.$

(iii) $u(t)$ white noise

For $|a_0| < 1$,

$$y(t) = \frac{b_0}{1 - a_0 z^{-1}} u(t-1) + \frac{1}{1 - a_0 z^{-1}} e(t)$$

If

$$E[u(t)] = 0, \qquad E[u^2(t)] = \sigma_u^2 \qquad \text{and} \qquad E[u(t)e(s)] = 0 \; (s,t = 1,2,\ldots,)$$

then

$$\mathbf{G}(\theta) = \operatorname{diag}\left\{ \frac{b_0^2}{1-a_0^2}\sigma_u^2 + \frac{\sigma_c^2}{1-a_0^2}, \sigma_u^2 \right\} \qquad \text{(nonsingular)}$$

Therefore $a^* = a_0, \; b^* = b_0.$

(iv) $u(t) = fy(t)$

$$\mathbf{G}(\theta) = \begin{bmatrix} 1 & f \\ f & f^2 \end{bmatrix} E[y^2(t-1)] \qquad |a_0 + b_0 f| < 1$$

The matrix is singular but (6.104a) yields the correct closed-loop pole, i.e.

$$a^* + b^* f = a_0 + b_0 f \qquad \text{(System identifiability)}$$

Constant linear feedback may therefore destroy parameter identifiability by constraining the components of $\mathbf{x}(t)$ to be linearly dependent.

(v) $u(t) = fy(t) + r(t)$

This example illustrates the use of an auxiliary 'dither' signal $r(t)$ to restore the parameter identifiability lost in (iv) by breaking up the linear dependence.

Rather than proceeding by direct substitution, partition $\mathbf{x}(t)$ as follows:

$$\mathbf{x}(t) = [y_1(t-1), u_1(t-1)]^T + [y_2(t-1), u_2(t-1)]^T$$

where

$$y(t) = y_1(t) + y_2(t), \qquad u(t) = u_1(t) + u_2(t)$$

$$u_1(t) = fy_1(t), u_2(t) = fy_2(t) + r(t) \qquad \text{(6.110a)}$$

$$y_1(t) = ay_1(t-1) + bu_1(t-1) + e(t) \qquad \text{(6.110b)}$$

$$y_2(t) = ay_2(t-1) + bu_2(t-1) \qquad \text{(6.110c)}$$

Equations (6.110) show that the quantities with suffices 1 and 2 are dependent only on the sequences $\{e(t)\}$ and $\{r(t)\}$ respectively. Hence, if these sequences are uncorrelated in the sense that

$$\bar{E}[r(t)e(s)] = 0 \qquad s,t = 1,2,\ldots,$$

then

$$\mathbf{G}(\boldsymbol{\theta}) = E[\mathbf{x}_1(t)\mathbf{x}_1^T(t)] + E[\mathbf{x}_2(t)\mathbf{x}_2^T(t)]$$

in an obvious notation. The first term on the right-hand side is singular (see (iv)) but the second is positive definite if $\{r(t)\}$ is chosen as white noise (as in (iii)). Then $\mathbf{G}(\boldsymbol{\theta})$ is also positive definite and parameter identifiability follows.

(vi) $u(t) = fy(t-1)$

Although the control law is still linear, the components of $\mathbf{x}(t)$ are no longer proportional:

$$\mathbf{G}(\boldsymbol{\theta}) = \begin{bmatrix} Ey^2(t-1) & fEy(t-1)y(t-2) \\ fEy(t-1)y(t-2) & f^2Ey^2(t-1) \end{bmatrix}$$

Therefore det $\mathbf{G} = [f \, var \, y(t)]^2 - [fR_{yy}(1)]^2 > 0$

(vii) $u(t) = f \, sign \, y(t)$

Finally, this example shows that the loss of identifiability in (iv) depends crucially on the *linear* nature of the dependence created between the components of $\mathbf{x}(t)$.

$$\mathbf{G}(\boldsymbol{\theta}) = \begin{bmatrix} Ey^2(t-1) & fE|y(t-1)| \\ fE|y(t-1)| & f^2 \end{bmatrix}$$

Therefore det $\mathbf{G} = f^2 \mathrm{var}|y(t)| \neq 0$. Hence the simple nonlinear control law guarantees consistency.

Simulations illustrating these analytical points concerning parameter identifiability are shown in Chapter 3 (Example 3.5).   ☐

Again, it must be emphasized that in (i)–(vii) the experimental conditions must be such that stability is guaranteed so that use of the ODE method is meaningful.

6.6.4 Bias in recursive least squares

The analysis of RLS used so far assumes that the system belongs to the class of models. The validity of the ODE method does not depend on this assumption and the method can be used to show that, in general, the failure of this assumption leads to parameter bias. A simple example will suffice here.

Example 6.12

$$\begin{array}{lll} \text{System:} & y(t) = ay(t{-}1) + e(t) + ce(t{-}1) & |a|,|c| < 1 \\ \text{Model :} & y(t) = \alpha y(t{-}1) \end{array}$$

Using RLS, in the usual notation

$$f(\alpha) = E[y(t{-}1)[y(t) - \alpha y(t{-}1)]]$$

where

$$y(t) = \frac{1 + cz^{-1}}{1 - az^{-1}} e(t)$$

Using complex integration

$$f(\alpha) = \sigma_e^2 \mathrm{res} \, \frac{(1+cz)\,(z-\alpha)\,(z+c)}{(1-az)\,(z-a)z}$$

$$= \sigma_e^2 \, \frac{(1+ac)\,(a+c) - \alpha a(1+2ac+c^2)}{(1-a^2)}$$

Then $f(\alpha^*) = 0$ yields

$$\alpha^* - a = c \, \frac{(1-a^2)}{(1+c^2+2ac)}$$

$$f'\,(\alpha^*) < 0$$

If $c = 0$, then $\alpha^* = a$, as expected. Otherwise, the parameter estimate is biassed.

It is interesting to note that, if a is known and α^* is observed asymptotically, then the value of c can be deduced. This should be compared with Example 3.7.   □

6.6.5 The self-tuning property

The previous example illustrated, via the ODE method, the concept of bias in RLS estimates. In certain self-tuning control algorithms a structural mismatch may exist between the system and model caused by correlated noise. In particular, adaptive control algorithms in which the use of RLS leads to system identifiability, despite the presence of correlated noise in the system, are said to satisfy the self-tuning property (see Chapter 7, Appendix 7.2). The proof of the self-tuning property for the pole-assignment regulator may be viewed as a consequence of local convergence (Result 2). The problem of computing the eigenvalues of $\mathbf{H}(\boldsymbol{\theta}^*)$, however, is not carried out. No general analytical expressions for the eigenvalues exist, but they may, of course, be computed numerically in any specific case and analytical expressions can be derived in simple cases. Consider the following illustrative example, an extension of Example 6.3:

Example 6.13

$$\text{System} \quad : \quad y(t) = bu(t-1) + e(t) + ce(t-1), \qquad Ee(t)^2 = 1$$

$$\textit{Model} \quad : \quad y(t) = \beta u(t-1)$$

$$\textit{Controller}: \quad u(t) = z_0 y(t)/\hat{\beta}(t)$$

This adaptive controller fails to yield the self-tuning property because it gives rise to a first-order closed loop characteristic polynomial which violates the conditions on the polynomial degrees (see Section 7.4.1). Using the ODE method, however, it is still possible to investigate local convergence properties of the algorithm. Briefly:

$$f(\beta) = Eu(t-1)[y(t) - \beta u(t-1)]$$

where $u(t) = z_0 y(t)/\beta$ leads to

$$y(t) = \frac{1+cz^{-1}}{1-\gamma z^{-1}} e(t) \qquad (\gamma = bz_0/\beta)$$

Provided $|\gamma| < 1$, then

$$f(\beta) = \gamma[c\gamma^2 + \gamma(1+c^2-2cz_0) + (c-z_0-z_0c^2)]/(1-\gamma^2)b \qquad (6.111)$$

yielding two possible zeroes of $f(\beta)$.
    If $c = 0$, then (cf. Example 6.3)

$$f(\beta) = \frac{\gamma(\gamma-z_0)}{b(1-\gamma^2)}$$

so that the correct closed loop pole is possible: $\gamma^* = z_0$, and therefore $\beta^* = b$ and

$$f'(b) = -\frac{1}{b^2}\frac{z_0^2}{1-z_0^2} < 0$$

as required for local convergence. For nonzero values of c, $\gamma^* \neq z_0$ and convergence may not occur at all. For example, if $c = 0.5$ and $z_0 = 0.4$, then the zeroes of (6.111) are $\gamma = 0$ (β infinite) and $\gamma = -1.7$, neither of which are acceptable solutions. Therefore, for this case, no convergence is possible.

$\square$

6.7 APPROXIMATE MAXIMUM LIKELIHOOD REVISITED

It was shown in Chapter 3 that the RLS algorithm could be extended to estimate the parameters of a noise polynomial C in the system

$$Ay(t) = Bu(t-1) + Ce(t)$$

where $u(t)$ may or may not be zero.
    Two algorithms were described for estimating the coefficients of C: the RELS algorithm and the AML algorithm. These are also known as Recursive Maximum likelihood algorithms, RML_1 and RML_2 respectively.
    In this section we consider local convergence properties of the AML/RML_2 algorithm.

6.7.1 The Approximate Maximum Likelihood algorithm

In general, to avoid the problem of bias (see Section 6.6.4, Example 6.12), we need to replace RLS with an estimation technique that can deal with correlated system noise.

The CARMA system:

$$Ay(t) = Bu(t-1) + Ce(t)$$

can be expressed in the form

$$y(t) = \mathbf{x}^T(t)\boldsymbol{\theta}_0 + e(t)$$

where, corresponding to the true system,

$$\boldsymbol{\theta}_0 = (-a_1,\ldots,-a_{n_a},b_0,\ldots,b_{n_b},c_1,\ldots,c_{n_c})^T$$

$$\mathbf{x}^T(t) = [y(t-1),\ldots,y(t-n_a),u(t-1),\ldots,u(t-1-n_b), e(t-1),\ldots,e(t-n_c)]$$

Note that $\mathbf{x}(t)$ is *not* wholly observable data and therefore cannot be used directly in an estimation algorithm.

The AML algorithm is given (Chapter 3) by a modification of RLS (cf. (6.75)):

$$\hat{\boldsymbol{\theta}}(t) = \boldsymbol{\theta}(t-1) + \frac{1}{t}\mathbf{R}(t)^{-1}\hat{\mathbf{x}}(t)\{y(t) - \mathbf{x}^T(t)\hat{\boldsymbol{\theta}}(t-1)\} \qquad (6.112a)$$

$$\mathbf{R}(t) = \mathbf{R}(t-1) + \frac{1}{t}\{\hat{\mathbf{x}}(t)\hat{\mathbf{x}}^T(t) - \mathbf{R}(t-1)\} \qquad (6.112b)$$

where $\hat{\mathbf{x}}(t)$ denotes $\mathbf{x}(t)$ with $e(t-i)$ replaced by the *residual* $\eta(t-i)$ for $i = 1,\ldots,n_c$, i.e.

$$\hat{\mathbf{x}}(t) = [y(t-1),\ldots, y(t-n_a), u(t-1),\ldots, u(t-1-n_b), \eta(t-1),\ldots, \eta(t-n_c)]^T$$
$$(6.112c)$$

where

$$\eta(t) = y(t) - \hat{\mathbf{x}}^T(t)\hat{\boldsymbol{\theta}}(t) \qquad (6.112d)$$

The associated ODEs are

$$\dot{\boldsymbol{\theta}} = \mathbf{R}^{-1}\mathbf{f}(\boldsymbol{\theta}) \qquad (6.113a)$$

$$\dot{\mathbf{R}} = \mathbf{G}(\boldsymbol{\theta},\mathbf{R}) \qquad (6.113b)$$

where

$$\mathbf{f}(\boldsymbol{\theta}) = \bar{E}\hat{\mathbf{x}}(t)\{y(t) - \hat{\mathbf{x}}^T(t)\boldsymbol{\theta}\} \qquad (6.113c)$$

$$\mathbf{G}(\boldsymbol{\theta},\mathbf{R}) = \mathbf{G}(\boldsymbol{\theta}) - \mathbf{R} \tag{6.113d}$$

$$\mathbf{G}(\boldsymbol{\theta}) = E[\hat{\mathbf{x}}(t)\hat{\mathbf{x}}^{\mathrm{T}}(t)] \quad \text{(symmetric)} \tag{6.113e}$$

Substituting $y(t)$ from the system equation

$$y(t) - \hat{\mathbf{x}}^{\mathrm{T}}(t)\boldsymbol{\theta} = \mathbf{x}^{\mathrm{T}}(t)\boldsymbol{\theta}_0 - \hat{\mathbf{x}}^{\mathrm{T}}(t)\boldsymbol{\theta} + e(t)$$

$$= -\hat{\mathbf{x}}^{\mathrm{T}}(t)(\boldsymbol{\theta} - \boldsymbol{\theta}_0) + [\mathbf{x}(t) - \hat{\mathbf{x}}(t)]^{\mathrm{T}}\boldsymbol{\theta}_0 + e(t)$$

i.e. term by term

$$\eta(t) = -\hat{\mathbf{x}}^{\mathrm{T}}(t)\,(\boldsymbol{\theta} - \boldsymbol{\theta}_0) + [C(z^{-1})-1]\,[e(t) - \eta(t)] + e(t)$$

so that

$$C(z^{-1})\eta(t) = \hat{\mathbf{x}}^{\mathrm{T}}(t)(\boldsymbol{\theta}_0 - \boldsymbol{\theta}) + C(z^{-1})e(t)$$

or

$$\eta(t) = \tilde{\mathbf{x}}^{\mathrm{T}}(t)\,(\boldsymbol{\theta}_0 - \boldsymbol{\theta}) + e(t) \tag{6.114a}$$

where

$$\tilde{\mathbf{x}}(t) = \frac{1}{C(z^{-1})}\,\hat{\mathbf{x}}(t) \tag{6.114b}$$

Then (6.113a) yields

$$\mathbf{f}(\boldsymbol{\theta}) = \tilde{\mathbf{G}}(\boldsymbol{\theta})\,(\boldsymbol{\theta}_0 - \boldsymbol{\theta}) \tag{6.115a}$$

where

$$\tilde{\mathbf{G}}(\boldsymbol{\theta}) = E[\hat{\mathbf{x}}(t)\tilde{\mathbf{x}}^{\mathrm{T}}(t)] \tag{6.115b}$$

This completes the construction of the ODEs associated with the AML algorithm.

6.7.2 Local Convergence of Approximate Maximum Likelihood

From (6.114a), a zero of $\mathbf{f}(\boldsymbol{\theta})$ is $\boldsymbol{\theta}^* = \boldsymbol{\theta}_0$ and therefore the true parameter values are a possible convergence point. It is therefore necessary to investigate the eigenvalues of the matrix

$$\mathbf{H}(\boldsymbol{\theta}_0) = -\mathbf{G}(\boldsymbol{\theta}_0)^{-1}\tilde{\mathbf{G}}(\boldsymbol{\theta}_0) \tag{6.116}$$

Of course, if $C(z^{-1})$ is unity then (6.116) reduces to $-\mathbf{I}_m$ as before (Section 6.6.2). It is not surprising that, in general, the properties of $\mathbf{H}$ are linked to those of C.

Theorem 6.1

If $C(z^{-1})$ is *positive real* (pr), i.e.

$$\text{Re } C(\exp(j\omega)) \geqslant 0 \qquad -\pi \leqslant \omega \leqslant \pi$$

then $\mathbf{H}(\boldsymbol{\theta})_0)$ has all its eigenvalues in the closed left half-plane.

Proof See Appendix 6.1.

The pr condition may seem rather strange, but it is intimately linked with the interpretation of our recursive estimation algorithm as a stochastic descent step (see Section 6.5.1).

Roughly, we can argue that, in the vicinity of the (assumed) minimum $\boldsymbol{\theta}_0$, the equation error variance $V(\boldsymbol{\theta})$ satisfies

$$V(\boldsymbol{\theta}_0) \simeq V(\boldsymbol{\theta}) + (\boldsymbol{\theta}_0 - \boldsymbol{\theta})^{\mathrm{T}} V'(\boldsymbol{\theta})$$

and therefore

$$(\boldsymbol{\theta}_0 - \boldsymbol{\theta})^{\mathrm{T}} V'(\boldsymbol{\theta}) \leqslant 0 \tag{6.117}$$

For the AML algorithm (6.112), the cost gradient has been replaced by

$$\hat{V}'(\boldsymbol{\theta}) = -\hat{\mathbf{x}}(t)\{y(t) - \hat{\mathbf{x}}^{\mathrm{T}}(t)\boldsymbol{\theta}\} \simeq -\hat{\mathbf{x}}(t)\eta(t) \tag{6.118}$$

Thus, in order to generate a descent (on the average) we demand that $\hat{\mathbf{x}}(t)$ and $\eta(t)$ are positively correlated.

From (6.117) and (6.118) and taking expectations

$$(\boldsymbol{\theta}_0 - \boldsymbol{\theta})^{\mathrm{T}} E[\hat{\mathbf{x}}(t)\tilde{\mathbf{x}}^{\mathrm{T}}(t)](\boldsymbol{\theta}_0 - \boldsymbol{\theta}) \geqslant 0$$

using (6.114a). We therefore are led to find conditions for the matrix $E[\hat{\mathbf{x}}(t)\tilde{\mathbf{x}}^{\mathrm{T}}(t)]$ (i.e. $\mathbf{G}(\boldsymbol{\theta}_0)$) to be nonnegative definite. Using (6.114b) and defining

$$\xi(t) = \boldsymbol{\alpha}^{\mathrm{T}}\tilde{\mathbf{x}}(t)$$

for arbitrary nonzero $\boldsymbol{\alpha}$, then

$$\boldsymbol{\alpha}^T E[\hat{\mathbf{x}}(t)\tilde{\mathbf{x}}^T(t)]\boldsymbol{\alpha} = E[\xi(t) \cdot C(z^{-1})\xi(t)] = \frac{1}{2\pi}\int_{-\pi}^{\pi} S_{\xi\xi}(\omega)[\text{Re } C(\exp(j\omega))]d\omega \geq 0$$

if C is pr, where $S_{\xi\xi}(\omega)$ denotes the spectral density of $\{\xi(t)\}$.

6.7.3 Positive real transfer functions

In general, a transfer function $D(z^{-1})$ is defined as *positive real* (pr) if

    (i) $D(z^{-1})$ is stable

    (ii) Re $D(\exp(j\omega)) \geq 0 \qquad -\pi \leq \omega \leq \pi$

and *strictly positive real* if the inequality is strict in (ii).

(a) Every first-order stable C is pr, i.e.

$$C(z) = 1 + c_1 z \qquad |c_1| < 1$$
$$\Rightarrow \text{Re } C(\exp(j\omega)) \geq 1 - |c_1| > 0$$

(b) If C is stable

$$C \text{ pr} \Leftrightarrow C^{-1} \text{ pr}$$

(c) $C_1, C_2 \ pr \Rightarrow C_1 + C_2 \ pr$
Therefore

$$C_2, C_1 - C_2 \text{ pr} \Rightarrow C_1 \text{ pr}$$

in particular if C_2 is simply a positive number.

(d) The stable polynomial

$$C(z^{-1}) = 1 + 1.5z^{-1} + 0.75z^{-2} \text{ is } not \text{ pr} \qquad\qquad (6.119)$$

Its zeroes are at $-0.0845 \pm j0.713$ within the unit circle but

$$\text{Re } C(\exp(j\omega)) < 0 \qquad\quad \text{if } 1.7837 < \omega < 2.4794$$

6.7.4   Counterexample to convergence

The non pr C in (6.119) can be used to demonstrate failure of convergence of both on-line estimation and adaptive control algorithms.

Example 6.14

Estimation of the parameters of the ARMA process

$$y(t) + 0.9y(t-1) + 0.95y(t-2) = e(t) + 1.5e(t-1) + 0.75e(t-2)$$

using AML yields eigenvalues

$$(0.162 \pm j1.383, -1, -1)$$

two of which lie in the right half-plane and therefore the algorithms will not give convergence to $\boldsymbol{\theta}_0$. This is illustrated in Figure 6.16.   □

Figure 6.16
Showing nonconvergence of parameters in Example 6.14.

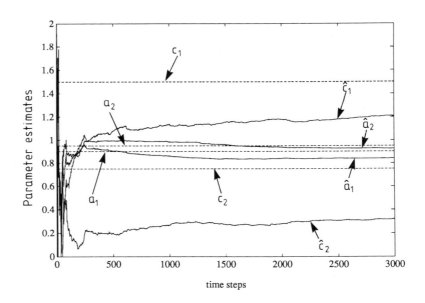

time steps

6.8   APPLICATION TO SELF-TUNING CONTROL

The discussion so far has concerned the use of ODE methods in general. While the application to self-tuning control has been mentioned, this has not been emphasized. The aim of this section is to show how the local convergence properties of a self-tuning control system may be analysed using the ODE method. Detailed discussion of the control design philosophies used here are left to Chapters 7 and 8. Specifically, consider the following (rather artificial) example of self-tuning regulation where two parameters must be estimated.

Example 6.15

It is required to regulate to zero the output of the system

$$y(t) = a_0 y(t-1) + b_0 u(t-1) + e(t) \tag{6.120}$$

The minimum variance regulator (identical to the pole assignment regulator with $\mathbf{T} = 1$ in this case) is given by

$$u(t) = -\frac{a_0}{b_0} y(t) \tag{6.121}$$

to yield closed loop behaviour

$$y(t) = e(t) \tag{6.122}$$

The system, however, is unknown and the following ARMAX model is assumed

$$y(t) = u(t-1) + d + \hat{e}(t) + c\hat{e}(t-1) \tag{6.123}$$

with $\boldsymbol{\theta} = (d,c)^{\mathrm{T}}$ is estimated on-line using AML. To achieve (6.122) using (6.123), the following regulator is implemented

$$u(t) = -cy(t) - d \tag{6.124}$$

where d,c are replaced by their current estimates $\hat{d}(t)$, $\hat{c}(t)$ at time t.
Freezing the estimates at nominal values yields

$$\mathbf{f}(\boldsymbol{\theta}) = E\left\{ \begin{bmatrix} 1 \\ \epsilon(t-1) \end{bmatrix} [y(t) - u(t-1) - d - c\epsilon(t-1)] \right\}$$

i.e.

$$f_1(d,c) = E\epsilon(t) \tag{6.125}$$

$$f_2(d,c) = E[\epsilon(t-1)\epsilon(t)] \tag{6.126}$$

Note that, once we have frozen the estimates, it is notationally irrelevant whether we use $\epsilon(t)$ or $\eta(t)$.

Substituting (6.124) in (6.120):

$$y(t) = -\frac{b_0 d}{1-z_0} + \frac{e(t)}{1-z_0 z^{-1}} = \epsilon(t) \tag{6.127a}$$

where

$$z_0 = a_0 - b_0 c \tag{6.127b}$$

Therefore

$$f_1(d,c) = -\frac{bd}{1-z_0} \tag{6.128a}$$

$$f_2(d,c) = \left[\frac{bd}{1-z_0}\right]^2 + \frac{z_0}{1-z_0^2} \tag{6.128b}$$

assuming $E[e^2(t)] = 1$. The stationarity conditions $\mathbf{f}(\boldsymbol{\theta}^*) = 0$ yield

$$d^* = 0, \qquad z_0^* = 0 \tag{6.129}$$

so that (6.127) implies (6.122) as required.

At $\boldsymbol{\theta}^*$, using (6.128),

$$\frac{d\mathbf{f}}{d\boldsymbol{\theta}} = -b\mathbf{I}_2$$

$$\mathbf{R} = E\left\{\begin{bmatrix} 1 \\ \epsilon(t-1) \end{bmatrix} [1, \epsilon(t-1)]\right\} = \mathbf{I}_2$$

Therefore the two eigenvalues of $\mathbf{H}(\boldsymbol{\theta}^*)$ take the value $-b$, implying that the convergence point (6.129) is locally stable provided that $b > 0$, i.e. the same sign as the assumed gain in the model (6.123).   □

6.9 SUMMARY

The chapter has covered the use of the Ordinary Differential Equation (ODE) method as a means of studying the local convergence properties of recursive estimators. The main points are:

- The ODE is a local convergence analysis tool. It assumes the recursive algorithm is stable
- A wide class of recursive algorithms can be analyzed. The presence or absence of adaptive or fixed coefficient control or signal processing is accommodated within the analysis
- The ODE replaces the time-varying stochastic recursive algorithm by a deterministic ODE with equivalent asymptotic convergence properties
- The ODEs associated with the RLS and AML algorithms have been developed
- The importance of the positive real condition for the C polynomial has been established for systems of the ARMA or ARMAX form

6.10 PROBLEMS

Problem 6.1

Consider the scalar recursion

$$x(t+1) = x(t)+\gamma(t+1)f(x(t))$$

where $\{\gamma(t)\}$ is a positive scalar sequence converging to zero. If a stable local convergence point x^* exists, show that the condition

$$\sum_{t=1}^{\infty} \gamma(t) = \infty$$

is necessary and sufficient for local convergence to x^*.

[*Hint*:   For $0<x<1$, $-\dfrac{x}{1-x} < \log(1-x) < -x.$]

Problem 6.2

The ARX system

$$y(t) = \mathbf{x}^T(t)\boldsymbol{\theta}_0+e(t)$$

in the usual notation, is stabilized using an adaptive control algorithm in which a model of the correct structure is assumed and the following (scalar gain) estimation step is employed:

$$\hat{\boldsymbol{\theta}}(t) = \hat{\boldsymbol{\theta}}(t-1)+\frac{1}{r(t)}\mathbf{x}(t)[y(t)-\mathbf{x}^T(t)\,\hat{\boldsymbol{\theta}}(t-1)]$$

$$r(t) = r(t-1)+\mathbf{x}^T(t)\mathbf{x}(t)$$

(a) Express the estimation algorithm in a form suitable for ODE analysis and write down the relevant ODEs.

(b) Write down the conditions for which $\boldsymbol{\theta}_0$ is the unique locally stable convergence point of the algorithm and hence show that all the local eigenvalues lie in the interval $(-1,0)$. Comment on this result with reference to RLS.

Problem 6.3

The discrete-time system

$$y(t) = ay(t-1)+u(t-1)+\xi(t)$$

is known to be open loop stable but the exact value of the constant parameter a is unknown.

The disturbance input $\{\xi(t)\}$ is known to be a zero mean stationary stochastic process and it is desired to estimate the parameter a without further modelling of the disturbance sequence. This is achieved indirectly by using the estimation model

$$y(t) = \alpha y(t-1)+\beta u(t-1)+d$$

where α is fixed at an arbitrary value whose magnitude is less than unity and on-line least squares estimation is employed to determine the constant parameters β, d. The input sequence $\{u(t)\}$ is chosen to be a white noise process independent of $\xi(t)$ with nonzero mean $\bar{u}$.

If local convergence of the estimates to β^*, d^* takes place, show that

$$a = \frac{d^*+\alpha\bar{u}}{d^*+\bar{u}}$$

and determine β^*.

Problem 6.4

Consider the system

$$y(t) = ay(t-1)+e(t)+ce(t-1), \qquad |a|<1$$

in the usual notation, where a,c are unknown. For estimation purposes, the system is represented by the model

$$y(t) = \alpha y(t-1)$$

Using the ODE approach, show that, if local convergence takes place, the asymptotic value attained by the RLS estimate of α has a bias given by

$$\alpha-a = c\sigma_e^2/\mathrm{var}\, y$$

in the usual notation.

Problem 6.5

Input/output data is generated by the system

$$y(t) = u(t-1) + bu(t-2) + e(t)$$

in the usual notation, where the white noise sequence $\{e(t)\}$ has zero mean and variance σ_e^2 and the input sequence $\{u(t)\}$ is a bounded deterministic function of time.
    A model is assumed of the form

$$y(t) = \alpha y(t-1) + u(t-1)$$

and α is estimated using RLS. Show that α^*, the only locally stable convergence point, is given by

$$\alpha^* = b \frac{\rho(0) + b\rho(1)}{(1+b^2)\rho(0) + 2b\rho(1) + \sigma_e^2}$$

provided that $\rho(0), \rho(1)$ exist and are finite, where

$$\rho(i) = \lim_{N \to \infty} \frac{1}{N} \sum_{t=1}^{N} u(t)u(t+i) \qquad (i=0,1)$$

Problem 6.6

Consider the MA(1) process

$$y(t) = e(t) + ce(t-1) \qquad |c| > 1$$

where $e(t)$ is a zero mean, unit variance discrete-time white noise process and c is unknown. Assuming a model with the correct structure, RELS estimation is employed to estimate the unknown parameter.
    Using the ODE approach, show that the only possible convergence point is c^{-1}.

Problem 6.7

The ARX system

$$y(t) = \mathbf{x}^T(t)\mathbf{\theta}_0 + e(t)$$

in the usual notation, is stabilized using an adaptive control algorithm in which a model of the correct structure is assumed and a modified version of RLS is used to estimate the unknown parameter vector $\mathbf{\theta}_0$. In the recursion for $\mathbf{P}(t)$, the data vector $\mathbf{x}(t)$ is replaced by $\mathbf{A}\mathbf{x}(t)$ where $\mathbf{A}$ is a nonsingular matrix.

(a)   Express the estimation algorithm in a form suitable for ODE analysis and write down the ODEs.

(b)   Write down the conditions for which $\boldsymbol{\theta}_0$ is the unique locally stable convergence point of the algorithm and hence show that, if $\mathbf{A} = \lambda\mathbf{I}$, the local eigenvalues are all equal to $-(1/\lambda^2)$.

Problem 6.8

It is required that the output y of the discrete-time system

$$y(t) = u(t-1)+u(t-2)+e(t)$$

in the usual notation, tracks the known set-point trajectory $\{r(t)\}$ in the sense that the control sequence $\{u(t)\}$ minimizes

$$\bar{E}[y(t)-r(t)]^2$$

for each t. The set point trajectory is deterministic and periodic, generated by repeating the sequence $\{0,0,0,1,1,1,1\}$.

The system is unknown and a self-tuning controller is implemented based on the incorrect model

$$y(t) = \beta u(t-2)$$

Calculate the quantities

$$\rho(i) = \lim_{N\to\infty}\frac{1}{N}\sum_{t=1}^{N} r(t)r(t+i) \qquad (i = 0,1)$$

and, using the ODE approach, show that $\hat{\beta}(t)$, the RLS estimate of β, can only converge to the value 1.75.

Problem 6.9

Show that the $m\times m$ matrix $\mathbf{A}$, expressed as a sum of m outer products, thus:

$$\mathbf{A} = \sum_{i=1}^{m} \mathbf{a}_i\mathbf{a}_i^{\mathrm{T}}$$

where $\mathbf{a}_1,\ldots,\mathbf{a}_m$ are m linearly independent m vectors, is symmetric and positive definite.

Show that the Levenberg–Marquardt regularization (6.94) can be carried out by a sequence of rank-one updates.

6.11   NOTES AND REFERENCES

All the basic material on the Robbins–Monro problem and the ODE approach to convergence analysis is to be found in the excellent book:

Ljung, L. and Soderstrom, T.
Theory and Practice of Recursive Identification, MIT Press, 1983.

As explained above the weakness of the ODE approach is the need to *assume* stability. For those who wish to look at the stability problem in depth, the following book is enthusiastically recommended:

Goodwin, G.C. and Sin, K.S.
Adaptive Filtering, Prediction and Control, Prentice Hall, 1984.

An additional analysis procedure for recursive algorithms is Averaging Analysis. This is covered in

Anderson, B.D.O., Bitmead, R.R., Johnson, C.R.Jr, Kokotovic, P.V. Kosut, R.L., Mareels, I.M.Y., Praly, L. and Riedle, B.D.
Stability of Adaptive Systems: Passivity and Averaging Analysis, MIT Press, 1986.

See also

Sastry, S. and Bodson, M.
Adaptive Control: Stability, Convergence and Robustness, Prentice Hall, 1989.

APPENDIX 6.1 LOCAL CONVERGENCE OF APPROXIMATE MAXIMUM LIKELIHOOD

Sufficient conditions for the matrix $\mathbf{H}(\boldsymbol{\theta}_0)$ given by (6.116) to have all its eigenvalues in the closed LHP is given by Theorem 6.1. The proof of this theorem follows immediately from two lemmas.

Lemma 6.1

If the symmetric matrix $(\tilde{\mathbf{G}} + \tilde{\mathbf{G}}^{\mathrm{T}})$ is nonnegative definite, then $\mathbf{H}$ has all its eigenvalues in the closed left half-plane.

Proof Consider the linear system $\dot{\mathbf{X}} = \mathbf{HX}$ and the Lyapunov function $V = \mathbf{X}^{\mathrm{T}}\mathbf{GX} > 0$. Then

$$\dot{V} = -\mathbf{X}^{T}(\tilde{\mathbf{G}} + \tilde{\mathbf{G}}^{\mathrm{T}})\,\mathbf{X} \leqslant 0 \qquad\qquad \square$$

Lemma 6.2

If $C(z^{-1})$ is *positive real*, then $(\tilde{\mathbf{G}} + \tilde{\mathbf{G}}^{\mathrm{T}})$ is nonnegative definite.

Proof For any nonzero vector $\boldsymbol{\alpha}$

$$\boldsymbol{\alpha}^{\mathrm{T}}(\tilde{\mathbf{G}} + \tilde{\mathbf{G}}^{\mathrm{T}})\,\boldsymbol{\alpha} = 2\bar{E}\{(\boldsymbol{\alpha}^{\mathrm{T}}\hat{\mathbf{x}}(t))(\boldsymbol{\alpha}^{\mathrm{T}}\tilde{\mathbf{x}}\,(t))\}$$

$$= 2\bar{E}\{\xi(t)\cdot C(z^{-1})\xi(t)\}$$

where $\xi(t) = \boldsymbol{\alpha}^T \tilde{\mathbf{x}}(t)$

$$= \frac{1}{\pi} \int_{-\pi}^{\pi} S_{\xi\xi}(\omega) C(\exp(j\omega)) d\omega$$

$$= \frac{1}{\pi} \int_{-\pi}^{\pi} S_{\xi\xi}(\omega) [\text{Re } C(\exp(j\omega))] d\omega$$

$$\geq 0 \qquad \text{if } C(z^{-1}) \text{ is pr}$$

where stationarity is assumed and $S_{\xi\xi}(\omega)$ is the spectral density of $\{\xi(t)\}$.   □

Part 2

Self-tuning Controllers

In this part we consider a number of feedback control algorithms which can be linked with the recursive estimators discussed in Part 1 in order to form a self-tuning or adaptive control system. Three forms of controller are considered:

  (i) pole assignment control

  (ii) minimum variance control

  (iii) multistage predictive control

The first, pole assignment control, is a classical approach to controller specification. The second, minimum variance control, uses an optimization framework. The third, predictive control, while using optimization is presented here in a way which links the classical pole assignment and optimization approaches.

7 Pole Assignment Control

7.1 OUTLINE AND LEARNING OBJECTIVES

The purpose of this chapter is to discuss a number of commonly used control algorithms associated with pole assignment self-tuning control. The aim of pole assignment control is to exactly match the closed loop characteristic equation of a feedback system to some desired form. This is used in controller design where the performance criterion for the control system can be expressed in the classical control terms of frequency response or transient response. Section 7.2 introduces pole assignment from this viewpoint. To further motivate the idea of pole assignment, Section 7.3 discusses three-term control in pole assignment terms.

The general algorithms for pole assignment synthesis are presented in Section 7.4, while Section 7.5 shows some algorithmic variations which indicate the special value of pole assignment design. Self-tuning versions of the pole assignment design procedures are covered in Section 7.6.

The learning objective is to understand the use of and implementation techniques associated with pole assignment and to see how they are practically implemented in a self-tuning framework.

7.2 POLE ASSIGNMENT — BASIC IDEAS

The design of a feedback controller has two main aims. The first is to modify in some way the dynamic response of a system. The second is to reduce the sensitivity of a system output to disturbances. Additionally, and linked to these aims, is the further objective of reducing the overall sensitivity of the closed loop system to parameter variations. For example, referring to Figure 7.1, the controller polynomials F,G,H are to be designed so as to ensure that the system output $y(t)$ tracks changes in the reference signal $r(t)$ in an acceptably fast way. In addition, the designer will usually require that in the

Figure 7.1 A closed loop control system.

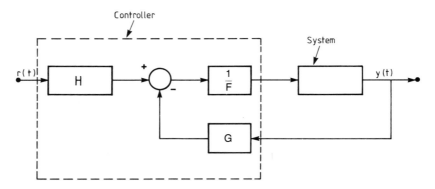

steady state (when $r(t)$ is constant) the output $y(t)$ is equal to the reference setpoint. To illustrate how this may be achieved with pole assignment design, consider the following first order example.

Example 7.1

Suppose we have a system with discrete time model given by

$$y(t) = \frac{z^{-1}b}{1-az^{-1}}u(t) \tag{7.1}$$

From Chapter 2 we know that this model arises from a continuous time process with transfer function

$$\frac{f}{1+\alpha s} \tag{7.2}$$

where α is the time constant of the system, f is the system gain, and α and f are related to the discrete time model parameters by

$$\left.\begin{aligned} a &= \exp(-\tau_s/\alpha) \\ b &= (1-a)f \end{aligned}\right\} \tag{7.3}$$

where τ_s is the sample time.

As noted above, a common control objective is to use a feedback controller to alter the speed of response of the system. Thus, for this example (Figure 7.2) the original system has a certain speed of response to a step change in $u(t)$ (associated with time constant α) but it is required to change this response rate to one associated with a chosen time constant β.

Figure 7.2
Shaping of step
response of a
first order
system by
feedback
(Example 7.1).

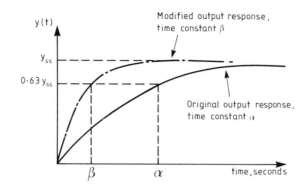

Figure 7.2
Shaping of step
response of a
first order
system by
feedback
(Example 7.1).

In terms of the system model (7.1), changing the speed of response corresponds to altering the value of the parameter a in the denominator by assigning it in some way to a value which corresponds to a faster speed of response. As noted previously, we also want to ensure that the output $y(t)$ comes into correspondence with the reference signal $r(t)$ in the steady state.

To satisfy these demands, we can propose the feedback controller structure

$$u(t) = -\,gy(t) + hr(t) \tag{7.4}$$

as in Figure 7.3. Combining controller (equation (7.4)) and model (equation (7.1)) gives

$$y(t) = \frac{bh}{1 - (a-bg)z^{-1}}\,r(t-1) \tag{7.5}$$

Note that the closed loop speed of response is now determined by $(a-bg)$ instead of a. In other words, the open loop system pole at $z = a$ has been

Figure 7.3
Closed loop
controller for
Example 7.1.

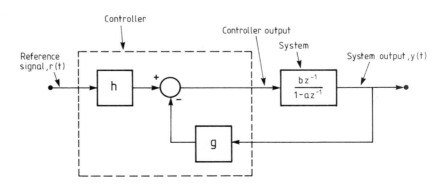

assigned to a closed loop position at $z = a - bg$. Moreover, we can specify the location of the pole as a design parameter. For example, suppose we require a single closed loop pole at $z = t_1$. Then the controller gain g is determined by the equation

$$t_1 = a - bg \tag{7.6}$$

i.e.

$$g = \frac{a - t_1}{b} \tag{7.7}$$

Equation (7.7) is the controller synthesis rule for the feedback controller gain g. For more complex systems the synthesis rules and the controller structure also become more complex, but the basic idea of a set of equations to be solved for the controller parameters is fundamental.

The remaining control objective is to ensure that at steady state the output $y(t)$ equals a (constant) reference setpoint $r(t)$. This objective is met by selecting the controller parameter h such that the closed loop transfer function is equal to unity at zero frequency. Zero frequency corresponds to the transfer function value when $z=1$ so we set

$$\left. \frac{bhz^{-1}}{1-(a-bg)z^{-1}} \right|_{z=1} = \frac{bh}{1-(a-bg)} = 1 \tag{7.8}$$

or

$$h = \frac{1-(a-bg)}{b} = \frac{1-t_1}{b} \tag{7.9}$$

Equation (7.9) is the design (synthesis) rule for selecting h so that output $y(t)$ equals input $r(t)$ under steady state conditions.

An alternative method which is used in process control applications to ensure steady state correspondence between output and setpoint is to incorporate integral action into the controller. This method will be outlined for the specific case of three term control in Section 7.3 and generally in Section 7.5.1.   □

The key idea of pole assignment is to shift the open loop poles to some desired set of closed loop poles. For many readers, however, the poles of a system may be an abstract concept so it is useful to give the technique simple time domain and frequency domain interpretations.

7.2.1 Time domain interpretation

The poles of a system determine the stability of a system and influence the nature of its transient response. If the discrete time pole $z=a$ in the model (7.1) is reassigned to a pole at $t_1 = a-bg$, then this is the same as altering the system time constant from α to β s, where (cf. equation (7.3))

$$\alpha = -\tau_s/\ell n(a) \tag{7.10}$$

$$\beta = -\tau_s/\ell n(t_1) \tag{7.11}$$

where $\ell n(\cdot)$ denotes the natural logarithm. If β is smaller than α, then the closed loop system will respond faster to sudden changes (see Figure 7.2).

The relationship between pole positions and the system transient response is often used to select the desired pole set in pole assignment synthesis. If a first order response is required then the desired pole set will be specified by the zero of the polynomial

$$\mathbf{T} = 1 - t_1 z^{-1} \tag{7.12}$$

where $t_1 = \exp(-\tau_s/\beta)$ and β is the desired system closed loop time constant.

If a second order response is required, then the desired pole set will consist of the two zeroes of the second order polynomial

$$\mathbf{T} = 1 + t_1 z^{-1} + t_2 z^{-2} \tag{7.13}$$

where

$$\left.\begin{array}{l} t_1 = -2\exp(-\xi\omega_n\tau_s)\cos\{\tau_s\omega_n(1-\xi^2)^{1/2}\} \\ t_2 = \exp(-2\xi\omega_n\tau_s) \end{array}\right\} \tag{7.14}$$

where ξ and ω_n are respectively the damping factor and natural frequency of the desired closed-loop second-order transient response.

In practice, the first and second order desired pole sets cover almost all applications. Moreover, because of the interpretation and relationships given above, we can use the method as a *transient response* assignment technique without direct reference to the poles of the system. This concept of transient response assignment is particularly powerful in the self-tuning context where the desired transient response shape becomes a user specified parameter.

7.2.2 Frequency domain interpretation

An alternative and useful approach is to interpret the pole assignment concept as a frequency response shaping method. The closed loop transfer function of the system in Figure 7.1 is

$$\frac{GB/FA}{1 + GB/FA}$$

The stability and closed loop transient performance are determined by the shape of the frequency response of GB/FA evaluated between zero frequency and infinity. These plots are usually termed Nyquist diagrams because of their association with Nyquist's stability criterion. The controller frequency response G/F is designed to shape GB/FA to ensure closed loop stability and to satisfy desired performance criteria. Pole assignment achieves a similar goal in the sense that G/F is chosen so that the object $FA + BG$ is shaped to *exactly* match a prespecified shape **T**. The classical procedure of shaping the system frequency response is a *design* method in that the system performance objectives are sufficiently general that many controller settings may satisfy the objective. The pole assignment approach is more strict. Specifically, we are asking for an *exact* performance objective and (usually) only one controller will fit our requirement. In this context, the procedure is termed a synthesis technique. Returning to the frequency response shaping discussion, one of the criteria used in selecting the desired pole set polynomial **T** given by

$$\mathbf{T} = 1 + t_1 z^{-1} + \ldots + t_{n_t} z^{-n_t} \tag{7.15}$$

is that it corresponds to a stable system and hence its frequency response locus avoids zero. We will return to this frequency response interpretation later (see Problem 7.2). Throughout the remainder of the chapter, however, we will use the transient response shaping approach.

7.3 THREE-TERM CONTROLLER DESIGN BY POLE ASSIGNMENT

The three-term or PID controller is a controller structure which combines proportional action, integral action and derivative action. Three-term control is the most commonly applied control algorithm and is frequently used to introduce newcomers to the basic concepts of control systems design. Because of this it is useful to use three-term control as a means of introducing the procedures used in pole assignment. A PID controller can be cast in the following discrete time form

$$u(t) = \frac{r(t)\,(g_0 + g_1 + g_2) - (g_0 + g_1 z^{-1} + g_2 z^{-2}) y(t)}{1 - z^{-1}} \qquad (7.16)$$

The coefficients g_0, g_1, g_2 are related to k_p, k_d, k_i, the proportional, derivative and integral gain settings by

$$k_p = -g_1 - 2g_2$$

$$k_d = g_2$$

$$k_i = g_0 + g_1 + g_2$$

In a general design setting where the underlying system may have complex dynamics a number of rules of thumb exist which assist in the selection of the controller coefficients k_p, k_d, k_i. In a synthesis situation, however, the requirements on the underlying system are quite strict. In particular, in order to synthesize exactly the PID controller coefficients we must assume that the system to be controlled has the special structure

$$y(t) = \frac{b_0 z^{-1}}{1 + a_1 z^{-1} + a_2 z^{-2}}\, u(t) \qquad (7.17)$$

The restriction on the system model form is to ensure that only one set of PID controller coefficients arise from the design.

The design process begins by combining the system model (7.17) with the controller equation (7.16) to obtain the closed loop equation relating $r(t)$ and $y(t)$. Thus

$$y(t) = \frac{b_0 z^{-1}(g_0 + g_1 + g_2)}{(1 - z^{-1})\,(1 + a_1 z^{-1} + a_2 z^{-2}) + b_0 z^{-1}(g_0 + g_1 z^{-1} + g_2 z^{-2})}\, r(t) \qquad (7.18)$$

We can now select the coefficients g_0, g_1, g_2 to give a desired closed loop performance. Suppose we have selected a desired closed loop rise time and damped nature frequency using the expressions (7.14) which relate pole positions to second-order transient-response characteristics. In this way we arrive at the corresponding desired closed loop **T** polynomial

$$\mathbf{T} = 1 + t_1 z^{-1} + t_2 z^{-2} \qquad (7.19)$$

We now determine the controller coefficients by equating the actual denominator of the closed loop equation (7.18) with the desired closed loop characteristic equation (7.19) thus:

$$(1-z^{-1})(1+a_1z^{-1}+a_2z^{-2}) + b_0z^{-1}(g_0+g_1z^{-1}+g_2z^{-2}) = 1+t_1z^{-1}+t_2z^{-2} \quad (7.20)$$

By equating coefficients of like powers of z^{-1}, the following solution for the controller settings is obtained

$$\left.\begin{array}{l} g_0 = \dfrac{t_1+(1-a_1)}{b_0} \\[3mm] g_1 = \dfrac{t_2+(a_1-a_2)}{b_0} \\[3mm] g_2 = \dfrac{a_2}{b_0} \end{array}\right\} \quad (7.21)$$

The steady state matching of $y(t)$ to $r(t)$ for a constant reference signal is ensured by the integral term in the PID controller, giving rise to the term $(1-z^{-1})$ in the denominator of equation (7.18). Specifically, at zero frequency the complex variable z is unity and the result follows immediately from (7.18).

It is important to note that the integral action in the PID controller gives steady state tracking even if the parameter values of the system or controller alter. The use of integral action in a controller will be reconsidered in Section 7.5.1 where the question of incremental control is addressed generally.

Example 7.2

The aim of this example is to illustrate via a computer simulation the influence of the pole locations on the transient response of a three-term controller. The example simulated is

$$y(t) = \frac{z^{-1}}{1 - 1.7z^{-1}+0.72z^{-2}} u(t)$$

Figure 7.4 shows the closed loop response of this system with three-term controllers (7.16) designed according to the synthesis rule defined by equation (7.21) and for various positions of closed loop poles. Note the correspondence between the closed loop transient behaviour and the damping factor and natural frequency specified by the designer via (7.14). Note that the normalized sample interval $\tau_s = 1$ has been used in this example.   □

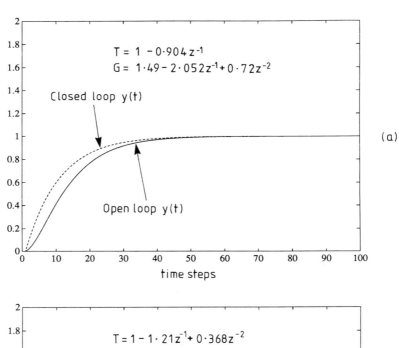

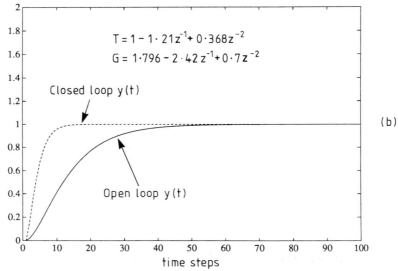

Figure 7.4 Three-term controller step response for various specified closed loop pole settings (Example 7.2) for a normalized sample rate of $\tau_s = 1$ s: (a) first order response, time constant $\beta = 10$ s; (b) second order response, $\xi = 1$, $\omega_n = 0.5$;

Figure 7.4 *(cont.)* (c) second order response, $\xi = 0.3$, $\omega_n = 0.2$; (d) second order response, $\xi = 0.4$, $\omega_n = 1$.

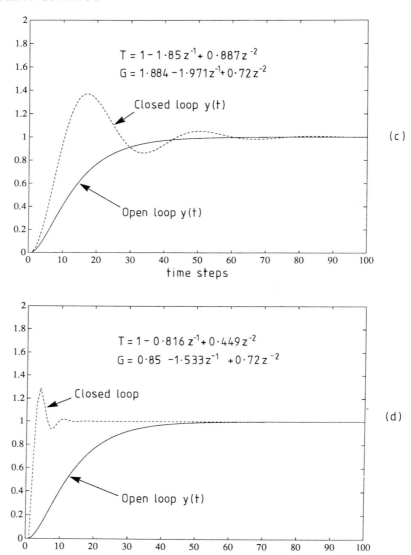

(c)

$$T = 1 - 1 \cdot 85\, z^{-1} + 0 \cdot 887 z^{-2}$$

$$G = 1 \cdot 884 - 1 \cdot 971 z^{-1} + 0 \cdot 72 z^{-2}$$

Closed loop y(t)

Open loop y(t)

time steps

(d)

$$T = 1 - 0 \cdot 816\, z^{-1} + 0 \cdot 449\, z^{-2}$$

$$G = 0 \cdot 85 - 1 \cdot 533 z^{-1} + 0 \cdot 72\, z^{-2}$$

Closed loop

Open loop y(t)

time steps

7.4 POLE ASSIGNMENT — GENERAL ALGORITHMS

In this section we will consider the use of pole assignment synthesis in general terms. The material first addresses the general problem of control of a process which must track a reference signal in the presence of random disturbance (Section 7.4.1). The special cases of servo-control and disturbance regulation are then considered separately in Sections 7.4.2 and 7.4.3 respectively. Finally, the solution of the synthesis equations for a pole assignment controller are considered.

7.4.1 Pole assignment control

In general, the control objective for a system will require the output $y(t)$ to follow a reference signal $r(t)$ in some predetermined way and to reject random disturbances which may corrupt the output. In this formulation the objective of servo following and disturbance regulation are combined. The pole assignment design for this combined objective is organized as follows. Consider a system defined by the equation:

$$Ay(t) = Bu(t-1) + Ce(t) \qquad (7.22)$$

where the controller is again of the form

$$Fu(t) = Hr(t) - Gy(t) \qquad (7.23)$$

Combining the controller and system equation yields the closed loop description

$$(FA+z^{-1}BG)y(t) = z^{-1}BHr(t) + CFe(t) \qquad (7.24)$$

The closed loop poles are then assigned to their desired locations, specified by **T**, by selecting F, G according to the polynomial identity

$$FA+z^{-1}BG = \mathbf{T}C \qquad (7.25)$$

where the polynomials F, G, H are given by

$$F = 1 + f_1 z^{-1} + \ldots + f_{n_f} z^{-n_f}$$

$$G = g_0 + g_1 z^{-1} + \ldots + g_{n_g} z^{-n_g}$$

$$H = h_0 + h_1 z^{-1} + \ldots + h_{n_h} z^{-n_h}$$

For a unique solution to (7.25) the degrees n_f, n_g should be selected as:

$$n_f = n_b$$

$$n_g = n_a - 1 \qquad (n_a \neq 0)$$

provided A, B have no common zeroes. In addition

$$n_t \leqslant n_a + n_b - n_c$$

Inserting equation (7.25) in the system equation (7.24), gives

$$y(t) = \frac{HB}{\mathbf{T}C} r(t-1) + \frac{F}{\mathbf{T}} e(t) \qquad (7.26)$$

where the noise polynomial C has been cancelled in the disturbance term. Note that this requires that C is inverse stable, a weak requirement (see Appendix).

The precompensator H is selected to achieve both low frequency gain matching and the cancellation of C from the servo pole set. The simplest choice is

$$H = C \left[\frac{\mathbf{T}}{B} \right]_{z=1} \tag{7.27}$$

yielding the closed loop equation

$$y(t) = \left[\frac{\mathbf{T}}{B} \right]_{z=1} \left[\frac{B}{\mathbf{T}} \right] r(t-1) + \frac{F}{T} e(t) \tag{7.28}$$

7.4.2 Servo control

In this section we consider the case of a noise-free process in which the object is to track a reference signal.

Consider a noise-free system modelled by

$$Ay(t) = Bu(t-1) \tag{7.29}$$

with a controller of the form

$$Fu(t) = -Gy(t) + Hr(t) \tag{7.30}$$

as before.

The closed loop system equation is (cf. (7.24))

$$y(t) = \frac{BH}{FA + z^{-1}BG} r(t-1) \tag{7.31}$$

If the desired closed loop pole set is again defined by the zeroes of

$$\mathbf{T} = 1 + t_1 z^{-1} + \ldots + t_{n_t} z^{-n_t}$$

then the controller coefficients which assign the actual pole set to the desired set are given by the solution to the polynomial equation

$$FA + z^{-1}BG = \mathbf{T} \tag{7.32}$$

As before the identity (7.32) can be expressed as a set of simultaneous equations by equating coefficients of like powers of z. A unique solution for the controller coefficients exists if the polynomials A and B have no common zeroes and the degrees n_f, n_g, n_t satisfy

$$n_f = n_b$$

$$n_g = n_a - 1 \qquad (n_a \neq 0) \tag{7.33}$$

$$n_t \leqslant n_a + n_b$$

This is now illustrated using a simple example. A fuller discussion of the identity and its solution is given in Section 7.4.4 and Appendix 7.1.

Example 7.3

Consider the identity for $n_a = 3$, $n_b = 2$, $n_t = 1$:

$$(1 + f_1 z^{-1} + f_2 z^{-2})(1 + a_1 z^{-1} + a_2 z^{-2} + a_3 z^{-3})$$

$$+ z^{-1}(b_0 + b_1 z^{-1} + b_2 z^{-2})(g_0 + g_1 z^{-1} + g_2 z^{-2}) = 1 + t_1 z^{-1}$$

Multiplying out and equating coefficients of z^{-i} for $i=1,\ldots,5$ leads to the equation set

$$
\begin{bmatrix}
1 & 0 & b_0 & 0 & 0 \\
a_1 & 1 & b_1 & b_0 & 0 \\
a_2 & a_1 & b_2 & b_1 & b_0 \\
a_3 & a_2 & 0 & b_2 & b_1 \\
0 & a_3 & 0 & 0 & b_2
\end{bmatrix}
\begin{bmatrix}
f_1 \\ f_2 \\ g_0 \\ g_1 \\ g_2
\end{bmatrix}
=
\begin{bmatrix}
t_1 - a_1 \\ -a_2 \\ -a_3 \\ 0 \\ 0
\end{bmatrix}
$$

or

$$\mathbf{A}\,\boldsymbol{\theta}_c = \mathbf{b} \tag{7.34}$$

The special banded structure of $\mathbf{A}$ is common to Sylvester matrices. Provided that A, B are coprime and of the correct degrees ($a_3 \neq 0 \neq b_2$), the matrix $\mathbf{A}$ is invertible and $\mathbf{A}^{-1}$ can be obtained by standard methods to yield the vector of controller parameters $\boldsymbol{\theta}_c$:

$$\boldsymbol{\theta}_c = \mathbf{A}^{-1}\mathbf{b} \tag{7.35}$$

The solution of equations of the form (7.34) is discussed further in Section 7.4.4 and Appendix 7.1.   □

Having satisfied the identity (7.32), the closed loop equation (7.31) takes the form

$$y(t) = \frac{BH}{\mathbf{T}} r(t-1) \qquad (7.36)$$

The closed loop poles are now at the desired locations specified by $\mathbf{T}$ and these in turn give the desired stability characteristics. However, we still have the problem of ensuring that the output $y(t)$ is equal to the reference $r(t)$ for a constant (or slowly changing) reference signal.

The simplest way to do this is to scale the reference signal by a constant amount as in the simple example at the beginning of this section. To do this we select H in (7.36) to be a constant h given by

$$H = h = \left.\frac{\mathbf{T}}{B}\right|_{z=1} \qquad (7.37)$$

This choice is made so that the closed loop transfer $BH/\mathbf{T}$ will be unity at zero frequency and thus $y(t) = r(t)$ for a constant reference setpoint.

An alternative procedure which is somewhat complex but leads to a better looking transient performance is to try to choose H to give desired system zeroes. The point of this is that the zeroes of a system (in this case the zeroes of BH) also influence the transient response shape. Thus cases arise where, although the denominator $\mathbf{T}$ of the closed loop equation corresponds to desired components of transient response, the BH term in the numerator of equation (7.36) influences the actual transient response of $y(t)$ to $r(t)$. The immediate answer to this would appear to be to cancel the B term by choosing

$$H = \frac{1}{B} \qquad (7.38)$$

but (as discussed in Chapter 2) the B polynomial may be inverse unstable, so cancellation must be *exact*, since any slight error in cancellation will leave an unstable component in the relationship between $r(t)$ and $y(t)$.

Example 7.4

Suppose that $B = 1-1.5z^{-1}$ and it is required that $\mathbf{T} = 1-0.7z^{-1}$ by pole assignment. Taking a sample interval of 1 s this corresponds to a closed loop time constant of

$$-\frac{1}{\ell n\,(0.7)} = 2.8 \text{ s.}$$

The closed loop transfer function after pole assignment is

$$y(t) = \frac{z^{-1}(1-1.5z^{-1})}{1-0.7z^{-1}} Hr(t) \tag{7.39}$$

In cancelling the B terms we note that the delay z^{-1} in (7.39) is not cancelled since this requires multiplying the numerator by z, which in turn means advancing $r(t)$ by one step. Pure time advances of an unmodelled or unpredictable signal are not possible. If, however, $r(t)$ follows a known trajectory over the system operating cycle then time advance is possible. An example of a system with a reference signal which is known over its entire time history is a batch chemical reactor where the reference signal $r(t)$ is the desired operating temperature over the duration of the batch reaction (see Chapter 9 for a discussion of programmed reference signals).

Assuming $r(t)$ is not known in advance then in equation (7.39), we might be tempted to ignore the delay and cancel the $(1-1.5z^{-1})$ term thus: select

$$H = \frac{1}{1-1.5z^{-1}} \tag{7.40}$$

Then

$$y(t) = \frac{z^{-1}(1-1.5z^{-1})}{(1-0.7z^{-1})(1-1.5z^{-1})} r(t) \tag{7.41}$$

$$= \frac{1}{1-0.7z^{-1}} r(t-1)$$

Suppose, however, that the value of H is slightly incorrect due to numerical errors, e.g.

$$H = \frac{1}{1-1.51z^{-1}} \tag{7.42}$$

The closed loop transfer function is now

$$y(t) = \frac{(1-1.5z^{-1})}{(1-0.7z^{-1})(1-1.51z^{-1})} r(t-1) \tag{7.43}$$

and the term $(1-1.51z^{-1})^{-1}$ will make the closed loop system unstable as can be seen by using the binomial expansion:

$$\frac{1}{1-1.51z^{-1}} = 1 + 1.51z^{-1} + (1.51)^2z^{-2} + (1.51)^3z^{-3} + \ldots$$

so that (7.43) yields (cf. (7.41))

$$y(t) = \frac{(1-1.5z^{-1})}{(1-0.7z^{-1})} (r(t-1) + 1.51\, r(t-2)$$

$$+ (1.51)^2\, r(t-3) + \ldots) \tag{7.44}$$

Unless $r(t)$ is specially chosen, this is clearly an unstable process.   □

The previous example illustrates the perils of cancelling a nonminimum phase B polynomial. The fact that pole assignment does not involve cancellation of B is a major advantage in digital control where many discrete systems are nonminimum phase. It is important to note that other self-tuning algorithms do demand B cancellation and are to be used with caution. (See Section 7.5.2 for further discussion and Chapter 8 on minimum variance control.)

7.4.3 Regulation

The previous section discussed pole assignment for the situation where stochastic disturbances are absent. In this section we consider the converse case where the servo signal is zero and we wish to regulate under the influence of such disturbances. This form of regulation problem occurs in control systems where the requirement is to maintain a steady plant output in the face of environmental disturbances.

The system is again modelled by

$$Ay(t) = Bu(t-1) + Ce(t) \tag{7.45}$$

and the required control algorithm is that of Section 7.4.1 with $r(t)$ set to zero.

The regulator equation is

$$Fu(t) = -Gy(t) \tag{7.46}$$

with pole assignment identity given by

$$FA + z^{-1}BG = \mathbf{T}C \tag{7.47}$$

The closed loop equation is

$$y(t) = \frac{F}{\mathbf{T}} e(t) \tag{7.48}$$

This is also an expression for the regulation error. Note that we have not minimized the regulation error, a task that will be achieved by the minimum variance regulator of Chapter 8. However, the closed loop pole set **T** has been defined so as to specify the loop bandwidth. If the output variance is still too large with this specified loop bandwidth then additional terms in the regulator can be added to reduce the variance of $y(t)$ without changing the closed loop pole positions (see Section 7.5.3).

7.4.4   Solution of the pole assignment identity

The pole assignment identity takes the general form

$$FA + z^{-1}BG = \mathbf{T}X$$

where X is either C or unity, corresponding to (7.25) and (7.32) respectively.

As noted above, the solution to this equation set is unique if A,B are coprime and the degrees of F,G are selected to satisfy

$$n_f = n_b$$
$$n_g = n_a - 1$$
$$n_t \leq n_a + n_b - n_x$$

In this form the equation can be solved for the unknown coefficients of F,G by matrix inversion (see Example 7.3). However, more effective algorithms exist for solving such polynomial identities (or Diophantine equations) and these methods allow for situations in which A,B may not be coprime. In the same spirit, the values used for n_a, n_b in modelling and recursively estimating A,B may be too large. This again can cause common factors in the estimated values of A,B and consequent ill-conditioning of the matrix inversion method.

Algorithms which are robust in these situations are discussed in Appendix 7.1 and used in Section 7.8.

7.5.   ALGORITHMIC MODIFICATIONS

The material in Section 7.4 covered the general forms of pole assignment for regulation and servo control. In this section we consider some modifications which are used in certain circumstances to deal with special requirements of the control system or to compensate for features of the underlying process.

7.5.1 Incremental control

Industrial process control schemes frequently specify an incremental control action whereby the control algorithm will output a signal sequence $\Delta u(t) = u(t) - u(t-1)$ while $u(t)$ is actually applied to the system. The effect of this is to introduce a digital integrator into the loop with the benefits that

(i) bumpless transfer between controllers is possible; and

(ii) automatic steady state reference setpoint tracking occurs despite the presence of unmodelled disturbances.

Disadvantages are

(i) destabilization of the control loop occurs; and

(ii) integral reset precautions are often needed.

Figure 7.5
Illustrating the concept of incremental control.

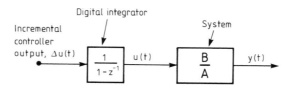

The effective open loop system when incremental control action is used is shown in Figure 7.5. Assume that the system to be controlled is such that its output is corrupted by a measurable disturbance $v(t)$ and a disturbance $\mathcal{D}(t)$. The system model used is (see Chapter 2):

$$Ay(t) = Bu(t-1) + Dv(t) + \mathcal{D}(t)$$

This can be expressed in the alternative incremental form

$$\bar{A}y(t) = B\Delta u(t-1) + \bar{D}v(t) + \bar{\mathcal{D}}(t) \qquad (7.49)$$

where

$$\bar{A} = (1-z^{-1})A = \Delta A \qquad \text{etc.}$$

Now assume a controller of the form

$$\Delta u(t) = -\frac{G}{F}(y(t) - r(t)) \qquad (7.50)$$

with G, F selected to satisfy the pole assignment identity

$$F\bar{A} + z^{-1}BG = \mathbf{T} \qquad (7.51)$$

Note that the degree of G must be increased by one to ensure that (7.51) has a unique solution (cf. (7.33)).

This gives the closed loop response

$$y(t) = \frac{GB}{\mathbf{T}}r(t-1) + \frac{F\bar{D}}{\mathbf{T}}v(t) + \frac{F}{\mathbf{T}}\bar{\mathcal{D}}(t) \qquad (7.52)$$

which can be rewritten as

$$y(t) = r(t) + \frac{(1-z^{-1})}{\mathbf{T}}\left\{FDv(t) - FAr(t) + F\mathcal{D}(t)\right\} \qquad (7.53)$$

The second term on the right-hand side can be interpreted as the composite closed loop disturbance. The disturbance term has its poles at the zeroes of $\mathbf{T}$ as required. The main point to note, however, is the term $(1-z^{-1})$ which differences all terms in the disturbance so that at low frequencies ($\omega \to 0$ and $z \to 1$) the disturbances will be exactly cancelled. In particular, constant offsets (corresponding to $\mathcal{D}(t) = d_0$) are removed exactly as are constant components of $v(t)$.

The effect of the incremental action is to exactly cancel disturbances at zero frequency, even if the parameters of the system change. Other disturbance rejection techniques rely upon the system parameters staying constant.

As was remarked earlier, the system zeroes also contribute to the transient response and situations arise where these additional contributions are unwelcome. In the incremental control case the contribution of the G polynomial to the servo response can be removed by using the modified control law

$$u(t) = \frac{1}{F}(gr(t) - G\,y(t))$$

where

$$g = \sum_{i=0}^{n_g} g_i$$

The resulting closed loop equation is

$$y(t) = \frac{gB}{\mathbf{T}} r(t-1) + \frac{F\bar{D}}{\mathbf{T}} v(t) + \frac{F}{\mathbf{T}} \tilde{\mathscr{D}}(t) \tag{7.54}$$

7.5.2   Time delays and nonminimum phase behaviour

In this section we consider the issues of system time delays and nonminimum phase behaviour. These are extremely important because they are aspects of the digital controller design problem which are easily handled in pole assignment design, but not in other procedures.

Cancellation of open loop zeroes

In Chapter 2 it was pointed out that the B polynomial in the discrete representation of a continuous time system may have some of its zeroes outside the unit disc. Systems in which this happens are said to be *nonminimum phase*. The problem with such nominimum phase systems is that they are inverse unstable (see Example 7.4). It is the possibility of this occurring in a digital model which prevents our cancelling the B polynomial in the pole assignment design algorithm. Consider the servo case only:

$$Ay(t) = Bu(t-1)$$

$$Fu(t) = -Gy(t) + Hr(t)$$

but with the modified pole assignment identity

$$AF + z^{-1}BG = \mathbf{T}B^{\dagger} \tag{7.55}$$

where $B = B^{+}B^{-}$ and B^{+} contains the inverse stable modes of B.
This leads to a closed loop configuration

$$y(t) = \frac{HB}{\mathbf{T}B^{\dagger}} r(t-1) = \frac{HB^{-}}{\mathbf{T}} r(t-1) \tag{7.56}$$

The modified identity (7.55) is simplified by noting that $B^{\dagger}$ must be a factor of F, i.e.

$$F = F_1 B^{\dagger} \tag{7.57}$$

so that we solve for F_1 and G using the identity

$$AF_1 + z^{-1}B^-G = \mathbf{T} \tag{7.58}$$

Treatment of time delays

Up to this point we have assumed that the system numerator has the form

$$z^{-1}B = z^{-1}(b_0 + b_1 z^{-1} + \ldots + b_{n_b} z^{-n_b}) \tag{7.59}$$

where b_0 is assumed to be nonzero. If the system has a pure time delay associated with it, we can simply assume within the pole assignment framework that some of the leading coefficients of B in equation (7.59) are zero. The pole assignment design procedure is special in that we are allowed to do this without causing problems at the synthesis stage. In particular, it does not influence the form of the synthesis procedure. This is not generally true of the optimal procedures to be considered in Chapters 8 and 9.

It is also important to recall that nonminimum phase behaviour may be introduced by a partial time delay. Again, this is an important justification for using pole assignment designs where unstable zeroes are not cancelled.

The fact that the leading coefficients of B can be zero is useful in circumstances where the time delay in the system is of unknown value. Within the pole assignment framework we need only assume the *minimum* time delay that may occur when proposing a system model. The most conservative choice for the system numerator is of the form (7.59). If, however, there are known to be $k-1$ zero coefficients ($b_0 = b_1 = \ldots = b_{k-2} = 0$), then it is useful to use this information and substitute $z^{-k}B$ for $z^{-1}B$ in the relevant equations. The pole assignment identity will now take the form

$$AF + z^{-k}BG = \mathbf{T} \tag{7.60}$$

where we can assume b_0 is nonzero. The identity (7.60) is solved in the usual way. Note that the usefulness of using $z^{-k}B$ is that it reduces the number of parameters to be estimated in a self-tuning framework. Thus, if the value of the integer time delay k is known, recursive estimation of the system parameters involves $k-1$ fewer unknown parameters. This is of significant help in a self-tuning pole assignment framework in which the controller synthesis stage is linked to recursive estimation of system model parameters.

The effect of variable time delays and other structural changes in the system on the performance of pole assignment controllers is most readily assessed in a self-tuning framework (see Section 7.8). At this point, however, it is instructive to look at an extended simulation example that compares a number of the controllers introduced earlier, but under conditions where the system is assumed known.

Example 7.5

Consider the second order open loop unstable nonminimum phase system

$$y(t) = 2y(t-1) - 1.25y(t-2) + 2u(t-1) - 5u(t-2)$$
$$+ 2u(t-3) + e(t) - 0.7e(t-1)$$

so that

$$A(z^{-1}) = 1 - 2z^{-1} - 1.25z^{-2}$$

giving unstable oscillatory poles. The system zeroes are given by

$$B(z^{-1}) = 2(1 - 0.5z^{-1})(1 - 2z^{-1})$$

giving two real zeroes (one of which is nonminimum phase). Also $k=1$ and

$$C(z^{-1}) = 1 - 0.7z^{-1}$$

It is required that the process output tracks a square wave setpoint (unit amplitude and period 200) in the face of a constant load disturbance of amplitude 0.5, starting at time step 400 and ending at time step 700. The demanded closed loop polynomial **T** is changed as follows:

(a) $0 \leqslant t \leqslant 200$, **T** $= 1 - 1.2z^{-1} + 0.8z^{-2}$

(b) $200 < t < 1200$, **T** $= 1 - 0.5z^{-1}$

The simulation (Figure 7.6) shows

(i) the effect of changing **T** from second order oscillatory to first order;

(ii) steady state tracking error when nonincremental control is used and the load disturbance is present ($400 \leqslant t \leqslant 700$);

(iii) the removal of steady state error when the incremental control is introduced ($t \geqslant 600$); and

(iv) the effect of cancelling $B^+ = 1 - 0.5z^{-1}$ in the incremental controller ($t \geqslant 800$).

The controller polynomials F, G, however, still contain the same number of parameters throughout and hence the synthesis procedure demands a constant computational effort. □

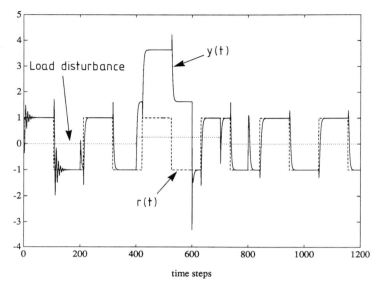

Figure 7.6 Performance of various pole assignment algorithms (Example 7.5). (a) At $t=20$ the desired closed loop pole set **T** changes from second order oscillatory to first order. (b) At $t=600$ incremental control is introduced in order to remove steady state error in $y(t)$ caused by the load disturbance. (c) At $t=800$ the inverse stable factor of B is cancelled in the control law.

7.5.3   Reduced variance regulation

It is clear from previous sections that the pole assignment strategy is inspired by a classical rather than an optimization approach to controller synthesis. It is therefore not surprising that, in achieving the specified closed-loop characteristics, the noise-rejecting properties of the control system may be impaired so that large output or input variances arise. In other words, there is nothing in the basic pole assignment approach requiring that such quantities are minimized or even 'small' in any sense.

The following example illustrates the problem in the case of output regulation to a zero setpoint.

Example 7.6

$$y(t) = 1.25y(t-1) + 0.5u(t-1) + 2u(t-2) + e(t) + 0.4e(t-1)$$

$$\mathbf{T}(z^{-1}) = 1 - 0.8z^{-1}; \qquad \sigma_e^2 = 1$$

Solving the polynomial identity (7.47) yields the regulator polynomials

$$F_0(z^{-1}) = 1 + 0.7086z^{-1}; \qquad G_0(z^{-1}) = 0.2829 \qquad (7.61)$$

and an output variance of 7.322, much larger than the minimum variance value of unity.   □

The solution (7.61) is obtained by demanding a unique solution to the identity and therefore making the standard assumptions (7.33) on the regulator polynomial degrees. Overparametrizing F and G, however, allows an infinity of possible solutions to the identity and the extra degrees of freedom thus created can be used for optimization purposes. Let F_0, G_0 denote the standard solution of (7.60) satisfying

$$n_{f_0} = n_b + k - 1, \qquad n_{g_0} = n_a - 1 \qquad (7.62)$$

then the general solution is given by

$$F(z^{-1}) = F_0(z^{-1}) + z^{-k}B(z^{-1})P(z^{-1}) \qquad (7.63a)$$

$$G(z^{-1}) = G_0(z^{-1}) - A(z^{-1})P(z^{-1}) \qquad (7.63b)$$

where P is an arbitrary polynomial. This implies

$$n_f = n_b + k + n_p \qquad (7.63c)$$

$$n_g = n_a + n_p \qquad (7.63d)$$

so that the standard case can be considered as corresponding to $n_p = -1$. For $n_p = 0$, there is a single degree of freedom, and so on.

Use of the extended polynomials F,G in the regulator

$$Fu(t) + Gy(t) = 0 \qquad (7.64)$$

is equivalent to implementing the control law

$$(F_0 + z^{-k}BP)u(t) + (G_0 - AP)y(t) = 0$$

i.e.

$$F_0u(t) + G_0y(t) = PCe(t) \qquad (7.65)$$

Hence the new regulator can be interpreted as a modification of the standard regulator in which the (reconstructed) noise signal is used as an extra input.

Note that this extra input can be interpreted as a disturbance feedforward signal (Figure 7.7). Again note that within the algorithm the noise feedforward is implicit within the general solution (7.63) to the diophantine equation.

The regulator (7.65) leads to the closed loop equation

$$\mathbf{T}y(t) = Fe(t) \tag{7.66}$$

as before. As a function of P, however, the output variance, denoted here by $\mathrm{var}(y)_{n_p}$, is given by

$$\mathrm{var}(y)_{n_p} = E\left[\frac{F_0 + z^{-k}BP}{\mathbf{T}}e(t)\right]^2 \tag{7.67}$$

This scalar function is quadratic in the coefficients of P and has a unique minimum given by

$$\frac{\partial[\mathrm{var}(y)_{n_p}]}{\partial p_i} = 0 \qquad i = 0, 1, \ldots, n_p \tag{7.68}$$

Figure 7.7
Interpretation of the reduced variance pole assignment regulator as a minimum degree pole assignment regulator F_0, G_0 with feedforward of the stochastic disturbance $e(t)$.

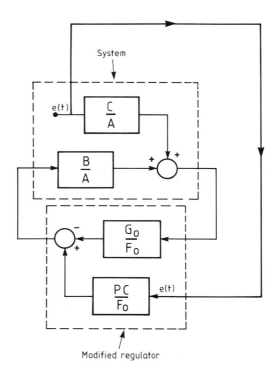

This is a set of n_p+1 linear equations for the optimal P coefficients and may be solved by a matrix inversion

$$\mathbf{p} = \mathbf{M}^{-1}\mathbf{d} \tag{7.69a}$$

where

$$\mathbf{p} = (p_0, p_1, \ldots, p_{n_p})^\mathrm{T} \tag{7.69b}$$

$\mathbf{M}$ is a matrix with entries M_{ij} $(i,j=0,\ldots,n_p)$ given by

$$M_{ij} = E\left\{\left[\frac{B}{\mathbf{T}}e(t)\right]\left[\frac{B}{\mathbf{T}}e(t+i-j)\right]\right\} \tag{7.69c}$$

$\mathbf{d}$ is a vector with entries d_i $(i=0,\ldots,n_p)$ given by

$$d_i = -E\left\{\left[\frac{F_0}{\mathbf{T}}e(t)\right]\left[\frac{B}{\mathbf{T}}e(t-k-i)\right]\right\} \tag{7.69d}$$

Note that $\mathbf{M}$ has a covariance structure, i.e. it is a symmetric positive definite Toeplitz matrix and has only n_p+1 distinct elements.

Example 7.7

Consider again the system described in Example 7.6 with the same desired closed-loop pole location. Employing P polynomials up to the second order yields the following table:

Table 7.1

n_p	−1	0	1	2
var(y)	7.322	2.566	2.486	2.482
var(u)	0.222	0.681	0.915	0.945

For $n_p=0$ the optimal solution is

$$p_0 = -0.542 \tag{7.70a}$$

$$F(z^{-1}) = 1 + 0.438z^{-1} - 1.083z^{-2} \tag{7.70b}$$

$$G(z^{-1}) = 0.824 - 0.677z^{-1} \tag{7.70c}$$

For this low-order example, a single degree of overparametrization achieves significant output variance reduction and little improvement results from further increases in the degree of P. This phenomenon occurs in many cases although it cannot be guaranteed in general.   □

The 'constant P' case is important because of its computational simplicity in avoiding a matrix inversion. The minimizing P is given by

$$p_0 = - E\left\{\left[\frac{F_0}{\mathbf{T}}e(t)\right]\left[\frac{B}{\mathbf{T}}e(t-k)\right]\right\} \bigg/ E\left[\frac{B}{\mathbf{T}}e(t)\right]^2 \tag{7.71}$$

On-line variance reduction

The exact calculation of the optimal P using (7.69) involves $2(n_p+1)$ complex integrations and the inversion of the (n_p+1)-dimensional M matrix. These computations can be avoided by recursively minimizing the cost function (7.67) while noting that its value remains unchanged if $\{e(t)\}$ is replaced by a *known* zero mean white noise sequence $\{w(t)\}$. Thus

$$\text{var}(y)_{n_p} = E\left[\frac{F_0+z^{-k}BP}{\mathbf{T}}w(t)\right]^2 \tag{7.72a}$$

$$= E\left[y_w(t) - \mathbf{x}_w^{\mathsf{T}}(t)\mathbf{p}\right]^2 \tag{7.72b}$$

where

$$\mathbf{T}y_w(t) = F_0 w(t) \tag{7.72c}$$

and $\mathbf{x}_w(t)$ is the column (n_p+1)-vector whose ith component is given by

$$\mathbf{T}[\mathbf{x}_w(t)]_i = - Bw(t-k-i) \qquad i=0,\dots,n_p \tag{7.72d}$$

By analogy with RLS estimation, a sequence of estimates $\{\hat{\mathbf{p}}(t)\}$ of the optimal $\mathbf{p}$ is generated as follows.

$$\hat{\mathbf{p}}(t) = \hat{\mathbf{p}}(t-1) + \mathbf{P}_w(t)\mathbf{x}_w(t)[y_w(t) - \mathbf{x}_w^{\mathsf{T}}(t)\hat{\mathbf{p}}(t-1)] \tag{7.73a}$$

$$\mathbf{P}_w(t)^{-1} = \mathbf{P}_w(t-1)^{-1} + \mathbf{x}_w(t)\mathbf{x}_w^{\mathsf{T}}(t) \tag{7.73b}$$

and a forgetting factor may be incorporated if required. The calculation of $\mathbf{P}_w(t)$ from (7.72b) is achieved by employing the standard rank-one updating procedure (Section 3.3).

Figure 7.8 On-line variance reduction for Example 7.6 ($\alpha = 0$). The variance reduction algorithm is introduced at $t = 500$. (a) The setpoint $r(t)$ and system output $y(t)$; and (b) control signal $u(t)$. Self-tuning variance reduction for Example 7.6 ($\alpha = 0$).

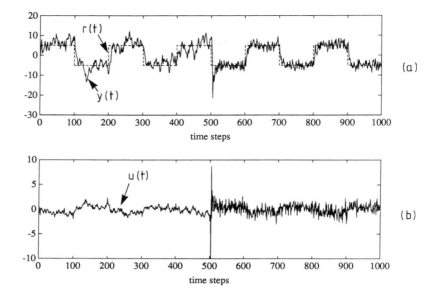

Figure 7.8 illustrates the behaviour of this algorithm for Example 7.6. Note the reduction in output variance when the algorithm is invoked at $t=500$ (Figure 7.8(a)) and the corresponding increase in control action (Figure 7.8(b)).

Affect on input variance

In Example 7.7, the reduction of output variance is achieved at the expense of an increase in the variance of the input signal. If required, this effect can be moderated by choosing P to minimize a modified cost function

$$J_{n_p}(\alpha) = \operatorname{var}(y)_{n_p} + \alpha \operatorname{var}(u)_{n_p} \qquad (\alpha \geqslant 0) \qquad (7.74)$$

reflecting a trade off between the two variances. For zero α, (7.74) leads to the smallest output variance for the given degree of overparametrization. In cases where this may lead to excessive controller action, nonzero values of α can be chosen.

Note that

$$\operatorname{var}(u)_{n_p} = E\left[\frac{G}{\mathbf{T}}e(t)\right]^2 \qquad (7.75a)$$

$$= E\left[\frac{G_0 - AP}{\mathbf{T}}e(t)\right]^2 \qquad (7.75b)$$

so that the cost function remains quadratic in $\mathbf{p}$ for positive α. Hence the optimal $\mathbf{p}$ is the solution of a set of linear equations as before.

In Example 7.7, choice of $\alpha=50$ leads to the following table:

Table 7.2

n_p	-1	0	1	2
var(y)	7.322	5.822	5.292	5.100
var(u)	0.222	0.236	0.239	0.240

The trade-off between var(y) and var(u) in comparison with the previous table is obvious.

Given the quadratic nature of the cost function (7.74), it can be minimized recursively by employing the same technique as for var(y)$_{n_p}$. Note that

$$J_{n_p}(\alpha) = E\left[\frac{F}{\mathbf{T}}e(t)\right]^2 + \alpha\, E\left[\frac{G}{\mathbf{T}}e(t)\right]^2 \tag{7.76a}$$

$$= E\left[\frac{F}{\mathbf{T}}w_1(t) + (\alpha)^{1/2}\frac{G}{\mathbf{T}}w_2(t)\right]^2 \tag{7.76b}$$

$$= E[y_w(t) - \mathbf{x}_w^{\mathsf{T}}(t)\mathbf{p}]^2 \tag{7.76c}$$

where $\{w_1(t)\}$, $\{w_2(t)\}$ are known, uncorrelated, zero mean white noise sequences having the same variance. Having cast the cost function into the form of expectation of a single squared term, the recursion is again given by (7.73) where

$$\mathbf{T}y_w(t) = F_0 w_1(t) + (\alpha)^{1/2}G_0 w_2(t) \tag{7.77a}$$

$$\mathbf{T}[\mathbf{x}_w(t)]_i = -\,Bw_1(t-k-i) + (\alpha)^{1/2}Aw_2(t-i) \qquad i=0,\ldots,n_p \tag{7.77b}$$

Note that (7.77a,b) coincide with (7.72c,d) when $\alpha = 0$.

Possible extensions

The overparametrization technique is easily extended to the combined servo/regulation problem and to the incremental form. In the latter case, the appropriate identity is

$$\Delta AF + z^{-k}BG = C\mathbf{T} \tag{7.78}$$

and the general solution takes the form (cf. 7.63):

$$F(z^{-1}) = F_0(z^{-1}) + z^{-k}B(z^{-1})P(z^{-1}) \tag{7.79a}$$

$$G(z^{-1}) = G_0(z^{-1}) - (1-z^{-1})A(z^{-1})P(z^{-1}) \tag{7.79b}$$

where

$$n_f = n_b + k + n_p \tag{7.79c}$$

$$n_g = n_a + 1 + n_p \tag{7.79d}$$

The cost function (7.74) takes the form

$$J_{n_p}(\alpha) = E\left[\frac{\nabla F}{\mathbf{T}} e(t)\right]^2 + \alpha E\left[\frac{G}{\mathbf{T}} e(t)\right]^2 \qquad (\alpha > 0) \tag{7.80}$$

and is again quadratic in **p**.

Further detailed development is left to the reader (see Section 7.10 Problems).

7.6 SELF-TUNING POLE ASSIGNMENT ALGORITHMS

The basic form of a self-tuning control system was outlined in Chapter 1. However, that was a long time ago and the outline was drawn in general terms. Accordingly, what we need to do here is to remind ourselves of what a self-tuning controller is in the special form appropriate for pole assignment design. Figure 7.9 shows the layout of a pole assignment self-tuner in which a recursive estimator (of the kind considered in Part 1, Chapter 3) is combined with a pole assignment synthesis rule in order to continuously update the controller coefficients. Certain performance requirements are fed into the synthesis block. These consist of the following controller design information:

(a) the desired closed loop pole set, specified by **T**;

(b) the form of the controller, e.g. whether it is a servo system, a regulator or a combination of both; and

(c) whether the controller is to be incremental or nonincremental in nature.

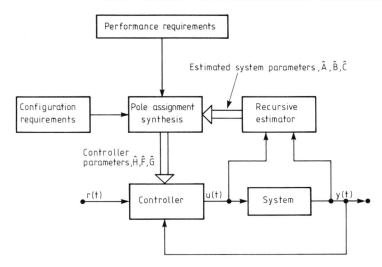

Figure 7.9
Self-tuning pole
assignment
system.

In addition, further information must be supplied concerning the configuration requirements of the self-tuner, including:

(d)   the sample rate to be used;

(e)   the degrees of the system model polynomials n_a, n_b, n_c, etc.;

(f)   the delay k in the system, if known.

Provided with this information, the self-tuning system can be set up to go through the following cycle of adaption.

Self-tuning cycle

At each sample interval t the following sequence of action is taken:

Step (i) Data capture
The system output $y(t)$, reference signal $r(t)$ and any other variables of importance are measured.

Step (ii) Estimator update
The date acquired in (i) is used together with past data and the previous control signal to update the parameter estimates in a model of the system using an appropriate recursive estimator.

Step (iii) Controller synthesis
The updated parameters from (ii) are used in a pole assignment identity to synthesize the parameters of the desired controller.
Step (iv) Control calculation
The controller parameters synthesized in (iii) are used in a controller to calculate and input the next control signal $u(t)$.   ☐

At the end of the cycle the control computer waits until the end of sample interval t and then repeats the cycle for interval $t+1$, and so on. The steps in the self-tuning cycle are computed sequentially. Figure 7.10 illustrates this in terms of a timing and sequence diagram. The total computation time must be less than the sample interval and is generally assumed to be much less. In any event the computation time introduces an additional time delay into the control loop. In view of the problems introduced by partial time delays, some users delay outputting the controller signal $u(t)$ until the end of the sample interval in which it was calculated. This ensures that the additional delay is one complete sample interval.

Figure 7.10
Timing and sequence diagram for self-tuning control.

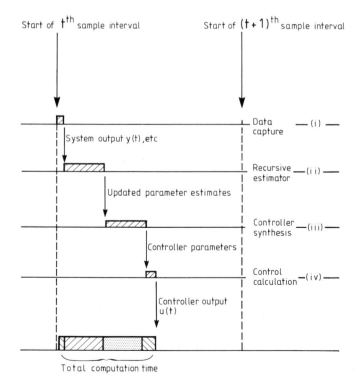

7.6.1   Servo self-tuner

In order to illustrate a specific self-tuning algorithm consider the case of a noise-free system given by the equation

$$Ay(t) = Bu(t-1) \tag{7.81}$$

It is required that the output $y(t)$ should follow a reference signal $r(t)$ with zero steady-state error and that the system closed loop dynamics should be governed by a pole set specified by **T**.

A basic self-tuning algorithm for this system is:

Algorithm: Pole Assignment Servo System

At each sample time t

Step (i) Read in the system output $y(t)$
Step (ii) Use this new measurement in a RLS algorithm to update the parameter estimates in the model

$$\hat{A}y(t) = \hat{B}u(t-1) \tag{7.82}$$

Step (iii) Synthesize the controller polynomials $\hat{F},\hat{G},\hat{H}$ by using the current estimates $\hat{A}, \hat{B}$ in the identity

$$\hat{F}\hat{A}+z^{-1}\hat{B}\hat{G} = \mathbf{T} \tag{7.83}$$

and set

$$\hat{H} = \left.\frac{\mathbf{T}}{\hat{B}}\right|_{z=1} \tag{7.84}$$

Step (iv) Generate and output the control $u(t)$ using $\hat{F},\hat{G},\hat{H}$ calculated in (iii) in the control equation

$$\hat{F}u(t) = -\hat{G}y(t) + \hat{H}r(t) \tag{7.85}$$

Step (v) Wait for sample time interval to elapse and then return to step (i).□

The above sequence is repeated at each sample interval until the controller parameters have converged to steady values. At this stage the controller coefficients can be fixed and the standard control sequence used as follows:

(i) read in the system output $y(t)$; and

(ii) generate and output the control signal $u(t)$ using equation (7.85).

Alternatively, the self-tuning sequence can be continued to allow it to adapt to parameter changes. If this second option is used, the recursive estimator must be organized so as to allow the estimator to track changing system dynamics (see Chapter 4).

The above algorithm starts at time step zero and may start with little or no knowledge of the true coefficients of A,B so that initially the parameter estimates (and hence the control action) may be poor. As time increases the parameter estimates should improve in accuracy and the controller converge to that which would have been obtained using off-line design with the correct values of A and B. Note that the possibly poor transient performance of the algorithm can be traced to our blind acceptance at *every* time step of the estimated $\hat{A}$, $\hat{B}$ as if they were correct. This 'certainty equivalence' approach is standard in self-tuning and is easy to understand and implement. More sophisticated approaches are possible (in which, for example, the covariance object $\mathbf{P}(t)$ is used in the controller synthesis to inject an element of initial 'caution') but these are not pursued here (see Chapter 4 for a more extensive discussion of algorithm management).

As an example of the above servo self-tuner, consider the following simulation.

Example 7.8   Self-tuning pole assignment servo

The following system

$$(1+0.5z^{-1}+0.7z^{-2})y(t) = (z^{-1}+0.2z^{-2})u(t) \tag{7.86}$$

was simulated with a self-tuning controller attached and the desired closed loop pole set specified by

$$\mathbf{T} = 1 - 0.6z^{-1} \tag{7.87}$$

Figure 7.11 shows the plots of the various system parameters and variables as the self-tuning procedure evolves. In particular Figure 7.11(a) shows the system output $y(t)$ and the square wave setpoint $r(t)$; Figure 7.11(b) shows the control signal $u(t)$. The system parameter estimates are shown in Figure 7.11(c) and the controller coefficients in Figure 7.11(d). The correct values for the controller parameters are $f_1 = 0.15$, $g_0 = 1.25$, $g_1 = 0.525$. Note that, in this case, the initial convergence has been artificially slowed down by using a small value $\mathbf{P}(0) = 5\mathbf{I}_4$ for the covariance initialization. In addition, random initial conditions were used for the parameter estimates.   □

7.6.2 Self-tuning regulation

The deterministic servo case considered in the previous section involved no disturbances, offsets or noise sources. Here we consider the complementary case of regulation against random noise with a zero set point. In this case the special nature of self-tuning will be shown to lead to simplifications in the system model. Specifically, if the system is given by

$$Ay(t) = Bu(t-1) + Ce(t) \tag{7.88}$$

in the usual notation, then a self-tuning pole assignment regulator leading to the correct closed loop equation

$$y(t) = \frac{F}{\mathbf{T}} e(t)$$

can be constructed, based on a system model which *assumes that C is unity.*
    The self-tuning algorithm takes the form:

Algorithm: Self-Tuning Pole Assignment Regulation

At each sample time t

Step (i) Sample the new system output $y(t)$
Step (ii) Update the polynomial estimates $\hat{A},\hat{B}$ using the RLS model

$$\hat{A}[X^{-1}y(t)] = \hat{B}[X^{-1}u(t-1)]+\hat{e}(t) \tag{7.89}$$

where

$$\hat{A} = 1 + \hat{a}_1 z^{-1} +\ldots+ \hat{a}_{n_a} z^{-n_a}$$

$$\hat{B} = \hat{b}_0+\ldots+\hat{b}_{n_b} z^{-n_b}$$

and X is a chosen inverse stable polynomial in z^{-1} (possibly unity) of degree n_x.

Figure 7.11
Illustrating the performance of the self-tuning pole assignment servo system of Example 7.8: (a) output and reference signals; (b) control signal; (c) estimated system parameters; and (d) controller parameters.

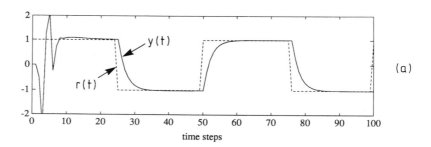

(a)

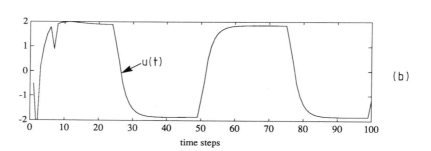

(b)

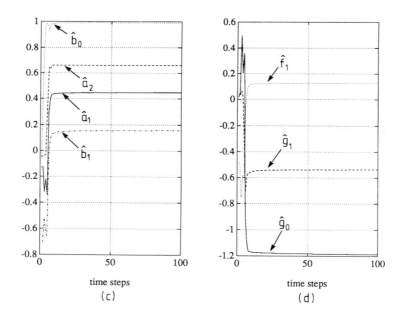

(c)

(d)

Step (iii) Synthesize the controller polynomials $\hat{F}, \hat{G}$ using

$$\hat{F}\hat{A} + z^{-1}\hat{B}\hat{G} = \mathbf{T}X \tag{7.90}$$

Step (iv) Apply control using

$$\hat{F}u(t) = -\hat{G}y(t) \tag{7.91}$$

Wait for sample interval to elapse and then return to step (i).   □

Note that the model (7.89) can be cast in the form

$$\hat{A}y(t) = \hat{B}u(t-1) + X\hat{e}(t) \tag{7.92}$$

which is equivalent to assuming $C = X$.

The most important point, however, about the above self-tuning cycle is that despite the presence of coloured noise the self-tuning estimator employs an RLS algorithm. The use of RLS in this way produces biased estimates of the system polynomials A,B. When the biased estimates are used in the 'wrong' identity (7.90), however, which also assumes $C = X$, then the final effect is to produce the correct controller polynomials F,G. Correct in this context means the values which would have been obtained if the correct A, B,C had been inserted in the pole assignment identity (7.47).

The convergence to a correct configuration from an incorrect assumption on the noise dynamics is termed a 'self-tuning property'. Such self-tuning properties exist for pole assignment and minimum variance controllers. A detailed proof that the property applies here is given in Appendix 7.2. For the moment we can make some observations and a heuristic justification of the property. The main observation is that the polynomial X filters the data $y(t)$, $u(t)$ and is useful in removing unwanted components of signal (e.g. sensor noise). Normally, however, X is set to unity.

Example 7.9   *Self-tuning pole-assignment regulation*

Consider the system of equation (7.86) extended to include a stochastic disturbance

$$(1+0.5z^{-1}+0.7z^{-2})y(t) = z^{-1}(1+0.2z^{-1})u(t)+(1-0.8z^{-1})e(t)$$

where $\{e(t)\}$ is a zero mean white noise sequence with unit variance.

The requirement is to self-tune a pole assignment regulator such that the closed loop pole set is given by (7.87) i.e. a single pole at $z=0.6$.

Figures 7.12(a) and (b) show the plots of the parameter estimates for the system model and the controller parameters. Note that the system model parameters are clearly biased while controller parameters converge to the correct values $f_1 = -0.05$, $g_0 = -1.85$, $g_1 = 0.175$. The remaining traces show (Figure 7.12(c)) the driving noise $e(t)$ and (Figure 7.12(d)) the difference between $e(t)$ and the residual $\eta(t)$ generated within the self-tuning estimator. As predicted by the self-tuning property this difference converges to zero (see Appendix 7.2 and the following paragraph). In this self-tuning experiment $X = 1$, $\mathbf{P}(0) = 5\mathbf{I}_4$ and a forgetting factor of $\lambda = 0.998$ was employed.   □

A heuristic justification for the self-tuning property is obtained by combining the model, identity and control equations (7.89), (7.90) and (7.91) to give

$$y(t) = \frac{\hat{F}}{\mathbf{T}} \hat{e}(t) \tag{7.93}$$

where X is cancelled as a common factor in numerator and denominator.

Likewise, combining the control law (equation (7.91)) with the system equation gives

$$y(t) = \frac{\hat{F}C}{A\hat{F} + z^{-1}B\hat{G}} e(t) \tag{7.94}$$

Now, *if* the closed loop system converges so that $\hat{e}(t)$ (and therefore also $\epsilon(t)$ and $\eta(t)$) becomes equal to $e(t)$, then by combining equations (7.93) and (7.94) we have

$$A\hat{F} + z^{-1}B\hat{G} = \mathbf{T}C \tag{7.95}$$

which is precisely the synthesis rule which we would use if we knew the correct values of A, B, C. Thus the self-tuning system has produced the correct controller coefficients F, G despite the fact that the model (equation (7.89)) does not correctly account for the noise dynamics.

The self-tuning algorithm will work, therefore, provided that $\eta(t)$ becomes equal to $e(t)$, the white noise signal which creates the system disturbance (e.g. see Figure 7.12(d)). At this point we simply note that, if the residual sequence satisfies this whiteness property, then the stationarity conditions for the convergence of the RLS estimator are satisfied (see Section 3.2).

The important aspect of the self-tuning property is that RLS estimation can be used rather than the more complex RELS or AML algorithms. A more general and rigorous proof of the self-tuning property is given in Appendix 7.2 and this reveals some important limitations of the property. For example, caution should be exercised when operating the reduced-variance regulator (Section 7.5.3) in a self-tuning mode. In this case, a possible explicit algorithm is as follows:

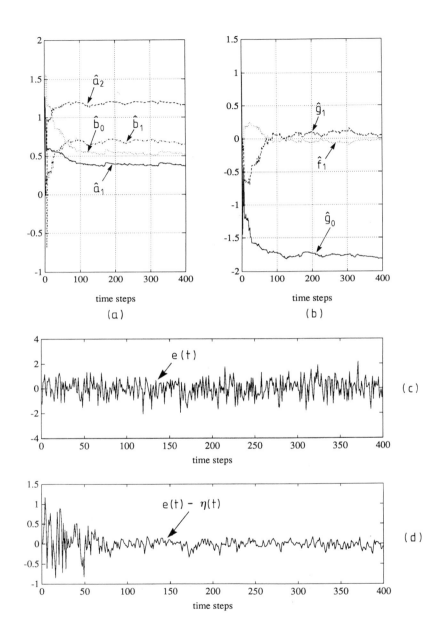

Figure 7.12
Illustrating the
performance of
the self-tuning
pole assignment
regulator system
of Example 7.9:
(a) estimated
system
parameters; (b)
controller
parameters; (c)
driving noise
$e(t)$; and (d)
$e(t) - \eta(t)$.

Algorithm: Reduced Variance Regulation

At time t, execute the following sequence:

Step (i) Sample the system output $y(t)$

Step (ii) Update the parameter estimates $\hat{A}$ and $\hat{B}$ using the RLS model

$$\hat{A}y(t) = \hat{B}u(t-1) + \hat{e}(t)$$

Step (iii) Synthesize the controller polynomials $\hat{F}_0, \hat{G}_0$ using

$$\hat{F}_0\hat{A} + z^{-1}\hat{B}\hat{G}_0 = \mathbf{T}$$

Step (iv) Solve for the optimal P using $\hat{A}_0, \hat{B}_0, \hat{F}_0, \hat{G}_0$

Step (v) Synthesize the regulator polynomials $\hat{F}, \hat{G}$ using

$$\hat{F} = \hat{F}_0 + z^{-1}P\hat{B}, \qquad \hat{G} = \hat{G}_0 - P\hat{A}$$

Step (vi) Apply control using

$$\hat{F}u(t) = - \hat{G}y(t)$$

Step (vii) Wait for sample time interval to elapse and then return to step (i).

☐

In practice this algorithm has been found to work well but only appears to be on firm theoretical ground for $n_p=0$. Indeed, if n_p is too large, the self-tuning property cannot hold (see Appendix 7.2). In such cases, C should be estimated.

Example 7.7 (continued)

Figure 7.8 (continued) shows the system and controller estimates corresponding to a self-tuning version of Figure 7.8. The system is that of Example 7.6 with variance reduction using scalar P introduced at $t = 500$. Note the change in controller coefficients which occurs at this point. The correct controller values are $f_1 = 0.438$, $f_2 = -1.083$, $g_0 = 0.867$, $g_1 = -0.677$. ☐

The following section works through a demonstration of the self-tuning property in the case of standard pole asignment regulation of a simple first order system.

Figure 7.8 *(cont.)* Compare Figure 7.8 a & b p. 270) Illustrating: (c) system parameter estimates; and (d) regulator parameter estimates.

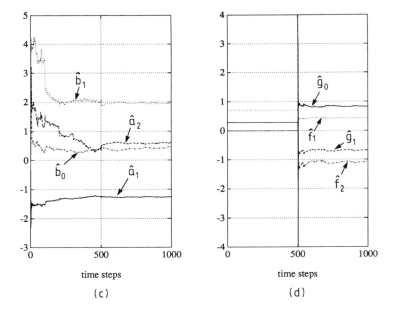

time steps

(c)

time steps

(d)

7.6.3 Self-tuning property – a simple example

Consider the first order system defined by

$$(1+az^{-1})y(t) = bz^{-1}u(t) + (1+cz^{-1})e(t) \qquad (|c| < 1) \qquad (7.96)$$

and assume that it is required to assign all closed loop poles to the origin so that **T** = 1. The controller is

$$u(t) = - gy(t) \qquad (7.97)$$

and, if a,b,c are *known*, the controller parameter g is obtained from

$$(1 + az^{-1}) + z^{-1}bg = 1 + cz^{-1} \qquad (7.98)$$

i.e.

$$g = (c - a)/b \qquad (7.99)$$

Substituting (7.97) into (7.96) then yields the closed loop equation

$$y(t) = e(t) \qquad (7.100)$$

If the system parameters are *unknown*, self-tuning pole assignment can be achieved based on the model

$$(1 + \hat{a}z^{-1})y(t) = \hat{b}z^{-1}u(t) + \hat{e}(t) \qquad (7.101)$$

The controller (7.97) is obtained by synthesis from

$$(1 + \hat{a}z^{-1}) + z^{-1}\hat{b}\hat{g} = 1 \qquad (7.102)$$

so that $\hat{g}$ (a function of the estimated model parameters) replaces g and is given by

$$\hat{g} = -\hat{a}/\hat{b} \qquad (7.103)$$

When the parameter estimates have converged, we now show that the controller based on (7.103) leads to the correct closed loop behaviour described by (7.100).

The RLS algorithm in conjunction with the model (7.101) and the control law leads to the stationarity condition

$$E[y(t-1)\eta(t)] = R_{y\eta}(1) = 0 \qquad (7.104)$$

to be satisfied at the convergence point (see Section 3.2).

Substituting the controller in (7.101) at convergence yields

$$y(t) = \eta(t) \qquad (7.105)$$

so that (7.104) can be expressed as

$$R_{\eta\eta}(1) = 0 \qquad (7.106)$$

Comparing (7.105) with (7.100), it is clear that we need to show that $\eta(t) = e(t)$, i.e. uncorrelated.

The data is generated by employing the control law together with the system (7.96). This yields

$$y(t) = \frac{1+cz^{-1}}{1+hz^{-1}} e(t) = \eta(t)$$

where $h = a+b\hat{g}$ and is assumed of magnitude less than unity. Then

$$R_{\eta\eta}(1) = E\left[z^{-1}\left[\frac{1+cz^{-1}}{1+hz^{-1}}\right]e(t) \cdot \left[\frac{1+cz^{-1}}{1+hz^{-1}}\right]e(t)\right]$$

$$= \sigma_e^2 \operatorname{res}\left[\frac{1+cz}{1+hz} \cdot \frac{z+c}{z+h}\right] = \sigma_e^2 \frac{(1-ch)(c-h)}{(1-h^2)}$$

where res[·] denotes the sum of the residues of [·] at its poles within the unit circle and $\sigma_e^2 = \operatorname{var}(e)$. Equation (7.106) then implies that $h = c$ so that $\hat{g}$ is given by (7.99) and the result follows. The local stability of this convergence point is discussed in Section 7.7.   □

This self-tuning result can be established in a more general setting (see Appendix 7.2). Note that the key figure of merit indicating correct convergence is the difference $\eta(t) - e(t)$ which should tend to zero. This can be checked easily in simulation, but in practice one must rely upon the qualitative nature of the control performance.

Note also that the choice $\mathbf{T} = 1$ is essential in order to satisfy the polynomial identity (7.95). Although a first order $\mathbf{T}$ appears to be achievable by insertion on the right-hand side of (7.102), the identity (7.95) would not be solvable because of the mismatch between the first order left-hand side and the second order right-hand side. Knowledge of n_c is therefore of importance for low order systems.

7.6.4   Self-tuning pole assignment controllers

In this section we describe the self-tuning pole assignment algorithm for a combined servo and regulation problem. The situation considered is a self-tuning version of the algorithm given in Section 7.4.1.

Algorithm: Pole Assignment Control

At each sample time t:

Step 1 Use RELS or AML to estimate $\hat{A}$, $\hat{B}$, $\hat{C}$ in the system model

$$\hat{A}y(t) = \hat{B}u(t-1) + \hat{C}\hat{e}(t) \qquad (7.107)$$

Step 2 Synthesize the controller coefficients using

$$\hat{A}\hat{F} + z^{-1}\hat{B}\hat{G} = \hat{C}T \qquad (7.108a)$$

$$\hat{H} = \frac{T(1)}{\hat{B}(1)}\hat{C} \qquad (7.108b)$$

Step 3 Apply control

$$\hat{F}u(t) = \hat{H}r(t) - \hat{G}y(t)$$

Step 4 Wait for tth time interval to elapse and return to step 1.   □

If the model (7.107) is not overparametrized, then, under weak conditions that appear to present no problems in practical applications, the algorithm converges asymptotically. Further, provided the reference signal is sufficiently exciting, the parameter estimates converge to their true values. The distinctive feature of the algorithm is that the C object must be estimated. If $\hat{C}^{-1}$ is unstable, then (7.018a) implies that the controller is destabilizing the model (and possibly the system). This can happen if C has zeroes close to the unit circle. In this situation, $\hat{C}$ must be monitored and, if necessary, prevented from misbehaving. This can be achieved by freezing $\hat{C}$ at a current 'good' value until a 'good' update is generated. Alternatively, $\hat{C}$ can be replaced by unity in (7.108) but this will change the resulting closed loop signal variances.

   If C is known (or a good prior estimate constructed off-line) then the data can be filtered by C^{-1} and RLS used to estimate $\hat{A}$, $\hat{B}$ (see equation (7.89)). In general, if $\hat{C}$ is arbitrarily assumed to be unity, then the achieved pole set will be incorrect and dependent on the reference signal (see Problem 7.12).

Example 7.10   *Self-tuning pole assignment control*

Consider the system used in Example 7.9 and given by:

$$(1+0.5z^{-1}+0.7z^{-2})y(t) = (z^{-1}+0.2z^{-2})u(t)+(1-0.8z^{-1})e(t)$$

The system was simulated using the self-tuning pole assignment controller just described. The desired pole set was given by

$$\mathbf{T} = 1-0.6z^{-1}$$

Figure 7.13 illustrates the behaviour of this algorithm. Note that, in this situation where the C polynomial is estimated, the estimated system parameters (Figure 7.13(c)) do not bias as in the pure regulator case (see Example 7.9) and the controller parameters (Figure 7.13(d)) converge to their correct values $f_1 = -0.05$, $g_0 = -1.85$, $g_1 = 0.175$.   □

Figure 7.13
Illustrating the performance of the self-tuning pole assignment controller system of Example 7.10: (a) output and reference signals; (b) control signal; (c) estimated system parameters; and (d) controller parameters.

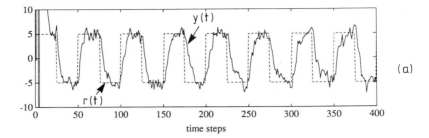

(a)

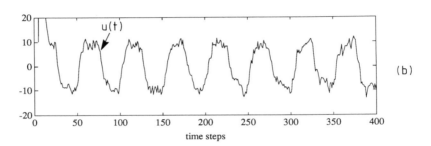

(b)

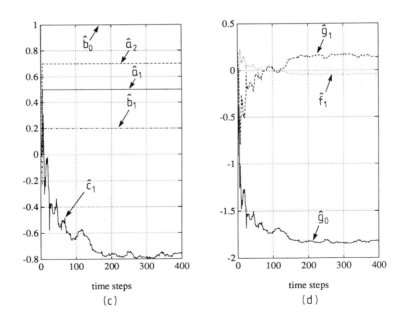

(c)   (d)

7.7   LOCAL CONVERGENCE — A SIMPLE EXAMPLE REVISITED

In order to emphasize the power of the ODE approach to convergence, we return to the simple first order regulator example of Section 7.6.3 to investigate whether the self-tuning property established there satisfies the required local stability conditions.

Recall that the system has structure

$$(1+az^{-1})y(t) = bz^{-1}u(t) + (1+cz^{-1})e(t) \tag{7.109}$$

while the model takes the RLS form

$$(1+\hat{a}z^{-1})y(t) = \hat{b}z^{-1}u(t) + \hat{e}(t) \tag{7.110}$$

The self-tuning regulator is

$$u(t) = -\hat{g}y(t) = -(\hat{a}/\hat{b})y(t) \tag{7.111}$$

which places the model poles at the origin.

In order to investigate possible local convergence points (a^*,b^*), we invoke the ODE results of Section 6.5 and form

$$\mathbf{f}(\boldsymbol{\theta}) = E\begin{bmatrix} -y(t-1) \\ u(t-1) \end{bmatrix}[(1+\hat{a}z^{-1})y(t)-z^{-1}\hat{b}u(t)] \tag{7.112}$$

Note that (7.110) and (7.111) allow (7.112) to be cast in the form

$$\mathbf{f}(\boldsymbol{\theta}) = -\begin{bmatrix} 1 \\ \hat{g} \end{bmatrix}E[y(t-1)\eta(t)]$$

so that

$$\mathbf{f}(\boldsymbol{\theta}^*) = 0 \rightarrow R_{y\eta}(1) = 0 \tag{7.113}$$

i.e. the local convergence condition implies the RLS stationarity condition (cf. 7.104). Note also, however, that the control law (7.111) has caused the persistent excitation condition to be violated by forcing the components of the regressor vector to be linearly dependent, so that the matrix

$$E\begin{bmatrix} -y(t-1) \\ u(t-1) \end{bmatrix}[-y(t-1)\ u(t-1)] = \begin{bmatrix} 1 & \hat{g} \\ \hat{g} & \hat{g}^2 \end{bmatrix}Ey^2(t-1) \tag{7.114}$$

is singular. The result is that (7.113) provides only one equation for the two unknowns a^*, b^* but this turns out to be sufficient for establishing the self-tuning property (i.e. system identifiability).

From (7.109) and (7.111) the closed loop system behaviour is given by

$$y(t) = \frac{1+cz^{-1}}{1+hz^{-1}} e(t) \tag{7.115a}$$

$$u(t) = -\hat{g}\, \frac{1+cz^{-1}}{1+hz^{-1}} e(t) \tag{7.115b}$$

$$h = a+b\hat{g} \tag{7.115c}$$

so that, assuming $|h|<1$ for stability,

$$f_1(\boldsymbol{\theta}) = -E\left\{ \frac{1+cz^{-1}}{1+hz^{-1}} e(t-1) \cdot \frac{1+cz^{-1}}{1+hz^{-1}} e(t) \right\}$$

$$= -\sigma_e^2\, \frac{(1-ch)(c-h)}{1-h^2} \tag{7.116a}$$

$$f_2(\boldsymbol{\theta}) = \hat{g}f_1(\boldsymbol{\theta}) \tag{7.116b}$$

Setting $\mathbf{f}(\boldsymbol{\theta}^*)$ to zero yields $h^*=c$, i.e.

$$g^* = -a^*/b^* = (c-a)/b \tag{7.117}$$

as required to reduce (7.115a) to

$$y(t) = e(t) \tag{7.118}$$

(cf. equation (7.48) with $F=1=\mathbf{T}$). As expected, this corroborates the results of our analysis in Section 7.6.3.

Note again, however, that the lack of excitation has prevented our determining the separate values of a^*, b^* and this causes a problem if we pursue the analysis and attempt to investigate local stability at $\boldsymbol{\theta}^*$ by calculating the eigenvalues of the matrix:

$$\mathbf{H}(\boldsymbol{\theta}^*) = (\mathbf{R}^*)^{-1}\mathbf{f}'(\boldsymbol{\theta}^*) \tag{7.119}$$

(see Section 6.5).

The matrix $\mathbf{R}^*$ is given by (7.114) evaluated at the convergence point, so that

$$\mathbf{R}^* = \sigma_e^2 \begin{bmatrix} 1 & g^* \\ g^* & (g^*)^2 \end{bmatrix} \tag{7.120}$$

This is singular and we replace it by the nonsingular matrix

$$\mathbf{R}^*(\delta) = \mathbf{R}^* + \delta\sigma_e^2\mathbf{I}_2 \tag{7.121}$$

where δ is small and will be allowed to tend to zero.
   From (7.116) and (7.117)

$$\mathbf{f}'(\mathbf{\theta}^*) = -\sigma_e^2 \begin{Bmatrix} b \\ b^* \end{Bmatrix} \begin{bmatrix} 1 & g^* \\ g^* & (g^*)^2 \end{bmatrix} \tag{7.122}$$

To calculate the eigenvalues of $\mathbf{H}$, we first find the two values of λ that satisfy

$$\det[\mathbf{f}'(\mathbf{\theta}^*) - \lambda\mathbf{R}^*(\delta)] = 0 \tag{7.123}$$

i.e.

$$\delta\lambda[\lambda(1 + (g^*)^2 + \delta) + (b/b^*)(1 + (g^*)^2] = 0 \tag{7.124}$$

If $\delta = 0$, equation (7.124) is trivially true. If we use a limiting argument, however, we generate the two eigenvalues

$$\lambda_1 = 0, \qquad \lambda_2 = -(b/b^*) \tag{7.125}$$

which lie in the closed left-half plane provided that b^* has the same sign as the true b. The validity of this condition is not settled by this analysis. The process by which the convergence takes place becomes important but this is beyond our present considerations.
   Finally, note that, if $\hat{b}$ is fixed and only $\hat{a}$ is estimated, the above analysis goes through in a similar way (see Problem 7.14).

7.8 APPLICATIONS

In this section we consider two rather different applications. Both are performed in a simulation environment but with different objectives. The first involves the behaviour of a self-tuning pole assignment controller and how it performs in difficult circumstances. The objective here is to illustrate the power of pole assignment self-tuning in the face of radical changes in the system to be controlled and the need for proper numerical algorithms in such circumstances.

7.8.1   Example 7.11: Self-tuning pole assignment control of a difficult system

Consider the situation in which the system to be controlled suffers structural changes from time to time in a troublesome way. Specifically, suppose it experiences changes in its time delay and order. In order to maintain control in such a situation the self-tuning estimator must be capable of tracking parameter changes, suggesting that a suitable parameter tracking algorithm (of the kind discussed in Chapter 4) should be used. Likewise, the controller design rule associated with the self-tuner should be capable of accommodating the differing system structures. The ability to handle time delay variation is intrinsic to pole assignment. However, a reduction in system degree will often cause the simple matrix inversion solution to the pole assignment identity to fail. In this case, the Kucera method for solving the identity (as discussed in Appendix 7.1) should be used.

The example is arranged to show the behaviour of a pole assignment incremental servo controller under a number of system changes (see Table 7.3). Using a sample interval of 1 s, a discrete time noise-free model is assumed with $n_a=2$, $n_b=2$ and the five parameters estimated using RLS. The estimates are then used in the incremental control algorithm of Section 7.5.1, where **T** is selected to be unity to correspond to closed loop deadbeat control. Throughout, Kucera's algorithm is used to solve the pole assignment identity (see Appendix 7.1) and directional forgetting is employed in order to track the system changes (see Section 4.5.8).

Figure 7.14 shows the performance of the algorithm in tracking a periodic three-level reference signal. The algorithm tracks whichever system is generating the input/output data. In particular, this includes System No. 4,

Table 7.3

System No.	Samples	Dynamics
1	1–79	$\dfrac{1}{1+10s+40s^2}$
2	80–159	$\dfrac{\exp(-2.7s)}{1+10s+40s^2}$
3	160–239	$\dfrac{\exp(-2.7s)}{1+10s}$
4	240–319	$\dfrac{1}{1+10s}$
5	320–400	$\dfrac{1}{10s(1+2.5s)}$

Figure 7.14
Performance of
pole assignment
of self-tuning
applied to the
difficult system
of Example 7.11.
The system
changes at 80
time step
intervals as
follows: System
1 second order,
no time delay;
System 2 second
order with time
delay; System 3
first order with
time delay;
System 4 first
order, no time
delay; System 5
second order
with integrator
and no time
delay.

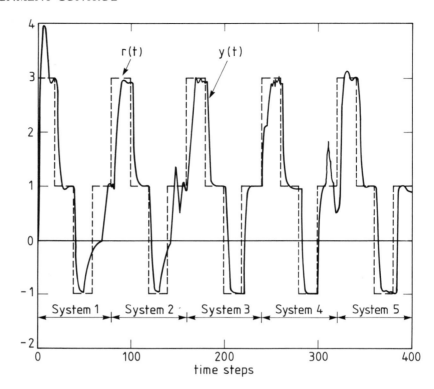

when the model is overparametrized and the solution of the identity involves approximate pole-zero cancellations.

7.8.2 Self-tuning control of a superheater attemperator

This study concerns the self-tuning control of a specific loop within a power station superheater system. In particular, it concerns the control of the attemperator loop in a multistage superheater. The role of the attemperator may be understood by reference to Figure 7.15 and the following description. A thermal power station of the kind considered here produces power by first generating steam within a boiler. The steam is then fed to a multistage superheater and hence to a turbine generator system in which the thermal energy of the superheated steam is converted to rotative energy (by the turbine) and hence to electricity (by the generator). The steam generated by the boiler is not suitable for use in a turbine. It is necessary to increase its temperature further in a series of superheaters. In order to regulate the temperature of steam leaving the first superheater stage an attemperator is used. The attemperator is a device consisting of an electrically operated spray

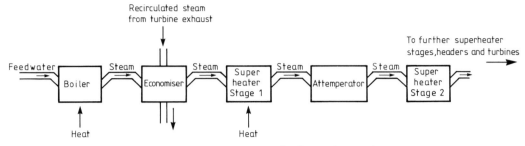

Figure 7.15   General scheme for a power station boiler and superheater system.

valve with which cold water is mixed with the superheated steam as it passes through a mixing chest (Figure 7.16). The control objective is to manipulate the water spray flow rate in order to regulate the temperature of the steam as it leaves the mixing chest. This temperature is denoted by the intermediate temperature $y(t)$.

The water flow rate to the attemperator spray is controlled by the position of a spray valve unit. The input to this, in the application considered here, is a 0 to 10 mA current signal which actuates a hydraulic valve position control circuit. The current signal is itself supplied by a mechanical 'servo-follower' of the type depicted in Figure 7.17. This is a form of mechanical integrator whereby a fixed control input $u(t)$ causes a cam to rotate at constant velocity. This in turn, depending upon the cam profile, produces a constant rate of change in the measured cam follower position. Mechanical cut-outs are provided at either end of the cam's angular travel. These provide a mechanical form of anti-wind-up for the integrator.

The overall attemperator system can be put in block diagram form, as shown in Figure 7.18. The figure gives the approximate linear transfer functions for

Figure 7.16
Scheme for a
steam
attemperator.

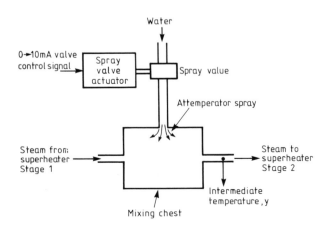

Figure 7.17
Scheme for a
mechanical servo
follower.

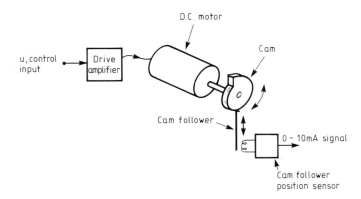

the system components. As indicated in the figure, the control objective is to regulate the intermediate temperature against fluctuations in upstream and downstream variation in temperature and steam flow. The control over the intermediate temperature should be fast but with predefined rate of change limits (in order to avoid thermal stressing of the mixing chest). The intermediate temperature should respond smoothly to reference setpoint changes and with little or no overshoot. In addition control actions should not require excessive spray valve action.

In order to test the use of self-tuning in this application a pole assignment self-tuner was applied to an analogue simulation of the 'servo-follower' and spray valve/mixing chest shown in Figure 7.18.

Because of the mechanical integration action of the servo-follower, the system is naturally in an incremental control form so that a discrete time model of the system could be written in incremental form. However, this was not done because the servo-follower is not an exact integrator and we wished to accommodate this in our model. Accordingly a model was employed of the form

$$y(t) = \frac{B}{A} u(t-1) \qquad (7.126)$$

Figure 7.18
Block diagram of
the steam
attemperator
system.

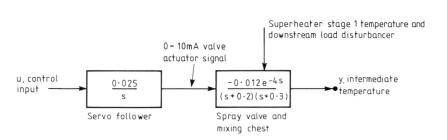

The pole assignment self-tuner was configured in the form shown in Figure 7.19. Note that the controller is driven by the error between setpoint $r(t)$ and the output $y(t)$. This is because of the incremental action embodied in the attemperator loop.

The setpoint $r(t)$ was rate limited to 0.05 V per sample interval. This was selected such that, using a sample interval of 4 s, the system output slew rate limit of 0.025 V/s would not be exceeded. Likewise saturation limits were imposed upon $u(t)$ to correspond to the saturation level of the servo follower drive system. With these management precautions and a pole set given by **T** $= 1 - 0.8z^{-1}$, the input/output results of Figure 7.20 were recorded. Note that the parameter and controller estimates corresponding to these results were not recorded, since the self-tuning programme used in this trial was based on an early minicomputer with limited memory. However, the fixed controller obtained from the trials and corresponding to the results shown is given by

$$u(t) = \frac{-(22.2 - 1.61z^{-1} - 1.9z^{-2})}{1 + 0.522z^{-1} + 0.377z^{-2} + 0.142z^{-3}} (r(t) - y(t)) \qquad (7.127)$$

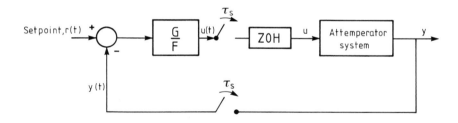

Figure 7.19 Self-tuning scheme for attemperator loop control tests.

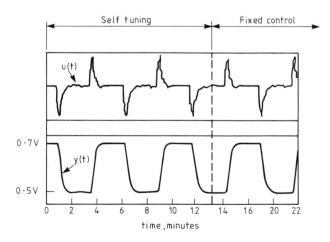

Figure 7.20 Input/output plots for self-tuning pole assignment control of a steam attemperator.

The motivation for this example is to illustrate the nature of physical limitations which often occur in a real system and how they are accommodated in the management software around a self-tuner. Note also that the engineering constraints are often expressed naturally in limits on rates of change with time, so that the pole assignment concept, through its relation with transient response performance, provides a natural format for embedding the constraints in the controller design criterion.

7.9 SUMMARY

The chapter has covered the topic of pole assignment synthesis control in both fixed coefficient and self-tuning forms. The main points are:

- Pole assignment allows the closed loop pole set of a system to be arbitrarily specified by the user
- The selected poles can be related to classical transient response or frequency response performance characteristics
- Unlike other methods (see Chapter 8) the method does not cancel open loop zeroes (except by choice)
- The method does not require prior knowledge of the system time delay, which may be variable
- Algorithmic forms have been given for servo design, regulator design and a controller design in which servo and regulator objectives are combined
- Further algorithmic variations allow the reduction of the regulation error and the inclusion of incremental action
- Some of the self-tuning algorithms display a special 'self-tuning property' and have the advantage that the C polynomial need not be estimated (Appendix 7.2)
- The solution of the pole assignment identity can usually be performed using simple matrix inversion but more reliably by other algorithms (Appendix 7.1)

7.10 PROBLEMS

Problem 7.1

For the following system and control rule:

$$Ay(t) = Bu(t-1)$$

$$Fu(t) = Hr(t) - Gy(t)$$

where $n_a = 3, n_b = 2$, show that the coefficients of F, G are given by the solution to

$$
\begin{bmatrix}
1 & 0 & b_0 & 0 & 0 \\
a_1 & 1 & b_1 & b_0 & 0 \\
a_2 & a_1 & b_2 & b_1 & b_0 \\
a_3 & a_2 & 0 & b_2 & b_1 \\
0 & a_3 & 0 & 0 & b_2
\end{bmatrix}
\begin{bmatrix}
f_1 \\
f_2 \\
g_0 \\
g_1 \\
g_2
\end{bmatrix}
=
\begin{bmatrix}
t_1 - a_1 \\
-a_2 \\
-a_3 \\
0 \\
0
\end{bmatrix}
$$

when $\mathbf{T} = 1 + t_1 z^{-1}$ specifies the desired closed loop pole set.

Problem 7.2

(a) Interpret the pole assignment identity as a means of matching the Nyquist locus of a system with a desired Nyquist locus.

(b) For the system

$$
y(t) = \frac{b_0 z^{-1}}{1 + a_1 z^{-1} + a_2 z^{-2}} u(t)
$$

with control $u(t) = (g_0 + g_1 z^{-1}) y(t)$ and desired pole set specified by $\mathbf{T} = 1 + t_1 z^{-1}$, show that an alternative but equivalent way of selecting g_0, g_1 is to demand that the actual and desired Nyquist responses and their first derivatives with respect to z are equal at zero frequency.

Problem 7.3

Many control systems use incremental control action in order to satisfy steady state accuracy requirements. This is done by forming the control signal $u(t)$ from

$$
F\Delta u(t) = F(u(t) - u(t-1)) = G(r(t) - y(t))
$$

and applying control action $u(t)$ to the system. Show how G, F can be derived using pole assignment and verify that $y(t) = r(t)$ at zero frequency.

Problem 7.4

Consider the CARMA system with $k = 1$ and

$$
B = 1 - 1.5z^{-1}, \qquad A = 1 - 0.9z^{-1}, \qquad C = 1 - 0.5z^{-1}
$$

Show that a control algorithm which attempts to cancel B in the closed loop numerator can be unstable if B changes to $(1 - 1.55z^{-1})$.

Problem 7.5

If a controller $Fu(t) = G(r(t)-y(t))$ is applied to the system $Ay(t) = Bu(t-1)$, show that the closed loop zeroes are given by GB. Illustrate how this can influence the closed loop transient response by proving that, for the system

$$y(t) = \frac{b}{(1-z^{-1})(1+a_1z^{-1})} u(t-1)$$

with controller $u(t) = (g_0+g_1z^{-1})(r(t)-y(t))$ and desired pole set specified by $\mathbf{T} = 1+t_1z^{-1}$, it is possible to achieve deadbeat control (i.e. $y(t) = r(t-1)$) with a nonzero choice of t_1.

How can the controller equation be modified to prevent the G polynomial modifying the system zeroes?

Problem 7.6

Given the identity $FA + z^{-1}BG = \mathbf{T}$, show that, if $F = F_0$, $G = G_0$ are particular solutions and the degrees of F, G are unrestricted, then an infinite number of solutions exist given by

$$F = F_0 + z^{-1}PB, \qquad G = G_0 - PA$$

where P is an arbitrary polynomial.

Problem 7.7

Given a system $Ay(t) = Bu(t-1)$ where

$$\frac{B}{A} = \frac{(1-bz^{-1})}{(1-a_1z^{-1})(1-a_2z^{-1})}$$

discuss the solution of

$$FA + z^{-1}BG = \mathbf{T}$$

with $F = 1+f_1z^{-1}$, $G = g_0 + g_1z^{-1}$ when $a_1 \rightarrow b$ in the cases

(a) $\mathbf{T}$ does not contain $(1-bz^{-1})$ as a factor

(b) $\mathbf{T} = (1-bz^{-1})\mathbf{T}_1$

Problem 7.8

Consider a system corrupted by an offset d and disturbance $v(t)$, thus

$$Ay(t) = Bu(t-1) + Dv(t) + d$$

Show how the incremental controller

$$F\Delta u(t) = G(r(t) - y(t))$$

influences the closed loop offset and disturbance rejection behaviour.

Problem 7.9

[*Note*: This question will require expertise from previous chapters.]

A system consisting of a biochemical production process has a product output rate $y(t)$ given by the nonlinear expression

$$y(t) = \beta \exp\{\alpha T\} (u(t))^p$$

where T is the process temperature and is measurable, $u(t)$ is the input feedrate and β, α, p are unknown constants.

(a) Find a transformation which allows the unknown constants β, α, p to be estimated using linear least squares.

(b) Hence suggest a feedback control system consisting of integral action alone which places the closed loop pole of the transformed system at a desired location. Show how this can be implemented as a self-tuning controller.

Problem 7.10

A deterministic system is described by the expression

$$Ay(t) = Bu(t-1)$$

where the polynomial B can be written as $B = B^+B^-$, B^+ has all its zeroes outside the unit disc and B^- has all its zeroes inside the unit disc.
    Derive the pole assignment synthesis rule for the controller polynomials F, G in the control law

$$Fu(t) = -Gy(t) + Hr(t)$$

to achieve the closed loop equation

$$\mathbf{T}y(t) = B^+H r(t-1)$$

Problem 7.11

The block diagram of P7.11 shows a feedback system which is to be used for self-tuning pole assignment. The system to be controlled is described by the discrete time representation

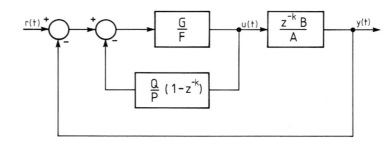

$$y(t) = \frac{z^{-k}B}{A} u(t)$$

in the usual notation, where k is a known integer time delay.

(a) Write down the closed loop transfer function for the system.

(b) Set the polynomials P and Q equal to A and B respectively and hence obtain a polynomial equation for F and G which will place the closed loop poles at the zeroes of the polynomial **T**, where

$$\mathbf{T} = 1 + t_1 z^{-1} + \ldots + t_{n_t} z^{-n_t}$$

(c) Compare the polynomial equation which you obtain with the normal pole assignment identity for this system. Indicate how the arrangement shown in Figure P7.11 might be advantageous in a self-tuning system if k is large.

Problem 7.12

The output $y(t)$, control signal $u(t)$ and zero mean disturbance $\xi(t)$ satisfy the dynamic relationship

$$y(t) = bu(t-1) + \xi(t)$$

where b is an unknown gain parameter.

It is required to design a controller that places the closed loop pole at $z = z_0$ and tracks a deterministic reference signal $\{r(t)\}$. A self-tuning servo controller

$$u(t) = gy(t) + hr(t)$$

is designed based on the RLS estimation of β in the model

$$y(t) = \beta u(t-1)$$

Find g,h in terms of z_0,β.
    Show that the estimate can converge to b in the cases:

(a) $\xi(t) = 0,$ $\bar{E}\left[\dfrac{r(t)}{1-z_0 z^{-1}}\right]^2 > 0$

(b) $\xi(t)$ = white noise $e(t)$

If $\xi(t) = e(t)+ce(t-1)$, show that any convergence point depends on the reference signal, but will give a pole close to z_0 provided that either $|c|$ is small or $\bar{E}[r(t)]^2 >> E[e(t)]^2$.

Problem 7.13

Investigate the validity of the self-tuning property in the case of self-tuning pole assignment regulation of the system

$$y(t) = Bu(t-1)+Ce(t)$$

based on RLS estimation of $\hat{B}$ in the model

$$y(t) = \hat{B}u(t-1)+\hat{e}(t)$$

Problem 7.14

It is desired to assign to $z = z_0$ the closed loop pole of the system

$$y(t) = bu(t-1)+e(t)$$

in the usual notation. The gain parameter b is unknown and an adaptive controller is constructed based on the over parametrized model

$$y(t) = \alpha y(t-1)+\beta u(t-1)$$

where β is fixed and α is estimated using RLS estimation.

Show that the only possible convergence point corresponds to the correct assignment of the pole, provided that the assumed gain has the correct sign, but that otherwise convergence cannot take place.

Problem 7.15

The standard pole assignment regulator of Section 7.4.3 can be implemented in an implicit self-tuning form by directly estimating the regulator coefficients using RLS with the model

$$\mathbf{T}y(t)-u(t-1) = F'(u(t-2)+\epsilon(t-1))+Gy(t-1)+\hat{e}(t)$$

where $F = 1+z^{-1}F'$, in the usual notation.

Show that, if the true F,G polynomials are coprime, then the only convergence points of the algorithm yield the desired closed loop pole set.

Calculate the local eigenvalues for the system

$$y(t)+ay(t-1) = bu(t-2)+e(t)$$

when $\mathbf{T} = 1$ and hence show that convergence can take place only if $|b| > |a| > 0$.

Problem 7.16

Develop the details of the *incremental* reduced variance pole assignment controller for the combined servo/regulation problem.

7.11 NOTES AND REFERENCES

The basic pole assignment approach to self-tuning is discussed in:

Wellstead, P.E., Edmunds, J.M., Prager, D.L. and Zanker, P.M.
  Pole zero assignment self-tuning regulator, *Int. J. Control,* **30**(1) 11–26, 1979.
Wellstead, P.E., Prager, D.L. and Zanker, P.M.
  A pole assignment self-tuning regulator, *Proc. IEE,* **126**(8), 781–7, 1979.
Astrom, K.J. and Wittenmark, B.
  Self-tuning controller based on pole-zero placement, *Proc. IEE,* **127**, 120–30, 1980.
Wellstead, P.E. and Sanoff, S.P.
  Extended self-tuning algorithm, *Int. J. Control,* **34**(3), 433–55, 1981.

A multivariable pole assignment self-tuner satisfying the self-tuning property is discussed in:

Prager, D.L. and Wellstead, P.E.
  Multivariable pole assignment regulator, *Proc. IEE, Part D,* **128**, 9–18, 1981.

In addition, see the original research material in the Ph.D. theses:

Edmunds, J.M.
  Digital adaptive pole shifting regulators. Ph.D. Thesis, Control Systems Centre, UMIST, Manchester, 1976.
Prager, D.L.
  Self-tuning control and system identification, Ph.D. Thesis, Control Systems Centre, UMIST, Manchester, 1980.
Sanoff, S.P.
  Extended pole shifting self tuners, Ph.D. Thesis, Control Systems Centre, UMIST, Manchester, 1982.

The reduced-variance algorithm was introduced in:

Zarrop, M.B. and Fischer, M.
  Reduced-variance pole assignment self-tuning regulation, *Int. J. Control,* **42**(5), 1013–33, 1985.

and the restricted nature of the self-tuning property for this algorithm along with discussion on a multivariable version appeared in:

Zarrop, M.B. and Davies, R.
  Multivariable reduced-variance pole assignment self-tuners, *Proc. Fifth IMA Conf. on Control Theory Strathclyde, UK, September 1988.*

General conditions under which the self-tuning property is valid are investigated in:

Casalino, G., Davoli, F., Minciardi, R. and Zappa, G.
  On implicit modelling theory: basic concepts and application to adaptive control, *Automatica,* **23**(2), 189–201, 1987.

Implicit pole assignment self-tuning algorithms, in which the controller parameters are generated directly by RLS estimation without the need to solve a polynomial identity, have been suggested by:

Elliott, H., Cristi, R. and Das, M.
  Global stability of adaptive pole placement algorithms, *IEEE Trans.* **Ac-30**(4), 348–56, 1985.
Gong, W.B.
  Correspondence on 'multivariable self-tuning regulators', *Proc. IEE, Part D*, **130**, 200, 1983.
Zarrop, M.B. Wellstead, P.E. and Partis, J.V.
  Implicit Algorithms for Pole Assignment Self-Tuning Controllers, CSC Report No. 589, Control Systems Centre, UMIST, 1983.

In simulation, however, the computational savings achieved by using implicit algorithms are often offset by slow convergence performance. Further, local convergence analysis using the ODE approach shows that, for stochastic systems, the algorithms are valid only for very restricted classes of systems (see Problem 7.15). This emphasizes the necessity of carrying out such convergence checks, often omitted in recent papers.

APPENDIX 7.1   SOLUTION OF THE POLE ASSIGNMENT IDENTITY

The pole assignment procedures discussed in Chapter 7 involve the solution of a set of linear equations for the controller polynomials F,G. These equations arise from a polynomial identity of the form

$$AF + z^{-1}BG = \mathbf{T} \qquad (A7.1.1)$$

where

(a) $A(0) = F(0) = \mathbf{T}(0) = 1$
(b) $\mathbf{T}(z^{-1}) = 0 \Rightarrow |z| < 1$

The case $n_a = 0$ is trivial. If $n_a \geqslant 1$, then the existence of at least one solution of (A7.1.1) requires three conditions:

(i) If A,B are not coprime, then any common zero must also be a zero of $\mathbf{T}$.

(ii) (A7.1.1) equates two polynomials and therefore

$$n_t \leqslant \max(n_a + n_f, n_b + n_g + 1) \qquad (A7.1.2)$$

(iii) The number of equations cannot exceed the number of unknowns and therefore

$$\max(n_a + n_f, n_b + n_g + 1) \leqslant n_f + n_g + 1 \qquad (A7.1.3)$$

If a strict inequality holds in (A7.1.3), then an infinite number of solutions are possible and this can be exploited as in the reduced variance algorithm (Section 7.5.3). For a unique solution, the equality must hold in (A7.1.3) and this leads to two possibilities:

$$n_f = n_b; \qquad n_g \geq n_a - 1; \qquad n_t \leq n_b + n_g + 1 \qquad \text{(A7.1.4a)}$$

$$n_f \geq n_b; \qquad n_g = n_a - 1; \qquad n_t \leq n_a + n_f \qquad \text{(A7.1.4b)}$$

From (A7.1.4) it is clear that the standard solution in which both n_f and n_g are of minimum degree:

$$n_f = n_b; \qquad n_g = n_a - 1 \qquad \text{(A7.1.5b)}$$

is only possible if

$$n_t \leq n_a + n_b \qquad \text{(A7.1.5c)}$$

In practice this does not usually cause a problem as only first or second degree T polynomials are considered. If, however, (A7.1.5b) is violated, then two solutions arise corresponding to choosing F or G of minimum degree.

Example

$$n_a = 1, n_b = 1, n_t = 3$$

Then the two cases yield, in the usual notation:
(1) $n_f = 1 = n_g$

$$\begin{bmatrix} 1 & b_0 & 0 \\ a_1 & b_0 & b_0 \\ 0 & 0 & b_0 \end{bmatrix} \begin{bmatrix} f_1 \\ g_0 \\ g_1 \end{bmatrix} = \begin{bmatrix} t_1 - a_1 \\ t_2 \\ t_3 \end{bmatrix} \qquad \text{(A7.1.6a)}$$

(2) $n_f = 2, n_g = 0$

$$\begin{bmatrix} 1 & 0 & b_0 \\ a_1 & 1 & b_0 \\ 0 & a_1 & 0 \end{bmatrix} \begin{bmatrix} f_1 \\ f_2 \\ g_0 \end{bmatrix} = \begin{bmatrix} t_1 - a_1 \\ t_2 \\ t_3 \end{bmatrix} \qquad \text{(A7.1.6b)}$$

If $n_t = 2$ ($t_3 = 0$), then (A7.1.5b) is satisfied and both sets of equations yield:

(3) $n_f = 1$, $n_g = 0$

$$\begin{bmatrix} 1 & b_0 \\ a_1 & b_0 \end{bmatrix} \begin{bmatrix} f_1 \\ g_0 \end{bmatrix} = \begin{bmatrix} t_1 - a_1 \\ t_2 \end{bmatrix} \tag{A7.1.7}$$

In (A7.1.6), (A7.1.7), the controller coefficients can be derived by inverting a matrix, provided that the matrix is nonsingular. The determinants of the three matrices are, in order:

$$b_0(b_0 - a_1 b_0), \qquad - a_1(b_0 - a_1 b_0), \qquad (b_0 - a_1 b_0)$$

and each vanishes if $-a_1$ (the zero of $A(z^{-1})$) coincides with $-b_0/b_0$ (the zero of $B(z^{-1})$), i.e. if A,B are not coprime.

In order to avoid this problem, it is preferable to solve the identity (A7.1.1) by Kucera's method which employs Euclid's algorithm to find the greatest common divisor g of A,B. If g divides **T**, it is cancelled throughout (A7.1.1) and a solution for F,G generated. If g does not divide **T** (the usual situation), then **T** is replaced by g**T**, provided that g is inverse stable so that conditions (b) and (i) are consistent. Then g can be cancelled and (A7.1.1) solved to give the minimum degree solution for either F or G.

Briefly, the essence of Kucera's method is as follows. For any polynomials A,B, the greatest common divisor g can be found together with two pairs of coprime polynomials P,Q and R,S such that

$$AP + BQ = g \tag{A7.1.8}$$

$$AR + BS = 0 \tag{A7.1.9}$$

which can be expressed in the matrix form

$$[g,0] = [A,B] \begin{bmatrix} P & R \\ Q & S \end{bmatrix} \tag{A7.1.10}$$

This implies that there is a sequence of linear operations on the columns of $[A,B]$ which transforms the array into $[g,0]$. Further, it immediately follows that these same operations applied to the unit matrix will result in the matrix

$$\begin{bmatrix} P & R \\ Q & S \end{bmatrix}$$

The algorithm is simpler to implement in terms of the reciprocal polynomials A^*, etc where

$$A^*(z) = z^{n_a} A(z^{-1})$$

Kucera's algorithm

Define the arrays

$$\mathcal{F} = [A^*, B^*] \tag{A7.1.11}$$

$$v = \begin{bmatrix} 1 & 0 \\ 0 & 1 \end{bmatrix} \tag{A7.1.12}$$

Step (i)   If one of the polynomials in $\mathcal{F}$ is zero, go to (iv).

Step (ii)   Let λ denote the ratio of the leading coefficient of the higher degree polynomial in $\mathcal{F}$ to that of the other polynomial and let $n(\geqslant 0)$ denote the difference in polynomial degrees.

Step (iii) Subtract λz^n times the lower degree polynomial from the other polynomial. Perform the same column operation on v; go to (i).

Step (iv) If the nonzero polynomial appears in the second column of $\mathcal{F}$, interchange the columns of both $\mathcal{F}$ and v. Stop.

The final entries in $\mathcal{F}$, v are given by

$$\mathcal{F} = [g^*, 0] \tag{A7.1.13}$$

$$v = \begin{bmatrix} P^* & R^* \\ Q^* & S^* \end{bmatrix} \tag{A7.1.14}$$

Example

Consider the transfer function

$$H(z^{-1}) = \frac{0.5\, z^{-1}}{1 + 0.1\, z^{-1}} \tag{A7.1.15}$$

The sequence of operations is as follows:

$\mathcal{F} = [z+0.1,\ 0.5]$ $v = \begin{bmatrix} 1 & 0 \\ 0 & 1 \end{bmatrix}$ Subtract 2z times column 2 from column 1.

$\mathcal{F} = [0.1,\ 0.5]$ $v = \begin{bmatrix} 1 & 0 \\ -2z & 1 \end{bmatrix}$ Subtract 5 times column 1 from column 2.

$\mathcal{F} = [0.1,\ 0]$ $v = \begin{bmatrix} 1 & -5 \\ -2z & 1+10z \end{bmatrix}$

Thus $g^* = 0.1$ and A,B are coprime. Note that the right-hand column of v is $[-B^*, A^*]^T/g^*$ satisfying (A7.1.9). In order to satisfy (A7.1.1) with $\mathbf{T}(z^{-1}) = 1 + 0.5z^{-1}$, the required controller polynomials are

$$F(z^{-1}) = 1, \qquad G(z^{-1}) = 0.8 \qquad \text{(A7.1.16)}$$

Now consider the case when (A7.1.15) is overparametrized and an exact stable pole-zero cancellation occurs:

$$H(z^{-1}) = \frac{0.5z^{-1}(1 + 0.9z^{-1})}{(1+0.1z^{-1})(1+0.9z^{-1})} \qquad \text{(A7.1.17)}$$

$$= \frac{z^{-1}(0.5 + 0.45z^{-1})}{1 + z^{-1} + 0.09z^{-2}}$$

Then the same sequence of column operations yields:

$$\mathcal{F} = [z^2+z+0.09, 0.5z+0.45] \qquad v = \begin{bmatrix} 1 & 0 \\ 0 & 1 \end{bmatrix}$$

$$\mathcal{F} = [0.1z+0.09, 0.5z+0.45] \qquad v = \begin{bmatrix} 1 & 0 \\ -2z & 1 \end{bmatrix}$$

$$\mathcal{F} = [0.1(z+0.9), 0] \qquad v = \begin{bmatrix} 1 & -5 \\ -2z & 1+10z \end{bmatrix}$$

Note that v follows the same sequence as before and the final right-hand column gives the minimal realization of H. The controller polynomials are therefore given by (A7.1.16).

If the plant (A7.1.15) is overparameterized and the parameters estimated as part of a self-tuning pole assignment control algorithm, it is unlikely that an exact pole-zero cancellation will occur and care must be taken to avoid ill-conditioning.

Consider

$$H(z^{-1}) = \frac{0.5z^{-1}(1+0.902z^{-1})}{(1+0.1z^{-1})(1+0.9z^{-1})}$$

The controller polynomials (A7.1.16) are now replaced by

$$F(z^{-1}) = 1 - 203.006z^{-1}, \qquad G(z^{-1}) = 405.012 + 40.511z^{-1} \quad \text{(A7.1.18)}$$

where the large coefficients arise from the inversion of a matrix which is almost singular. In general, poor control action would result. The use of Kucera's algorithm, however, allows detection of the approximate pole-zero cancellation. After two column transformations:

$$\mathcal{F} = [0.098z + 0.09, -0.0082], \qquad v = \begin{bmatrix} 1 & -5.102 \\ -2z & 1+10.204z \end{bmatrix}$$

If we accept -0.0082 as close enough to zero (compared to $(0.098, 0.09)$) to terminate the algorithm, then a pole-zero cancellation is imposed and the right-hand column of v gives the model

$$H(z^{-1}) = \frac{5.102z^{-1}}{z^{-1}+10.204} = \frac{0.5z^{-1}}{1+0.098z^{-1}} \qquad (A7.1.19)$$

leading to

$$F(z^{-1}) = 1, \qquad G(z^{-1}) = 0.804 \qquad (A7.1.20)$$

(cf. A7.1.16). Using the controller (A7.1.20) on the 'true' plant (A7.1.15) yields $\mathbf{T}(z^{-1}) = 1 + 0.502z^{-1}$, an acceptable error in pole position.

The above discussion indicates that the test for a zero polynomial in $\mathcal{F}$ required in Step (i) of the algorithm should be implemented by comparing the norms of the two vectors of coefficients. If their ratio is less than some chosen 'epsilon', then the algorithm is terminated.

APPENDIX 7.2   SELF-TUNING PROPERTY

Consider the system

$$A_0 y(t) = B_0 u(t-1) + C_0 e(t) \qquad (A7.2.1)$$

in the usual notation, where the polynomial coefficients are unknown. A self-tuning pole assignment regulation algorithm is implemented based on the RLS estimation of A, B coefficients in the model

$$Ay(t) = Bu(t-1) + X\hat{e}(t) \qquad (A7.2.2)$$

where X is a known inverse stable polynomial and the degrees of A, B are n_α, n_β respectively ($n_\alpha \neq 0$).

Given the estimates $\hat{A}$, $\hat{B}$ at time t, the control $u(t)$ is generated from

$$\hat{F}u(t) + \hat{G}y(t) = 0 \tag{A7.2.3}$$

$$\hat{A}\hat{F} + z^{-1}\hat{B}\hat{G} = X\mathbf{T} \tag{A7.2.4}$$

If

(a) $\hat{F} \to F, \hat{G} \to G$ asymptotically

the self-tuning property holds if F, G satisfy the identity

$$A_0F + z^{-1}B_0G = C_0\mathbf{T} \tag{A7.2.5}$$

The analysis that follows provides a set of sufficient conditions for (A7.2.5) to be satisfied at all convergence points.

Remarks

(i) Conditions under which (a) and other convergence and stability properties hold are beyond the scope of this discussion and are not investigated here.

(ii) No prior assumptions are made concerning the various polynomial degrees. In this way the framework covers the standard regulator and the 'extended' regulator (Section 7.6.2), controller overparametrization (Section 7.5.3) and includes the possibility of system/model mismatch.

Assuming

(b) $\hat{A} \to A, \hat{B} \to B$ asymptotically

the convergence point (A,B) must satisfy the stationarity conditions for RLS:

$$E[X^{-1}y(t-i)\cdot\eta(t)] = 0 \qquad i = 1, \ldots, n_\alpha \tag{A7.2.6a}$$

$$E[X^{-1}u(t-i)\cdot\eta(t)] = 0 \qquad i = 1, \ldots, n_\beta+1 \tag{A7.2.6b}$$

together with the condition

(c) A, B are coprime

to satisfy (A7.2.4).

The asymptotic closed-loop behaviour of the system is given by substituting (A7.2.3) into (A7.2.1) using (A7.2.4):

$$y(t) = \frac{FC_0}{H} e(t); \qquad u(t) = -\frac{GC_0}{H} e(t) \qquad (A7.2.7)$$

where

$$H = A_0 F + z^{-1} B_0 G \qquad (A7.2.8)$$

and is assumed inverse stable.

From (A7.2.2), (A7.2.7)

$$\eta(t) = \frac{C_0 \mathbf{T}}{H} e(t) \qquad (A7.2.9)$$

$$y(t) = \frac{F}{\mathbf{T}} \eta(t); \qquad u(t) = -\frac{G}{\mathbf{T}} \eta(t) \qquad (A7.2.10)$$

Substituting in (A7.2.6)

$$E\left[\eta(t) \frac{F}{X\mathbf{T}} \eta(t-i)\right] = 0 \qquad i = 1, \ldots, n_\alpha \qquad (A7.2.11a)$$

$$E\left[\eta(t) \frac{G}{X\mathbf{T}} \eta(t-i)\right] = 0 \qquad i = 1, \ldots, n_\beta + 1 \qquad (A7.2.11b)$$

Note that (A7.2.5) states that H should be identical to $C_0\mathbf{T}$ and therefore the task is to use (A7.2.11) to establish (see A7.2.9)) that $\{\eta(t)\}$ is an uncorrelated sequence.

Define the stationary sequence

$$\xi(i) = E[\eta(t)\cdot(X\mathbf{T})^{-1}\eta(t-i)] \qquad i \geq 1 \qquad (A7.2.12)$$

Then (A7.2.11) can be cast in the form

$$\xi(i) + f_1\xi(i+1) + \ldots + f_{n_f}\xi(i+n_f) = 0 \qquad i=1, \ldots, n_\alpha \qquad (A7.2.13)$$

$$g_0\xi(i) + g_1\xi(i+1) + \ldots + g_{n_g}\xi(i+n_g) = 0 \qquad i=1, \ldots, n_\beta+1$$

Allowing for possible controller overparametrization, from (A7.2.4), (A7.2.8)

$$n_f = n_\beta + n_\pi + 1; \qquad n_g = n_\alpha + n_\pi \qquad \text{(A7.2.14a)}$$

$$n_t + n_x \leqslant n_\alpha + n_\beta + n_\pi + 1 \qquad \text{(A7.2.14b)}$$

$$n_h = \max(n_a + n_\beta, n_\alpha + n_b) + n_\pi + 1 \qquad \text{(A7.2.14c)}$$

where

$$n_\pi \geqslant -1.$$

Thus (A7.2.13) consists of $(n_\alpha + n_\beta + 1)$ homogeneous linear equations for $\{\xi(i), i = 1, 2, \ldots, n_\alpha + n_\beta + n_\pi + 1\}$ and it follows that if

(d) $\qquad\qquad\qquad F, G$ are coprime

(e) $\qquad\qquad\qquad n_\pi = 0 \qquad$ or $\qquad -1$

then

$$\xi(i) = 0 \qquad i = 1, 2, \ldots, \mu(= n_\alpha + n_\beta + n_\pi + 1) \qquad \text{(A7.2.15)}$$

In order to extend (A7.2.15) by induction, write (A7.2.9) in the form

$$\eta(t) = C_0 \mathbf{T} e(t) - h_1 \eta(t-1) - \ldots - h_{n_h} \eta(t-n_h) \qquad \text{(A7.2.16)}$$

Multiplying by $(X\mathbf{T})^{-1}\eta(t-\mu-1)$ and taking expectations yields

$$\xi(\mu+1) = E[C_0 \mathbf{T} e(t) \cdot (X\mathbf{T})^{-1}\eta(t-\mu-1)] - h_1\xi(\mu) - \ldots - h_{n_h}\xi(\mu+1-n_h) \qquad \text{(A7.2.17)}$$

Hence $\xi(\mu+1)$ is zero provided that

$$\mu + 1 > \max(n_t + n_c, n_h) \qquad \text{(A7.2.18)}$$

or, using (A7.2.14),

(f) $\qquad\qquad\qquad n_t \leqslant n_\alpha + n_\beta + n_\pi + 1 - \max(n_c, n_x)$

(g) $\qquad\qquad\qquad n_\alpha \geqslant n_a \ ; n_\beta \geqslant n_b$

Note that (g) implies that the model cannot be underparameterized.
  By induction

$$\xi(i) = 0 \qquad i \geqslant 1 \qquad \text{(A7.2.19)}$$

Writing

$$XT = \gamma_0 + \gamma_1 z^{-1} + \ldots + \gamma_s z^{-s}$$

then

$$E[\eta(t)\eta(t-i)] = \gamma_0 \xi(i) + \gamma_1 \xi(i+1) + \ldots + \gamma_s \xi(i+s)$$
$$= 0 \qquad i \geqslant 1$$

Therefore $\{\eta(t)\}$ is an uncorrelated sequence and (A7.2.5) follows immediately.

Discussion

Conditions (a)–(g) have been dictated by the method of proof which is standard. They are sufficient conditions for the self-tuning property to hold and may not be necessary. More important, they may not be consistent. For example, model overparametrization (allowed by (g)) may lead to convergence problems, thus violating (a) and/or (b). Similarly, standard conditions for the existence of solutions of (A7.2.5) (such as coprimeness of A_0, B_0) have not been demanded above but are clearly required to avoid convergence problems. From (A7.2.5), assuming that $n_a \neq 0$,

(h) A_0, B_0 are coprime

$$n_f = n_b + n_p + 1; \qquad n_g = n_a + n_p \qquad (A7.2.20)$$

Comparing (A7.2.14a) with (A7.2.20), condition (g) can be replaced by

(g)' $n_\alpha - n_a = n_\beta - n_b \geqslant 0$

Also

$$n_t + n_c \leqslant n_a + n_b + n_p + 1 \qquad (A7.2.21)$$

This can be combined with (f) to yield

(f) $n_t \leqslant n_\beta + n_\pi + 1 + \min(n_\alpha - n_x, n_a - n_c)$

This completes the general analysis.

In the case of the *extended regulator* (Section 7.6.2) without overparametrization and assuming $n_x \leqslant n_c$, the conditions on the polynomial degrees reduce to

$$n_t \leqslant n_a + n_b - n_c \qquad (A7.2.22)$$

the standard result. This includes the *standard regulator* ($n_x=0$).

For the *reduced variance regulator* (Section 7.5.3), condition (e) is crucial as it constrains the degree of controller overparameterization allowed. In the case of correct model structure, this implies at most a constant P.

Although (e) is only a sufficient condition, a simple algebraic argument can be used to show that too much controller overparametrization can cause problems. From (A7.2.4) and (A7.2.5), assuming n_x is zero,

$$\frac{C_0 A - A_0}{z^{-1}(B_0 - C_0 B)} = \frac{G}{F} \qquad (A7.2.23)$$

For simplicity, assume that $n_\alpha = n_a$, $n_\beta = n_b$. Then condition (d) is certainly violated if the degree of $C_0 A - A_0$ is less than $n_g + 1$ i.e.

$$n_p \geqslant n_c \geqslant 1 \qquad (A7.2.24)$$

taking into account that $C_0 A - A_0$ and $z^{-1}(B_0 - CB_0)$ have a common factor of z^{-1}. Further, any common factors of F, G must appear in **T**, but this is impossible if

$$n_g + 1 - \deg(C_0 A - A_0) > \deg (\mathbf{T}) \qquad (A7.2.25)$$

i.e.

$$n_p \geqslant n_t + n_c \qquad (A7.2.26)$$

Therefore, if (A7.2.26) is satisfied the self-tuning property cannot hold.

The case $n_a = 0$ can be dealt with separately and the details are left to the reader (see Problem 7.13).

8 Minimum Variance Control

8.1 OUTLINE AND LEARNING OBJECTIVES

The control design rule considered in this chapter is based upon optimization methods. In particular, the minimum variance control strategy is discussed (Section 8.2), whereby at each time step a control signal is applied which minimizes the variance of the output of a system. In its basic form, minimum variance control is intended to solve the regulator problem for which the reference signal is a constant setpoint. It is possible, however, to generalize the approach to cover a general reference signal $r(t)$ and disturbances. The generalized minimum variance controller is considered in Section 8.3 and the limitations of the method are outlined in Section 8.4. The self-tuning version of generalized minimum variance control is introduced in Section 8.5. This is complemented by the algorithm in Section 8.6 which shows how the cost function associated with generalized minimum variance can be selected to give pole assignment properties. The convergence of self-tuning minimum variance controllers is discussed in Section 8.7. The chapter closes with an extended discussion of self-tuning minimum variance control applied to active suspension systems (Section 8.8).

The learning objective is to understand simple optimal self-tuning controllers and their implementation.

8.2. MINIMUM VARIANCE CONTROL

8.2.1 Introduction

The aim of the standard minimum variance controller is to regulate the output of a stochastic system to a constant (zero) set point. This goal can be expressed in optimization terms:

For each time t, choose the control $u(t)$ to minimize the output variance

$$J = E[y^2(t+k)] \tag{8.1}$$

where k is the time delay.

Note that the cost junction J involves k because $u(t)$ will only affect $y(s)$ for $s \geqslant t+k$. If the controller leads to closed-loop stability so that the output is a stationary process (a reasonable requirement), then J will have the same minimum value for each t (asymptotically).

8.2.2   A simple example

Consider the following example of a first order process.

Example 8.1

In difference equation form

$$y(t) = ay(t-1) + bu(t-1) + e(t) + ce(t-1) \tag{8.2}$$

where $\{e(t)\}$ is zero mean white noise of variance σ_e^2.
   In this case $k = 1$ and therefore

$$y(t+1) = ay(t) + bu(t) + e(t+1) + ce(t) \tag{8.3}$$

Defining

$$\hat{y}(t+1|t) = ay(t) + bu(t) + ce(t) \tag{8.4}$$

(the reason for this notation will be clarified later) and noting that, whatever the choice of controller, $u(t)$ cannot physically be a function of $y(t+1)$, then $\hat{y}(t+1|t)$ is functionally independent of $e(t+1)$. Further, from equations (8.2) and (8.4), we see that $\hat{y}(t+1|t)$ is a linear or nonlinear function of $e(t)$, $e(t-1)$, . . ., according to whether $u(t)$ is a linear or nonlinear controller. The significance of these properties emerges when we form the cost function J:

$$J = E[y^2(t+1)] = E[\hat{y}(t+1|t) + e(t+1)]^2$$
$$= E[\hat{y}(t+1|t)]^2 + E[e(t+1)]^2 + 2E[\hat{y}(t+1|t)e(t+1)] \tag{8.5}$$

The last term on the right-hand side vanishes for

   (i) any linear controller

   (ii) any nonlinear controller, provided $\{e(t)\}$ is an independent sequence (not just uncorrelated).

Condition (ii) is satisfied if we make the common assumption of gaussian white noise.

Then

$$J = E[\hat{y}(t+1|t)]^2 + \sigma_e^2 \tag{8.6}$$

and therefore J is minimized if $u(t)$ can be chosen to satisfy

$$\hat{y}(t+1|t) = ay(t) + bu(t) + ce(t) = 0 \tag{8.7}$$

This yields an implementable control law only if $e(t)$ can be expressed as a function of available data. This is achieved by inverting the process equation (8.2) to read:

$$e(t) = y(t) - ay(t-1) - bu(t-1) - ce(t-1) \tag{8.8}$$

or, in transfer function terms,

$$e(t) = \frac{1}{1+cz^{-1}} [(1-az^{-1})y(t) - bz^{-1}u(t)] \tag{8.9}$$

The recursion (8.8) always requires unknown initial values of the noise signal (unless c is zero) so that this reconstruction of $e(t)$ is only valid asymptotically and provided $|c|<1$. Fortunately, this last condition is a weak one for stationary stochastic processes (see Appendix).

Substituting equation (8.9) into $\hat{y}(t+1|t)$ yields:

$$\hat{y}(t+1|t) = \frac{1}{1+cz^{-1}} [(a+c)y(t) + bu(t)] \tag{8.10}$$

and setting $\hat{y}(t+1|t)$ to zero yields the Minimum Variance (MV) regulator:

$$u(t) = -\frac{(a+c)}{b} y(t) \tag{8.11}$$

Note that equation (8.3) can be written in the form

$$y(t+1) = \hat{y}(t+1|t) + e(t+1) \tag{8.12}$$

so that closed loop behaviour under (8.11) is given by

$$y(t+1) = e(t+1) \tag{8.13}$$

Thus the minimum achievable output variance is σ_e^2. This is not achieved if the time delay k is greater than unity as we will see later.

Clearly the control law (8.11) exploits the noise structure of the process (8.2). It is of interest to investigate the deterioration in control performance if this structure is neglected or incorrectly specified.

Suppose that c is (incorrectly) assumed to be zero, i.e. the assumed process model is:

$$y(t) = ay(t-1) + bu(t-1) + \hat{e}(t) \tag{8.14}$$

Setting c to zero in the controller (8.11) yields:

$$u(t) = -\frac{a}{b} y(t) \tag{8.15}$$

Substituting (8.15) into the true process model (8.2) yields:

$$y(t+1) = e(t+1) + ce(t) \tag{8.16}$$

so that

$$J = (1+c^2)\sigma_e^2 < 2\sigma_e^2 \tag{8.17}$$

The resulting output variance can therefore take a value up to double the desired minimum.

Figure 8.1 shows a simulation of the system (8.2) with parameter values

$$a = -0.8, \qquad b = 1, \qquad c = 0.98, \qquad \sigma_e^2 = 1$$

switching from no control to incorrect MV regulation (equation (8.15)) to true MV regulation (equation (8.11)). In particular, Figure 8.1(a) shows the output $y(t)$ and Figure 8.1(b) shows the noise process $e(t)$. For the period $400 \leq t \leq 799$ the control is incorrect minimum variance (equation (8.15)), for the period $800 \leq t < 1200$ the control is true MV. Note that the output variance reduces as the control changes and that in the final phase (MV) the output $y(t)$ is the same as the corrupting noise $e(t)$, as predicted by equation (8.13).   □

8.2.3 Output prediction

Before passing to a more general framework, we return to equation (8.12). Note that $y(t+1)$ is the sum of two independent terms, the first of which is a function of data up to time t only, and that

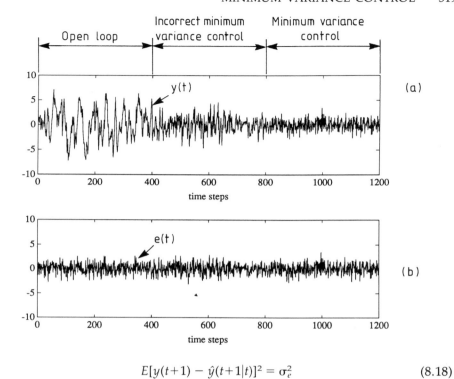

Figure 8.1 Simple MV regulator (Example 8.1). Comparing the open loop system response ($0 \leqslant t \leqslant 399$), incorrect MV control by neglecting C ($400 < t \leqslant 799$) and MV control ($800 \leqslant t \leqslant 1200$).

$$E[y(t+1) - \hat{y}(t+1|t)]^2 = \sigma_e^2 \qquad (8.18)$$

which is the minimum achievable output variance. It is clear that $e(t+1)$ cannot be reconstructed or estimated from the data available at time t. Thus $\hat{y}(t+1|t)$ can be interpreted as the 'best' estimate (or prediction) of $y(t+1)$ based on data up to time t (and hence the notation). It represents the closest we can approximate $y(t+1)$ in the sense of minimizing the left-hand side of equation (8.18). For Example 8.1, the output predictor is given by equation (8.10).

This discussion allows us to give a simple interpretation to the MV regulator derived in the last section:

'If you want to force $y(t+1)$ to zero, choose the control $u(t)$ to set to zero the best prediction of $y(t+1)$.'

This philosophy carries over to a more general framework and the MV controller can be said to be the simplest *predictive controller* (see Chapter 9). Optimal predictors will be discussed further in Chapter 11.

8.2.4 A more general framework

We now examine how the cost function (8.1) is minimized in a more general case. Consider the system described by the CARMA model

$$Ay(t) = z^{-k}Bu(t) + Ce(t) \tag{8.19}$$

so that

$$y(t+k) = \frac{B}{A}u(t) + \frac{C}{A}e(t+k) \tag{8.20}$$

Now define polynomials F, G of the form

$$\left.\begin{aligned} F &= 1+f_1z^{-1}+\ldots+f_{k-1}z^{-(k-1)} \\ G &= g_0+g_1z^{-1}+\ldots+g_{n_g}z^{-n_g} \\ n_g &= \max(n_a-1, n_c-k) \end{aligned}\right\} \tag{8.21}$$

to satisfy the polynomial equation

$$C = AF + z^{-k}G \tag{8.22}$$

so that F represents the first k terms in the expansion of C/A.

Making use of this identity in the system equation gives:

$$y(t+k) = \left[\frac{B}{A}u(t) + \frac{G}{A}e(t)\right] + Fe(t+k) \tag{8.23}$$

If t is the current time sample then the second term in the right-hand side of equation (8.23) will involved $e(t)$, $e(t-1)$, ... These may be found from the system equation (as for the simple example) using equation (8.19) in the form:

$$e(t) = \frac{A}{C}y(t) - z^{-k}\frac{B}{C}u(t) \tag{8.24}$$

Substituting this equation into equation (8.23) gives

$$y(t+k) = \left[\frac{B}{A}u(t) + \frac{G}{C}y(t) - z^{-k}\frac{BG}{AC}u(t)\right] + Fe(t+k)$$

which can be simplified using the identity (8.22) to yield

$$y(t+k) = \left[\frac{BF}{C}u(t) + \frac{G}{C}y(t)\right] + Fe(t+k) \tag{8.25}$$

This expression for y(t+k) is the sum of two independent terms. We can interpret the first term:

$$\hat{y}(t+k|t) = \left[\frac{BF}{C} u(t) + \frac{G}{C} y(t)\right]$$ (8.26)

as the best prediction of $y(t+k)$ based on data up to time t.

Then:

$$Fe(t+k) = y(t+k) - \hat{y}(t+k|t)$$ (8.27)

is the output prediction error arising from the noise sources $e(t+1)$, $e(t+2)$, ..., $e(t+k)$. These sources cannot be eliminated by the control signal $u(t)$.

Substituting equations (8.25) and (8.26) into the cost function (8.1) yields:

$$J = E[y^2(t+k)] = E[\hat{y}(t+k|t) + Fe(t+k)]^2$$

$$= E[\hat{y}(t+k|t)]^2 + (1 + f_1^2 + \ldots + f_{k-1}^2)\,\sigma_e^2$$ (8.28)

Clearly J is minimized by setting the predicted output to zero as before:

$$\hat{y}(t+k|t) = 0$$ (8.29)

to yield the control law

$$BFu(t) + Gy(t) = 0$$ (8.30)

and output signal

$$y(t) = Fe(t)$$ (8.31)

corresponding to the minimum output variance

$$J_{min} = (1 + f_1^2 + \ldots + f_{k-1}^2)\sigma_e^2$$ (8.32)

Equation (8.30) is the minimum variance control strategy, shown in Figure 8.2. The design procedure is:

(i) Solve for F, G using

$$C = FA + z^{-k}G$$ (8.22)bis

Figure 8.2 A
MV regulator.

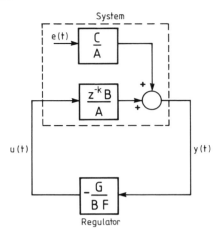

(ii) Apply control

$$u(t) = -\frac{G}{BF}y(t) \tag{8.33}$$

Note that:

(a) a polynomial identity (equation (8.22)) must be solved as in pole assignment;

(b) the time delay k must be known;

(c) if $k=1$, then $F=1$ and the polynomial identity yields

$$G(z^{-1}) = z[C(z^{-1}) - A(z^{-1})] \tag{8.34}$$

(d) the controller attempts to cancel explicitly the forward path zeroes, some of which may lie outside the unit circle (i.e. B may be inverse unstable).

Comment (d) indicates that the MV strategy is not a robust technique. This problem can be resolved within the optimization framework using the Generalized Minimum Variance (GMV) strategy discussed in Section 8.3.

Example 8.2

Consider the system

$$y(t) = 2y(t-1) + 2u(t-2) + e(t) + 0.5e(t-1) \tag{8.35}$$

where $\sigma_e^2=2$. In this case $n_f=1$, $n_g=0$ so that the identity (8.22) takes the form

$$(1+0.5z^{-1}) = (1-2z^{-1})(1+f_1z^{-1}) +z^{-2}g_0 \qquad (8.36)$$

leading to $f_1 = 2.5$, $g_0 = 5$.
    The MV controller is therefore

$$u(t) = - \frac{5}{2(1+2.5z^{-1})} y(t)$$

i.e.

$$u(t) = -2.5y(t)-2.5u(t-1) \qquad (8.37)$$

leading to closed loop behaviour

$$y(t) = e(t)+2.5e(t-1) \qquad (8.38)$$

The minimum output variance is therefore

$$J_{min} = 2[1+(2.5)^2] = 14.5 \qquad (8.39)$$

Note that J_{min} is about seven times as large as it would have been if the time delay had been unity. It is clear that a large value of f_1 is responsible and that increasingly large F coefficients will arise from the expansion of C/A if the system is open loop unstable. For such systems longer time delays lead to rapidly increasing values for the minimum achievable output variance. □

It is also important to emphasize comment (b) above, that the minimum variance regulator requires a knowledge of the system time delay k in order to construct the prediction $\hat{y}(t+k|t)$ and hence the control law. The following example illustrates the possible result of assuming an incorrect time delay.

Example 8.3

Consider the system

$$y(t) = 0.32y(t-1)+2u(t-2)+e(t)+0.5e(t-1)$$

where $\sigma_e^2 = 2$.

The associated minimum variance regulator has F, G polynomials

$$F = 1 + 0.82z^{-1}$$

$$G = 0.2624$$

and the minimum variance controller is

$$u(t) = -\frac{0.2624}{2(1+0.82z^{-1})}\, y(t)$$

Figure 8.3 shows the input/output behaviour using the minimum variance regulator. At time step 400, however, the system time delay is changed from $k=2$ to 3 while the regulator is unaltered. This causes the closed loop system poles to move to -1.014, $0.257 \pm j0.439$. Note that the system now has an unstable pole. Accordingly, the closed loop system becomes unstable with steadily increasing output variance. The sensitivity of the minimum variance regulator to the time delay is a significant disadvantage of the technique. Ways of avoiding this problem while preserving the optimization framework are considered later in the chapter.   □

8.2.5   A servo controller

The minimum variance regulator in the form just discussed is only aimed at regulating the system output $y(t)$ about zero. However, many industrial

Figure 8.3 Illustrating the output behaviour of the MV regulator of Example 8.3. For time 0 to 400 the correct time delay of $k = 2$ is present in the system. At time step 400 the delay is increased to $k = 3$ with consequent instability.

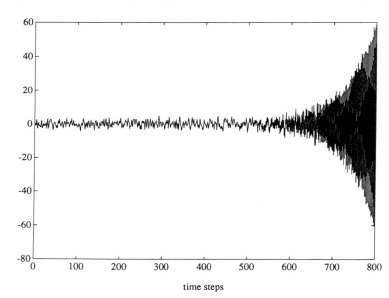

time steps

situations exist in which the requirement is to regulate $y(t)$ against random disturbances but at a constant nonzero value. The MV regulator can be adapted in a number of ways in order to allow a nonzero setpoint $r(t)$. The most obvious way is to apply the MV regulator to the signal $y(t)-r(t)$. Under MV control the system output should then be $r(t)$ plus the regulation error $Fe(t)$.

As an alternative to this we now indicate a popular and easy way in which a nonzero reference signal $r(t)$ can be incorporated into the MV scheme. The idea is to modify the algorithm as indicated in Figure 8.4 by the addition of a digital integrator. In this configuration the control law generates the control increment

$$\Delta u(t) = \frac{G}{BF}(r(t)-y(t)) \tag{8.40}$$

and $u(t)$ is derived from the increment by the digital integrator thus

$$u(t) = \Delta u(t) + u(t-1) \tag{8.41}$$

The controller polynomials are here obtained by solving the modified identity

$$C = FA\Delta + z^{-k}G \tag{8.42}$$

$$n_f = k-1; \qquad n_g = \max(n_a, n_c-k)$$

The closed loop equation then becomes

$$y(t) = \frac{G}{C}r(t-k) + \Delta Fe(t) \tag{8.43}$$

Figure 8.4 MV Regulator with incremental action and nonzero reference signal $r(t)$.

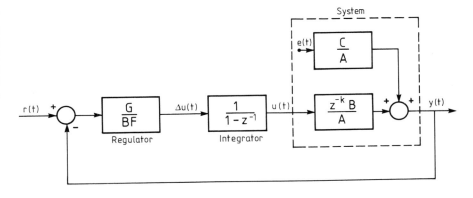

Note that:

(i) there is a k step delay between a change in the reference signal $r(t)$ and its appearance at the output $y(t)$;

(ii) the transient response from $r(t)$ to $y(t)$ is determined by the transfer function G/C; and

(iii) the regulation is no longer minimum variance because of the modified polynomial F arising from the expansion of $C/A\Delta$ rather than C/A.

The incremental MV controller is popular where $r(t)$ is constant over long periods or changes rather slowly. The reason for this can be seen by considering the component $y_r(t)$ of output contributed by $r(t)$:

$$y_r(t) = \frac{G}{C}r(t-k) \tag{8.44}$$

Using equation (8.42) this becomes

$$y_r(t) = r(t) - \frac{FA}{C}\Delta r(t) \tag{8.45}$$

Now if $r(t)$ is a constant then $\Delta r(t)$ is zero and the output equals the reference as required. Moreover, if $r(t)$ varies slowly then $y_r(t) \approx r(t)$, so that slowly changing references can be tracked.

Example 8.2 (continued)

The modified identity (8.42) now requires $n_f=1$, $n_g=1$ so that for the system given by equation (8.35) the following partition applies:

$$(1+0.5z^{-1}) = (1-2z^{-1})(1-z^{-1})(1+f_1z^{-1})+z^{-2}(g_0+g_1z^{-1}) \tag{8.46}$$

with solution $f_1 = 3.5$, $g_0 = 8.5$, $g_1 = -7$.

The steady-state closed-loop behaviour (for constant reference signal) is given by equation (8.43)

$$y(t) = \Delta Fe(t) = e(t)+2.5e(t-1)-3.5e(t-2) \tag{8.47}$$

(cf. equation (8.38)) so that the output variance is

$$\text{var } y(t) = 2[1+(2.5)^2+(3.5)^2] = 39 \tag{8.48}$$

(cf. equation (8.39)).

Clearly there has been a considerable trade-off between the magnitude of the output variance and the ability of the controller to track the set point. This is shown in Figure 8.5 which compares this algorithm (Figure 8.5(a)) with the use of the standard MV controller (8.37) in which $y(t)$ is replaced by $y(t) - r(t)$ (Figure 8.5(b)). The setpoint $r(t)$ in this case is a square wave of amplitude 4 and period 100 time steps. Note that the incremental algorithm tracks the setpoint whereas the standard MV controller does not. The variance of the incremental regulator, however, is significantly greater than the standard MV regulator.   □

The incremental MV algorithm is *ad hoc* but simple. It owes much of its popularity to the preference in industrial process control loops for incremental control action. As Figure 8.5 reveals, however, the regulation performance can deteriorate significantly as a result of its use *if* the true system structure is of the CARMA form. If a CARIMA model is more suitable, then it is not difficult to show that the incremental MV algorithm is optimal in the sense that it minimizes

$$E[y(t+k) - r(t)]^2$$

A more systematic way of accounting for a general reference signal and of dealing with the nonrobustness of ordinary MV is the generalized method which follows in Section 8.3.

Figure 8.5
Comparison of
incremental and
nonincremental
MV control
(Example 8.2):
(a) shows the
incremental MV
regulator; and
(b) the standard
MV regulator
with $y(t)$
replaced by
$y(t) - r(t)$.

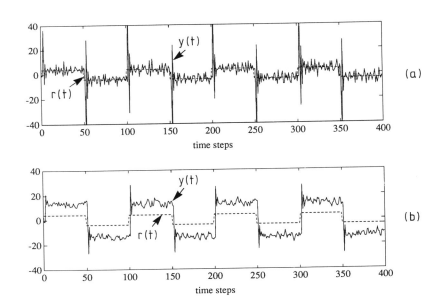

8.2.6 Detuned minimum variance

A useful variant on the basic MV regulator is the detuned MV algorithm. This is based on the observation that in some cases the basic form of the MV regulator is satisfactory except that the output variance minimization results in a very large control signal $u(t)$. One solution to this is to detune the MV regulator. The detuning algorithm considered here is presented specifically as a bridge between pole assignment control and MV control. It also serves as an introduction to the general MV control considered in Section 8.3.

The basic idea of detuned MV is to minimize the cost function

$$J = E[(\mathbf{T}y(t+k))^2] \tag{8.49}$$

where $\mathbf{T}$ is a polynomial of degree n_t.

The cost (8.49) is minimized by a regulator given by

$$u(t) = -\frac{G}{BF}y(t) \tag{8.50}$$

where F,G are obtained as the solution to the identity

$$C\mathbf{T} = FA + z^{-k}G \tag{8.51}$$

and $n_f = k-1$, $n_g = \max(n_a, n_c + n_t - k)$

With this design the closed loop regulation error for $y(t)$ is given by

$$y(t) = \frac{F}{\mathbf{T}}e(t) \tag{8.52}$$

Thus the detuning algorithm can be seen as a form of pole assignment algorithm in which the regulator assigns poles to the zeroes of $\mathbf{T}$. In particular, by manipulation of the coefficients of $\mathbf{T}$, a trade-off between the variance of $y(t)$ and $u(t)$ can be introduced. In practice it is usually sufficient to use a first order $\mathbf{T}$ polynomial:

$$\mathbf{T} = 1 - t_1 z^{-1} \tag{8.53}$$

By increasing t_1 from zero (corresponding to MV), a value can usually be reached which gives satisfactory output regulation without excessive control action (see Section 8.10 Problems).

As in the standard MV algorithm, the control law (8.50) requires the cancellation of all system zeroes. For nonminimum phase systems, therefore, we face the same robustness problem. Before resolving this problem by an extension of the detuning concept, it is of interest to consider an alternative approach that has a claim to optimality.

8.2.7 Peterka's algorithm

For general B, a more robust MV strategy (in the sense of avoiding the attempted cancellation of nonminimum phase zeroes) is given by the regulator

$$Fu(t)+Gy(t) = 0 \tag{8.54a}$$

where

$$AF+z^{-k}BG = CB^* \tag{8.54b}$$

and B^* is the polynomial with $B^*(0)$ set to unity and whose zeroes are the stable zeroes of B together with the reciprocal of the unstable zeroes of B.

The main disadvantage here is the need to factorize B into its stable and unstable components. For example, if

$$B(z^{-1}) = 2+3z^{-1}-2z^{-2}$$

$$= 2(1-0.5z^{-1})(1+2z^{-1})$$

then

$$B^*(z^{-1}) = (1-0.5z^{-1})(1+\tfrac{1}{2}z^{-1}) = 1-0.25z^{-2}$$

Note that the polynomial identity (8.54b) expresses a pole assignment operation and the closed loop behaviour of the system is given by

$$y(t) = \frac{F}{B^*} e(t) \tag{8.55a}$$

$$u(t) = -\frac{G}{B^*} e(t) \tag{8.55b}$$

Note also that, for a minimum phase system, the design yields the standard MV regulator (see Section 8.10 Problems).

Example 8.4

Consider the nonminimum phase system

$$y(t) = 2y(t-1)+u(t-2)+2u(t-3)+e(t)$$

with $\sigma_e^2=1$. To construct the MV regulator, the relevant polynomial equation is

$$1 = (1-2z^{-1})(1+f_1z^{-1})+z^{-2}g_0$$

yielding $f_1=2$, $g_0=4$ and the regulator

$$u(t) = -4u(t-1) - 4u(t-2) - 4y(t)$$

The closed loop output is

$$y(t) = e(t)+2e(t-1)$$

so that

$$\text{var } y(t) = 5$$

To construct Peterka's regulator, the unstable zero at -2 is reflected in the unit circle so that

$$B^*(z^{-1}) = 1+0.5z^{-1}$$

and the polynomial identity (8.54b) takes the form

$$1+0.5z^{-1} = (1-2z^{-1})(1+f_1z^{-1}+f_2z^{-2})+z^{-2}(1+2z^{-1})g_0$$

with the solution $f_1 = 2.5 = f_2$, $g_0 = 2.5$. The regulator takes the form

$$u(t) = -2.5[y(t)+u(t-1)+u(t-2)]$$

yielding the closed loop output

$$y(t) = -0.5y(t-1)+e(t)+2.5e(t-1)+2.5e(t-2)$$

with variance

$$\text{var } y(t) = 11.5$$

As expected we have paid for robustness through a higher variance level. □

When B has stable zeroes, some lying close to the unit circle, then the robustness problem again arises but Peterka's algorithm cannot help us. In this case an *ad hoc* modification due to Astrom is useful (see Section 8.11 Notes and references; also see Problem 8.10).

8.3   GENERALIZED MINIMUM VARIANCE CONTROL

The detuning idea discussed in Section 8.2.6 is a valuable link between the basic MV algorithm and general MV. Specifically, equation (8.49) introduces the concept of a modified cost function based on filtering the process output. This idea is now extended and generalized.

The MV control strategy can be extended to accommodate a servo-input $r(t)$ and to overcome the problem associated with attempting to cancel a nonminimum phase B. The approach introduces a system pseudo-output $\phi(t)$ defined by

$$\phi(t+k) = Py(t+k) + Qu(t) - Rr(t) \tag{8.56}$$

which can be interpreted as a generalized system output (Figure 8.6) based on the original system with feedforward added, a filtering action on the output, and the setpoint (filtered by R) subtracted out. The role of the

Figure 8.6
Generalized
system output.

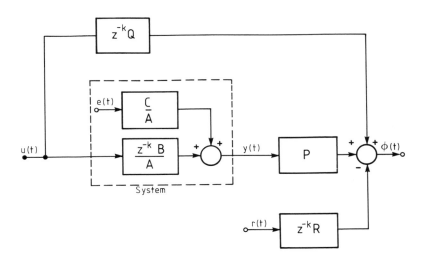

feedforward term is to shift the system open loop zeroes from B to $PB + QA$, since substituting the system equation in Equation (8.56) gives:

$$\phi(t+k) = \frac{PB + QA}{A} u(t) + \frac{PC}{A} e(t+k) - Rr(t) \tag{8.57}$$

The cost function to be minimized is the variance of the pseudo-output:

$$J = E[\phi^2(t+k)] \tag{8.58}$$

and the derivation procedure, as before, involves splitting $\phi(t+k)$ into two parts, one of which can be set to zero by control action $u(t)$ and a second part which is a function of $e(t+1),\ldots,e(t+k)$ and cannot be modified by control action at time step t.

Begin by defining $\mathbf{E}$ and G via the identity

$$PC = \mathbf{E}A + z^{-k}G \tag{8.59a}$$

where, assuming without loss of generality that $\phi(t)$ is scaled so that $P(0)$ is unity,

$$\mathbf{E} = 1 + e_1 z^{-1} + \ldots + e_{k-1} z^{-(k-1)} \tag{8.59b}$$

$$G = g_0 + g_1 z^{-1} + \ldots + g_{n_g} z^{-n_g} \tag{8.59c}$$

$$n_g = \max(n_a - 1, n_p + n_c - k) \tag{8.59d}$$

Multiply through the system equation

$$Ay(t+k) = Bu(t) + Ce(t+k) \tag{8.60}$$

by $\mathbf{E}$ and substitute for $\mathbf{E}A$ from equation (8.59a):

$$PCy(t+k) = B\mathbf{E}u(t) + Gy(t) + C\mathbf{E}e(t+k)$$

Adding $QCu(t) - CRr(t)$ to both sides gives

$$C[Py(t+k) + Qu(t) - Rr(t)] = (B\mathbf{E} + QC)u(t) + Gy(t) - CRr(t) + C\mathbf{E}eP(t+k)$$

which can be cast in the form

$$\phi(t+k) = \frac{1}{C} [(B\mathbf{E} + QC)u(t) + Gy(t) - CRr(t)] + \mathbf{E}e(t+k) \tag{8.61}$$

The error term $\mathbf{E}e(t+k)$ is uncorrelated with the remainder of the right-hand side and the analysis proceeds exactly as for the MV algorithm (Section 8.2.4).

The cost function is therefore minimized by setting the first term on the right-hand side to zero. The generalized MV (GMV) controller is therefore given by

$$(B\mathbf{E} + QC)u(t) = - Gy(t) + CRr(t) \qquad (8.62)$$

or

$$Fu(t) + Gy(t) + Hr(t) = 0 \qquad (8.63a)$$

where

$$F = B\mathbf{E} + QC \qquad (8.63b)$$
$$H = -CR \qquad (8.63c)$$

This equation should be compared in form to the pole assignment structure in Chapter 7.

Substituting 8.63 in the process equation (8.60) yields the closed loop equation

$$y(t) = \frac{z^{-k}BR}{PB + QA} r(t) + \frac{(B\mathbf{E} + QC)}{(PB + QA)} e(t) \qquad (8.64)$$

Note:

(i) If $Q = 0$, both transfer functions in equation (8.64) have a common factor B in the numerator and denominator, and the open loop zeroes are again cancelled.

(ii) The influence of P is equivalent to filtering all data by the transfer function $1/P$. (Compare P with the detuning polynomial $\mathbf{T}$ in Section 8.2.6.)

(iii) For zero steady state tracking error (on average), equation (8.64) yields the condition

$$\left. \frac{BR}{PB + QA} \right|_{z=1} = 1 \qquad (8.65)$$

This is most easily satisfied by choosing

$$R = P(1) \qquad \text{and} \qquad Q(1) = 0 \qquad (8.66)$$

(iv) The algorithm is equivalent to a pole assignment algorithm if the poles defined by the zeroes of a chosen polynomial T are used to fix P, Q through the identity

$$PB + QA = T \qquad (8.67)$$

First, using equations (8.59a), (8.63) and (8.67) the closed loop characteristic polynomial is

$$AF + z^{-k}BG = B(AE + z^{-k}G) + AQC = C(BP + AQ) = CT \qquad (7.47)\text{bis}$$

as required. Secondly, multiplying the identity 8.67 by E and using equations (8.59a) and (8.63) yields

$$PF - z^{-k}GQ = TE \qquad (8.68)$$

This identity relates E and the controller polynomials F and G directly to the P,Q cost function polynomials and can be used in self-tuning to adjust P and Q on-line to achieve desired pole positions (see Section 8.6). The solution of the identities (8.67) and (8.68) are discussed further in the Problems section.

Example 8.4 (continued)

Consider again the nonminimum phase system

$$y(t) = 2y(t-1) + u(t-2) + 2u(t-3) + e(t)$$

where $\sigma_e^2 = 0.5$. To construct the MV servo controller (8.40), the relevant identity is (8.42), i.e.

$$1 = (1 - 2z^{-1})(1 - z^{-1})(1 + f_1 z^{-1}) + z^{-2}(g_0 + g_1 z^{-1})$$

yielding $f_1 = 3$, $g_0 = 7$, $g_1 = -6$ and the controller

$$\Delta u(t) = \frac{7 - 6z^{-1}}{(1 + 2z^{-1})(1 + 3z^{-1})} [r(t) - y(t)]$$

The performance of this controller is shown in Figure 8.7(a) for a square wave reference and demonstrates the destabilizing effect of attempting to cancel the nonminimum phase zero at $z = -2$ outside the unit circle.

A suitable GMV controller can be constructed by choosing P,Q both equal to unity so that

$$PB + QA = 2$$

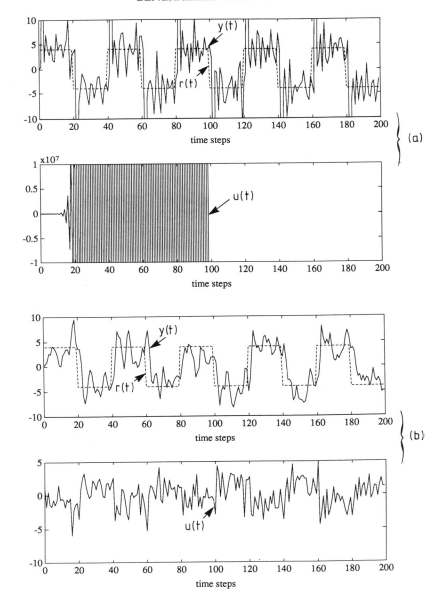

Figure 8.7 (a) Performance of MV Incremental Controller for nonminimum phase system (Example 8.4). (b) Performance of GMV Controller for nonminimum phase system (Example 8.4).

and the closed loop pole is assigned to the origin. The partition (8.59a) takes the form

$$1 = (1 - 2z^{-1})(1 + e_1 z^{-1}) + z^{-2} g_0$$

yielding $e_1 = 2$, $g_0 = 4$ so that

$$BE + QC = 2(1 + 2z^{-1} + 2z^{-2})$$

$$G = 4$$

Synthesis of the controller (8.62) is completed by choosing

$$H = -\tfrac{2}{3}$$

so that (8.65) is satisfied and steady state tracking achieved.
This leads to the closed loop equation

$$y(t) = \tfrac{1}{3}(1 + 2z^{-1})r(t-2) + (1 + 2z^{-1} + 2z^{-2})e(t)$$

This is precisely the nonincremental pole assignment controller with $\mathbf{T} = 1$ (see Section 7.4). Its performance is shown in Figure 8.7(b).   □

8.4   TIME DELAYS AND MINIMUM VARIANCE CONTROL

In MV control it is crucial that the time delay integer k be correctly selected. Incorrect values of k can destabilize the control or at least cause the regulation error to be much larger than need be. Consider the two possibilities separately.

(a)   If the assumed delay k_m is less than the actual system delay k, then the control will attempt to cancel output noise components before the signal can be transmitted through the system delay. The result is an unrealizable controller which can generate large feedback gains which ultimately destabilize the system.

(b)   Alternatively, if k_m is larger than k, then the variability of the noise will not be reduced to its lowest possible value, since the fastest transmission rate through the system is not being used.

The GMV algorithm, *provided Q is present*, need not suffer from Problem (a). If the assumed delay k_m is used in the generalized system (Figure 8.8) then the feedforward path provides the necessary route by which the control signal can influence the generalized output $\phi(t)$.

8.5   SELF-TUNING MINIMUM VARIANCE CONTROL

We now show how the generalized MV approach leads to a self-tuning algorithm in which the controller parameters are directly estimated by RLS.

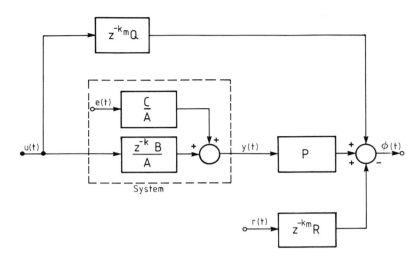

Figure 8.8
Using an
assumed
minimum
system delay k_m
in the
generalized
system $(k_m \leqslant k)$.

First, recall that the generalized system output, given by

$$\phi(t+k) = Py(t+k) + Qu(t) - Rr(t) \qquad (8.56)\text{bis}$$

can be expressed as the sum of two uncorrelated parts, only the first of which can be influenced by control action:

$$\phi(t+k) = \frac{1}{C}[Fu(t) + Gy(t) + Hr(t)] + \mathbf{E}e(t+k) \qquad (8.69)$$

where the polynomials F, G and H are obtained from the identity (8.59a) and equations (8.63).

The generalized MV controller is

$$Fu(t) + Gy(t) + Hr(t) = 0 \qquad (8.63a)\text{bis}$$

so that

$$\phi(t) = \mathbf{E}e(t) \qquad (8.70)$$

in closed loop. We propose the following self-tuning algorithm for a system represented by the standard CARMA model (8.19).

Algorithm: ***Generalized Minimum Variance***

At time step t

Step (i) Form the pseudo-output $\phi(t)$

$$\phi(t) = Py(t) + Qu(t-k) - Rr(t-k)$$

Step (ii) Estimate $\hat{F}$, $\hat{G}$, $\hat{H}$ from

$$\phi(t) = \hat{F}u(t-k) + \hat{G}y(t-k) + \hat{H}r(t-k) + \hat{e}(t) \tag{8.71}$$

Step (iii) Apply control $u(t)$ using

$$\hat{F}u(t) = -\hat{G}y(t) - \hat{H}r(t) \tag{8.72}$$

Step (iv) Wait for the sample time interval to elapse and return to step (i). □

Note that the above algorithm does not require solution of an identity and direct estimation of the noise polynomial C is avoided.

We can justify the above algorithm by showing that, if $\hat{F}$, $\hat{G}$, $\hat{H}$ converge to constant values, then the correct controller configuration is a possible outcome. To see this, recast equation (8.61) in the form

$$\phi(t) = Fu(t-k) + Gy(t-k) + Hr(t-k) + \mathbf{E}e(t) + (1-C)[\phi(t) - \mathbf{E}e(t)] \tag{8.73}$$

Under GMV control, equation (8.70) holds so that equation (8.73) yields

$$\phi(t) = Fu(t-k) + Gy(t-k) + Hr(t-k) + \mathbf{E}e(t) \tag{8.74}$$

where the error term is uncorrelated with the remaining terms on the right-hand side. This is consistent with the form of equation (8.71) and the RLS stationarity conditions (Section 3.2).

Note that this argument only establishes that the assumption of convergence to the correct control configuration does not lead to a contradiction.

Remarks

The important issues to note at this point are:

(i) The above self-tuning argument is a steady state one. It tells us that convergence to the correct configuration is a possibility, but not in what manner the parameters converge (if at all).

(ii) The algorithm calculates *directly* the control coefficients; no intermediate control synthesis stage is required.

(iii) There is apparently no need to model the noise dynamics C by using a RML algorithm. In fact, C *is* calculated since it occurs as a factor of $H(=-RC)$. If R is set to unity we will directly estimate C, the noise dynamics.

(iv) Usually the choice $R=P(1)$, $Q(1)=0$ is made in Step (i) to form $\phi(t)$. This choice guarantees zero steady-state tracking error (see Section 8.3).

Example 8.4 (continued)

If the model coefficients of our nonminimum phase system are unknown, the above self-tuning controller can be implemented. With the choice

$$P = 1 = Q, \qquad R = \tfrac{2}{3}$$

as before, Figure 8.9 shows how the self-tuner performs. In particular, Figure 8.9(a) shows the input/output response when the setpoint $r(t)$ is a square wave of period 40 times steps and amplitude 4. The corresponding control signal is shown in Figure 8.9(b) and the controller coefficients in Figure 8.9(c). The correct values for these coefficients are $h_0=\tfrac{2}{3}$, $f_0=2$, $f_1=4$, $f_2=4$, $g_0=4$. The self-tuning estimator in this case was initialized with $\mathbf{P}(0)=5\mathbf{I}_5$ and a constant forgetting factor $\lambda=0.998$ was applied. $\qquad\square$

8.6   GENERALIZED MINIMUM VARIANCE WITH POLE ASSIGNMENT

A key problem with the GMV approach is the need to make a prior selection of the cost polynomials P,Q,R. A modified approach and one which links nicely to pole assignment is to 'self-tune' P and Q so that a chosen closed-loop pole set is obtained. The approach involves making use of the relationship (8.68) between P,Q and F,G within the self-tuner. For simplicity, R and Q are assumed to satisfy (8.66) in order to achieve zero steady state error.

Algorithm: Generalized Minimum Variance with Pole Assignment

At time step t

Step (i) Form the pseudo-output $\phi(t)$

$$\phi(t) = P_t y(t) + \Delta Q_t u(t-k) - P_t(1)r(t-k)$$

Figure 8.9 Self-tuning the GMV Controller (Example 8.4): (a) shows the input/output behaviour for a square wave setpoint; (b) the corresponding control signals and (c) shows the estimated controller parameters.

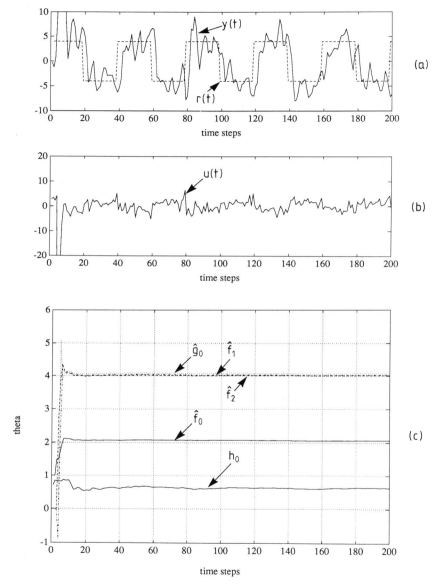

Step (ii) Estimate $\hat{F}_t$, $\hat{G}_t$, $\hat{H}_t$ from

$$\phi(t) = \hat{F}_t u(t-k) + \hat{G}_t(t-k) + \hat{H}r(t-k) + \hat{e}(t)$$

Step (iii) Apply control $u(t)$ from

$$\hat{F}_t u(t) = -\hat{G}_t y(t) - \hat{H}_t r(t)$$

Step (iv) Recompute the P and Q (and **E**) required to assign the closed loop poles at the zeroes of **T** using:

$$P_{t+1}\hat{F}_t - z^{-k}\hat{G}_t \Delta Q_{t+1} = \mathbf{T}\mathbf{E}_{t+1} \tag{8.75}$$

Step (v) Set $P_t = P_{t+1}$, $Q_t = Q_{t+1}$ and return to (i).   □

In Step (iv) the updated P, Q are determined by solving an identity using estimated controller coefficients. If the procedure converges, then the steady state solution satisfies

$$PF - z^{-k}G\Delta Q = \mathbf{T}\mathbf{E} \tag{8.68)bis}$$

implying that P,Q have 'self-tuned' to give the desired closed loop pole set.

Note that, if we make the standard normalizations $P(0) = 1 = \mathbf{T}(0)$, then the identity (8.67) yields

$$b_0 + q_0 = 1$$

The definition (8.63a) of the controller polynomial F then implies that $F(0)$ is unity and therefore this coefficient need not be estimated in Step (ii). Alternatively, if f_0 is estimated, then $(f_0)_t$ should be introduced as a scaling factor on the right-hand side of the identity (8.75). In this case, h_0 can be fixed instead by noting that

$$h_0 = [-RC]_{z=0} = -P_t(1)$$

at each time t. This is used later in the simulations. Finally, note that the forced coupling of the P,Q polynomials to the controller polynomials F,G can lead to instability if the algorithm is initialized carelessly. There is no guarantee of convergence from arbitrary initial values of P,Q.

Example 8.4 (continued)

In order to check the performance of the GMV pole assignment self-tuner for our first order example, it is useful to calculate the values of the various polynomials assuming that the system is known.

To solve the standard pole assignment identity (7.47), it is sufficient to assume the controller polynomial degrees $n_f = 2$, $n_g = 0$. It is then possible to solve the identity (8.68) for P,Q, given F,G if we choose $n_p = 1$, $n_q = 0$ (see Problem 8.8).

Selecting the polynomial

$$\mathbf{T} = 1 - 0.5z^{-1}$$

leads to

$$F = 1 + 1.5z^{-1} + 1.5z^{-2}$$

$$G = 1.5$$

$$P = 1 - \frac{5}{7}z^{-1}, \qquad Q = \frac{5}{7}, \qquad \mathbf{E} = 1 + \frac{9}{7}z^{-1}$$

$$H = -R = -P(1) = -\frac{2}{7}$$

using (7.47) and (8.68)

The values of the polynomial coefficients calculated above (scaled by the estimated coefficient f_0) are the required convergence values generated by the self-tuner for the chosen $\mathbf{T}$. Figure 8.10 shows the trajectories of the various estimates and the otuput behaviour. This performance should be compared with that of the explicit algorithm discussed in Section 7.4.   □

To summarize: the main point of this approach is that it links pole assignment with the optimization framework that gives rise to the MV and GMV control algorithms. In practice, however, if pole assignment is the chosen design philosophy, it is preferable to go directly to the explicit framework discussed in Chapter 7.

8.7   CONVERGENCE — A SIMPLE MINIMUM VARIANCE EXAMPLE

As a simple example of local convergence analysis, we now derive the *self-tuning property* for the adaptive *minimum variance control* of the simple first order system discussed earlier.

Example 8.1 (continued)

Consider the *system*

$$y(t) = ay(t-1) + bu(t-1) + e(t) + ce(t-1) \qquad |c| < 1 \qquad (8.76)$$

where $\{e(t)\}$ is a zero mean, unit variance, white noise sequence. The MV control law is

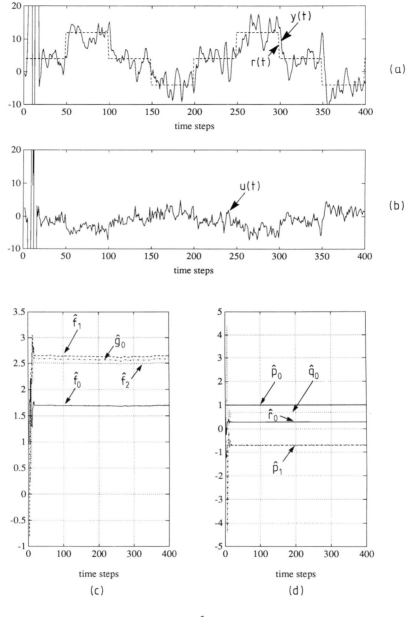

Figure 8.10 Self-tuning GMV PA control (Example 8.4) showing: (a) output and reference signal behaviour; (b) control signal behaviour; (c) estimated control parameters; and (d) estimated P, Q parameters.

$$u(t) = -\frac{1}{b}(c+a)\,y(t) \tag{8.77}$$

leading to closed loop behaviour

$$y(t) = e(t) \tag{8.78}$$

If the parameters a,b,c are unknown, however, the control law (8.77) cannot be implemented.

Consider the *estimation model*

$$y(t) = \alpha y(t-1) + \beta u(t-1) + \hat{e}(t) \tag{8.79}$$

where β is fixed and not estimated.

Corresponding to (8.79), the *MV controller* is

$$u(t) = -\frac{\hat{\alpha}(t)}{\beta} y(t) \tag{8.80}$$

where $\hat{\alpha}(t)$ is the estimate of α at time t.

If *RLS estimation* is used

$$\hat{\alpha}(t) = \hat{\alpha}(t-1) + \frac{1}{t} R(t)^{-1} y(t-1)[y(t) - \beta u(t-1)$$

$$- \hat{\alpha}(t-1)y(t-1)] \tag{8.81a}$$

$$R(t) = R(t-1) + \frac{1}{t} [y^2(t-1) - R(t-1)] \tag{8.81b}$$

To derive the ODEs, fix the estimate $\hat{\alpha}$ at some nominal value α. Then

$$f(\alpha) = \bar{E}[y(t-1) [y(t) - \beta u(t-1) - \alpha y(t-1)]] \tag{8.82}$$

where $y(t)$ is generated by (8.76) under the 'frozen' version of (8.80), i.e.

$$u(t) = -\frac{\alpha}{\beta} y(t) \tag{8.83}$$

so that

$$f(\alpha) = \bar{E}[y(t-1)y(t)] \tag{8.84}$$

where

$$y(t) = \frac{1+cz^{-1}}{1-\gamma z^{-1}} e(t) \tag{8.85a}$$

$$\gamma = a - b\alpha/\beta \tag{8.85b}$$

Assuming that $|\gamma| < 1$ for stability, (8.85) represents a stationary process and therefore $\bar{E}$ can be replaced by E in (8.84). Then

$$f(\alpha) = \text{res}\left\{\frac{(z+c)\,(1+cz)}{(z-\gamma)\,(1-\gamma z)}\right\}$$

$$= \frac{(\gamma+c)\,(1+c\gamma)}{1-\gamma^2} \tag{8.86}$$

If $\hat{\alpha}(t) \rightarrow \alpha^*$, then $f(\alpha^*) = 0$ and, since $|c| < 1$, $\gamma^* = -c$, i.e.

$$\alpha^* = \frac{\beta}{b}\,(c+a) \tag{8.87}$$

leading to (8.77) and (8.78), the correct input/output behaviour.

To check the eigenvalue condition for local convergence, from (8.81b)

$$G(\alpha,R) = E\,y^2(t-1) - R \tag{8.88}$$

Therefore

$$R^* = [E\,y^2(t-1)]_{\alpha=\alpha^*} = 1 \tag{8.89}$$

using (8.87) and (8.78).

From (8.86)

$$f'(\alpha) = -\frac{b}{\beta}\frac{d}{d\gamma}\,f(\alpha) = -\frac{b}{\beta} \qquad \text{for } \gamma = -c$$

Therefore

$$(R^*)^{-1}\,f'(\alpha^*) = \left[-\frac{b}{\beta}\right] < 0$$

if β has the correct sign (cf. Section 7.7).

8.8   APPLICATION — REGULATION OF ACTIVE SUSPENSION SYSTEMS

As we have already remarked, MV type regulation is a natural choice of design algorithm when the application clearly requires the minimization of some physically meaningful cost function. As an example of a situation in which this occurs consider the active control of vehicle suspension systems. Here the term active is used to mean feedback control. Conventional suspension systems are termed passive since they only include springs and hydraulic dampers. Feedback control of suspension must involve an active component, usually a hydraulic device, which is used to actuate the system. Although active suspensions take a number of forms, Figure 8.11 shows a simplified active vehicle suspension system, in which the conventional suspension spring, k, and hydraulic damper, b, are augmented by an actuator which can apply a force F in parallel. The mass m_b represents the vehicle body mass and m_w represents the wheel and axle assembly effective mass. The spring coefficient k_t represents compliance of the tyre.

The variables x_b, x_w, x_g represent the respective displacements of the vehicle body, wheel/axle assembly and the road surface. The control requirement for the active suspension system has two components. First, it must maintain the vehicle level at a required height above the road. Second, it should attempt to reduce the influence of road noise upon the body. The levelling control will operate so as to maintain the body height x_b at a fixed value or setpoint. The road noise isolation control will operate so as to minimize fluctuations in the body velocity $v_b(= \dot{x}_b)$. In this application study we will restrict our attention to the road noise isolation problem.

The task of road noise isolation is a regulator problem. Figure 8.12 illustrates this situation in which the objective is to minimize in some sense the variability of the body velocity v_b in response to the disturbance caused by the ground velocity v_g by using the actuator force F as control input. Clearly

Figure 8.11
Simplified model
of a vehicle
suspension
system.

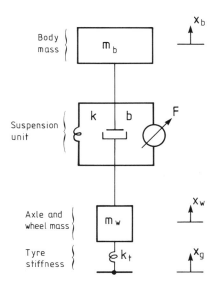

Figure 8.12
Road noise
isolation as a
regulation
problem.

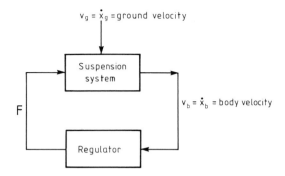

a MV type controller is a strong candidate for this application since an obvious cost function for minimization is the variance of the body velocity vibrations, i.e.

$$V = E\{v_b^2(t)\} \tag{8.90}$$

Moreover, the problem is one of stochastic regulation since the road velocity disturbances v_g can be modelled by a random process. The dynamics of the road velocity noise varies as the vehicle speed changes and as the road surface alters. These changes can be rather large, as for example if a vehicle were to drive from a smooth road surface to an irregular pavè surface. The changes in the stochastic description of v_g mean that a self-tuning regulator is a natural choice for the road noise isolation control law, the aim being to obtain a regulator which will retune as the road surface noise changes.

The simplified model of Figure 8.11 can be shown to have the following continuous time transfer function in terms of the complex variable s:

$$v_b(s) = \frac{B(s)}{A(s)} F(s) + \frac{C(s)}{A(s)} v_g(s) \tag{8.91}$$

where

$$B(s) = (m_w s^2 + k_t)s$$

$$C(s) = (bs+k)k_t$$

and

$$A(s) = (m_b s^2 + bs + k)(m_w s^2 + bs + k_t + k) - (bs+k)^2$$

In Chapter 2 a correspondence was set up between the order of a continuous time model and a discrete time model. Thus it is possible to predict that the discrete time model corresponding to equation (8.91) will be fourth order.

Assuming that the actuator force $F(s)$ is applied via a hydraulic actuator with first order dynamics and with input $u(s)$, the resulting discrete-time model associated with (8.91) is

$$v_b(t) = \frac{\bar{B}}{A} u(t-1) + \frac{C}{A} e_g(t) \tag{8.92}$$

where

$$A = 1 + a_1 z^{-1} + \ldots + a_5 z^{-5}$$

$$\bar{B} = (1 - z^{-1})(b_0 + b_1 z^{-1} + \ldots + b_3 z^{-3})$$

$$C = 1 + c_1 z^{-1} + \ldots + c_3 z^{-3}$$

and $e_g(t)$ is an equivalent discrete time noise source which replaces the continuous time road noise signal $v_g(s)$. In the simplified analysis presented here the source $e_g(t)$ is assumed to be white noise. In fact, as indicated earlier, the road noise spectrum will change as the road surface changes.

An additional point to note is that the $\bar{B}$ polynomial in equation (8.92) contains a factor $(1 - z^{-1})$. This is associated with the rate control, implied in the continuous time model by the factor s in the $B(s)$ polynomial (equation (8.91)). The natural occurrence of this $(1 - z^{-1})$ factor means that the discrete time model can be put into incremental form. Thus

$$v_b(t) = \frac{B}{A} \Delta u(t-1) + \frac{C}{A} e_g(t) \tag{8.93}$$

where

$$B = b_0 + b_1 z^{-1} + b_2 z^{-2} + b_3 z^{-3}$$

It is important to distinguish between the introduction of incremental action in a controller (see Section 8.2.5) and the incorporation of the factor $1 - z^{-1}$ in $\bar{B}$ into the data vector, as in equation (8.93). Specifically, in equation (8.93) we are removing a zero of $\bar{B}$ on the unit disk from the model in order to enable minimum variance control to be applied and to reduce the number of B coefficients by one. The introduction of integral action (Section 8.2.5) *increases* the degree of the A polynomial and is used to achieve steady state tracking of a constant reference.

A fourth order model of the type (8.93) requires the estimation of 12 parameters. The computational load associated with this may be too high for a fast sample rate system. In such circumstances, it is useful to check whether a lower order model might be feasible. In this application a reasonable approach to simplification is to check the system model when the tyre is much stiffer than the suspension. This gives the transfer function

$$v_b(s) = \frac{s}{m_b s^2 + bs + k} F(s) + \frac{bs + k}{m_b s^2 + bs + k} v_g(s) \tag{8.94}$$

The corresponding discrete time model would have (including first order hydraulic actuator dynamics) the following degrees: $n_a = 3$, $n_b = 2$, $n_c = 1$. During the specification of the self-tuning system and during subsequent validation trials it will be useful to check whether this reduced order model gives reasonable control action.

The next task in designing a self-tuning regulator for the suspension system is to select a sample interval for the digital controller. In this feasibility study we will neglect the practical implications for microprocessor speed and base the decision on the suspension system bandwidth. Again, for simplicity in this initial study, we will assume that $e_g(t)$ is a white noise discrete time process.

For a typical medium sized car the passive suspension components in Figure 8.11 give a step response of the form shown in Figure 8.13. The upper trace shows the step response of the car body velocity in response to the unit step in hydraulic value actuator input which is shown in the lower trace. Using

Figure 8.13 (a) Response of the suspension of a typical medium-sized car to a unit step change (b) in suspension actuator value.

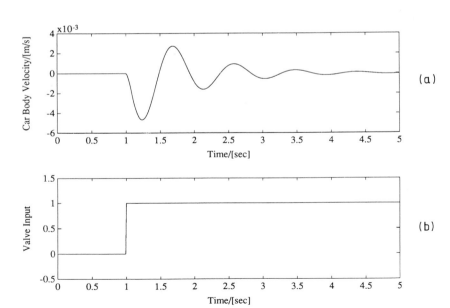

the sample rate selection ideas suggested in Chapter 4, an appropriate sample rate with which to commence control trials would be 0.1 s. Note, however, that this is only an initial guide. During the installation of a self-tuning system we would adjust the sample interval to achieve, in practical terms, the best controller behaviour. This would include a consideration of the desired bandwidth of the active suspension as well as the choice based on step response characterization. In MV control an important implication of sample rate selection is the corresponding locations of the discrete time zeroes. In particular, the zeroes of B can be moved by changing the sample rate. Thus one technique used to avoid nonminimum phase B is to experimentally adjust the sample interval in an attempt to locate the zeroes of B at reasonable positions inside the unit circle. This may not lead to an acceptable sampling interval but is worth investigating in case it proves helpful, as the next paragraph will demonstrate.

For example, Figure 8.14 shows the open-loop pole/zero locations for the discrete time model of the suspension system (B/A from equation (8.93)), as evaluated at various sample intervals for a typical passenger vehicle. The following observations from Figure 8.14 are relevant to the sample interval selection task for the suspension control problem.

(a) The sample interval $\tau_s = 0.01$ s (Figure 8.14(a)) is probably too fast for this application. This is because the dominant complex pole pair are close to the $+1$ point in the z-plane. The complex zeroes on the unit circle mean that MV could not be used because of the requirement of an inverse stable B.

(b) The sample interval $\tau_s = 0.05$ s (Figure 8.14(b)) is inappropriate for MV because of the real negative zero which lies outside the unit circle.

(c) The sample interval $\tau_s = 0.1$ s (Figure 8.14(c)) is acceptable for MV control, since all the zeroes are inside the unit circle and the poles are at reasonable positions for digital control purposes (recall the discussion in Chapter 4 on sample rate selection).

(d) The sample interval $\tau_s = 1$ s (Figure 8.14(d)) is too slow. The symptom is the clustering of the poles and zeroes around the z-plane origin.

On the basis of the above discussion a choice of sample interval $\tau_s = 0.1$ s would be practically reasonable and technically appropriate for minimum variance regulation.

Further information is contained in Figure 8.14 which relates to the model simplification discussion raised previously. The complex zero pair in Figure 8.14(c) is almost cancelled by a neighbouring complex pole pair. The close

Figure 8.14
Pole-zero plot
for discrete time
model of
simplified
suspension
system at
various
sampling
intervals: (a)
$\tau_s = 0.01$ s; (b)
$\tau_s = 0.05$ s;

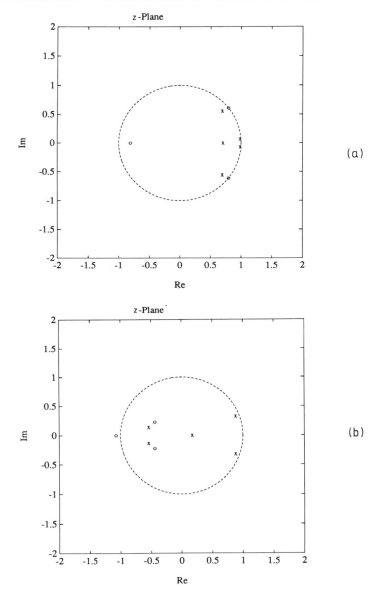

(a)

(b)

proximity of these pole/zero pairs suggests that they can be neglected in the self-tuning model. Physically, they are associated with the stiffness of the tyre being less significant than the suspension spring stiffness as discussed earlier in this section. Thus both the evidence of the pole/zero map and physical arguments can be used to suggest that a third order suspension model is adequate.

Figure 8.14
(cont.)
(c) $\tau_s = 0.1$ s;
and (d) $\tau_s = 1$ s.

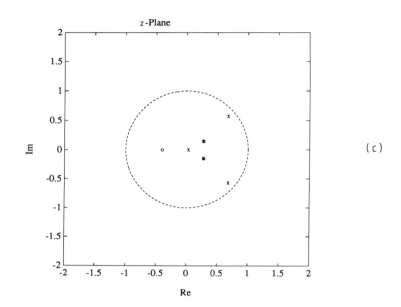

(c)

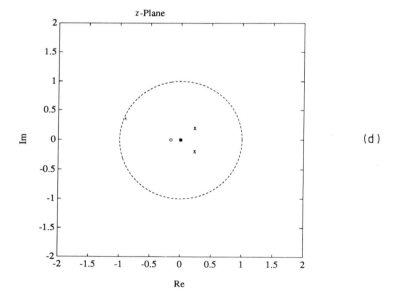

(d)

With the sample rate selected as 0.1 s, all poles and zeroes for both fifth and third order models are inside the unit circle. Hence MV control can be applied. If nonminimum phase behaviour *had* been present the following options would have been considered:

(a) to use Peterka's MV algorithm to avoid cancelling the nonminimum phase modes (Section 8.2.7);

(b) to use GMV to obtain a minimum phase generalized output;

(c) to use a multistep algorithm of the type discussed in Chapter 9;

(d) to change from an optimal control specification to a pole assignment design rule (Chapter 7) since this completely avoids the problem of inverse unstable B objects and yet can allow variance reduction (Section 7.5.3).

A further alternative is to assume a discrete time model that has a delay larger than actually exists. For MV control, if the assumed delay is large enough, it can be shown that a locally stable controller exists.

For the purposes of this application we will apply a straightforward MV algorithm. A reduced order model is used with the incremental form of equation (8.93) and sample interval 0.1 s. Specifically we have

$$v_b(t) = \frac{b_0 + b_1 z^{-1}}{1 + a_1 z^{-1} + a_2 z^{-2} + a_3 z^{-3}} \Delta u(t-1) + \frac{1 + c_1 z^{-1}}{1 + a_1 z^{-1} + a_2 z^{-2} a_3 z^{-3}} e_g(t)$$

Note that this model is simplified in two senses: first, by neglecting the pole/zero pair associated with the tyre dynamics; second, by the incorporation of the factor $1 - z^{-1}$ from $\bar{B}$ into the incremental actuator signal $\Delta u(t-1)$.

The MV control is computed for $\Delta u(t)$, and the actuator input at time t formed from

$$u(t) = \Delta u(t) + u(t-1)$$

The self-tuning MV controller was applied to a continuous time simulation of the simplified suspension system shown in Figure 8.11 with parameters selected to correspond to a typical medium-sized passenger vehicle. An implicit MV algorithm was applied, with an estimation model

$$v_b(t) = (f_0 + f_1 z^{-1}) \Delta u(t-1) + (g_0 + g_1 z^{-1} + g_2 z^{-2}) v_b(t-1) + \hat{e}(t) \tag{8.96}$$

and corresponding self-tuning control law

$$\Delta u(t) = - \frac{(g_0 + g_1 z^{-1} + g_2 z^{-2})}{f_0 + f_1 z^{-1}} v_b(t) \tag{8.97}$$

The suspension system was operated in open loop (as a passive system) for the first 50 s of operation. During this time a small square wave dither was applied to the actuator to provide excitation to the recursive estimator.

After 50 s the control action generated by the MV self-tuner was applied. Figure 8.15 shows the resulting behaviour over the first 300 s of operation. Figure 8.15(c) shows the output variable, the car body velocity. Note the distinct drop in body velocity variability as the self-tuner is introduced at 50 s. Traces 8.15(a) and (b) show the corresponding car body displacement and ground displacement. The reduction in body displacement is less noticeable as the self-tuner is introduced, but nonetheless a reduction in body movement can be discerned. The price paid for this reduction in body velocity variance is shown in the large fluctuation in the suspension strut displacement (Figure 8.15(d)) defined by

$$\text{strut displacement} = x_s(t) = x_b(t) - x_w(t) \tag{8.98}$$

The parameter estimates associated with this self-tuning controller are plotted in Figures 8.16(c) and (d). Also shown in this figure are the incremental actuation signal $\Delta u(t)$ applied to the control valve (Figure 8.16(a)) and the control coefficient factor (Figure 8.16(b)). This latter object is worth explaining since it is a general trick which can be used to introduce in a graceful way the feedback action of a self-tuning controller. Specifically, the control signal $\Delta u(t)$ from the MV controller is multiplied by the control coefficient factor $\lambda_{ccf}(t)$ before being applied to the actuator. Thus

$$u(t) = \lambda_{ccf}(t) \left\{ \frac{1}{1 - z^{-1}} \right\} \Delta u(t) \tag{8.99}$$

The control coefficient factor is defined by

$$\lambda_{ccf}(t) = \begin{matrix} 1 - \exp((t-50)/10); & t \geqslant 50 \\ 0; & t < 50 \end{matrix}$$

The effect is to introduce the control action gently, thus avoiding large transient fluctuations in the control and output signals.

Finally, Figure 8.17 shows some diagnostic information which is often used to assess the successful operation of a minimum variance self-tuning controller. The trace of the covariance matrix $\mathbf{P}(t)$ is a good indicator of the recursive

Figure 8.15
Self-tuning
MV regulation of
an active
suspension
system. The
regulator
minimizes the
variance of the
car body vertical
velocity (c).
The input
disturbance is
the ground
displacement
(b). The strut
displacement (d)
is a measure of
the suspension
strut's movement
from its nominal
position. The
dotted lines at
±0.1 m indicate
the physical
limits on strut
travel.

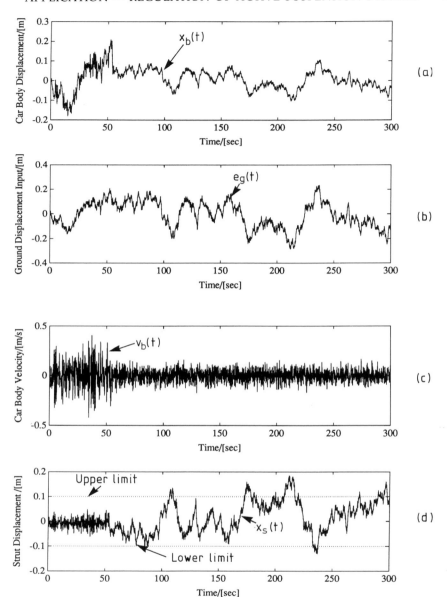

estimator performance. The figure shows the trace decreasing gradually during open loop operation (0 to 50 s) and decreasing rapidly as the self-tuning controller is invoked at the 50th second. This is reasonable behaviour for an RLS algorithm with no forgetting factors or other covariance management techniques applied.

Figure 8.16
Self-tuning MV
regulation of an
active
suspension
system. The
regulator
coefficients (c,d)
are obtained by
an implicit
algorithm. The
incremental
value position
(a) is the
regulator output
and the control
coefficient $\lambda_{ccf}(t)$
(b) is used to
phase in the
feedback control
over an interval.

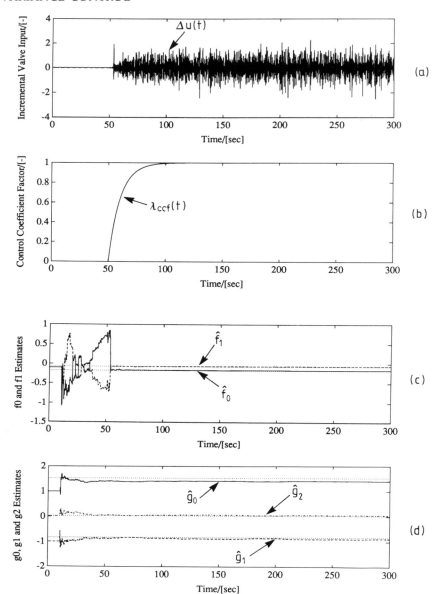

Figure 8.16
Self-tuning MV regulation of an active suspension system. The regulator coefficients (c,d) are obtained by an implicit algorithm. The incremental value position (a) is the regulator output and the control coefficient $\lambda_{ccf}(t)$ (b) is used to phase in the feedback control over an interval.

The difference between the residual and the driving noise $e_g(t)$ can also be used in MV regulators (for $k=1$) to assess performance (Figure 8.17(b)). The two signals should be identical for a correctly converged algorithm. A further means of validating a MV self-tuner is to compare the cumulative loss under self-tuning with the theoretical cumulative loss.

Cumulative loss under self-tuning (CLST) is given by

Figure 8.17
Diagnostic
information for
the self-tuning
suspension
controller. (a)
The trace of the
covariance
matrix $\mathbf{P}(t)$; (b)
the difference
between the RLS
residual and the
ground noise;
and (c) the
cumulative loss
under self-
tuning (CLST)
and the
theoretical
cumulative loss
(TCL).

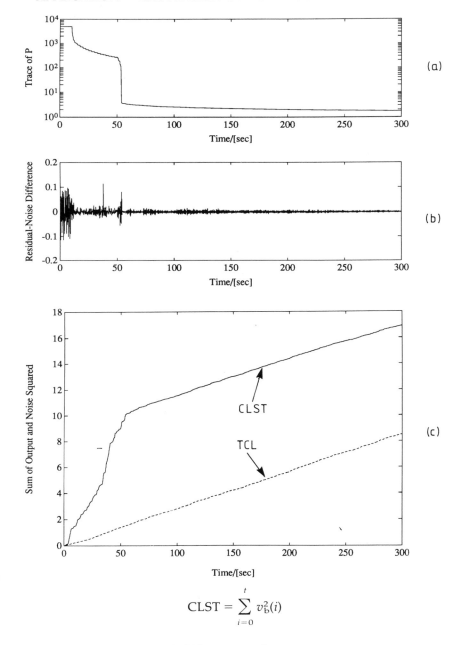

$$\text{CLST} = \sum_{i=0}^{t} v_{\text{b}}^2(i)$$

Theoretical cumulative loss (TCL) is given by

$$\text{TCL} = \sum_{i=0}^{t} e_{\text{g}}^2(i)$$

When the self-tuner has converged the CLST trace should run parallel to that for TCL. The initial large increase in CLST is due to nonoptimal behaviour during open loop operation and tuning-in (see Figure 8.17, lowest trace).

The MV regulator obtained here would require further development before real application. One problem is that MV control calls for large strut displacements (see Figure 8.15(d)). Indeed, the peak strut displacement exceeds the permitted limits of strut travel which are indicated as dotted lines at ±0.1 m on Figure 8.15(d). The strut deflections can be reduced by weighting the strut movement in the MV cost function. A possible cost is:

$$V = E\{v_b^2(t) + \lambda x_s^2(t)\} \tag{8.100}$$

where λ is a scalar weighting coefficient.

The cost function (8.100) is a form of GMV cost leading to the so-called *lamda controller*. The coefficient λ is increased to achieve a trade-off between output variance and strut displacement variance. Figure 8.18 shows the variances of body velocity and strut deflection plotted against λ. Notice that $\lambda = 2$ achieves a large reduction in strut variance with only a modest increase in body velocity variance. The car body velocity and strut deflection for a self-tuning controller minimizing (8.100) for $\lambda = 2$ are shown in Figure 8.19.

Further development of this application would involve controller bandwidth investigations and tests with varying road noise spectra in order to establish the adaptive performance of the controller.

Figure 8.18 Variation of the body velocity variance and strut displacement variance with strut weighting coefficient λ.

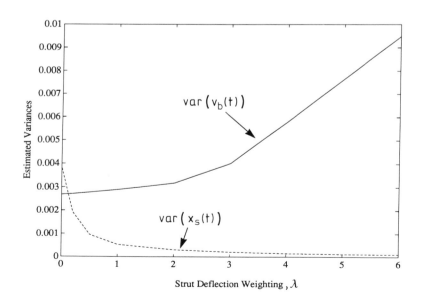

Figure 8.19
Self-tuning
suspension
regulation with
a strut
weighting of
$\lambda = 2$. Note the
much reduced
strut
displacement
(compared with
Figure 8.15(d)),
and slightly
increased car
body velocity
variability.

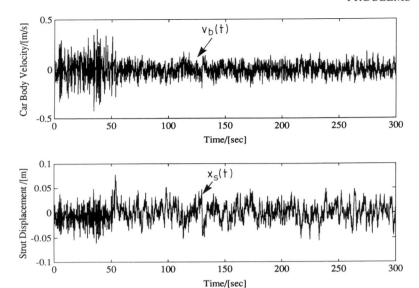

8.9 SUMMARY

In this chapter the following material has been presented.

- The basic MV control law
- The extension of this basic law to a GMV framework
- The self-tuning implementation of MV controllers
- The connection between pole assignment and MV control
- An appliction of MV strategies to noise reduction

8.10 PROBLEMS

Problem 8.1

Consider the system given by

$$Ay(t) = z^{-k}Bu(t) + Ce(t)$$

with

$$A = 1 - 1.8z^{-1} + 0.81z^{-2} = (1-0.9z^{-1})^2$$

$$B = 1 + 0.6z^{-1}$$

$$C = 1 - 1.1z^{-1} + 0.3z^{-2} = (1-0.5z^{-1})(1-0.6z^{-1})$$

For $k = 1$, find the minimum variance controller parameters g_0, g_1 in

$$u(t) = -\frac{(g_0 + g_1 z^{-1})}{B} y(t)$$

For $k = 2$, find the minimum variance controller parameters f_1, g_0, g_1 in

$$u(t) = -\frac{(g_0 + g_1 z^{-1})}{(1 + f_1 z^{-1})B} y(t)$$

What is the variance of $y(t)$ in each case if $e(t)$ has unit variance?

Problem 8.2

The GMV controller minimizes

$$V = E\{[Py(t+k) + Qu(t) - Rr(t)]^2\}$$

for systems given by

$$Ay(t) = z^{-k}Bu(t) + Ce(t)$$

using the feedback rule

$$Fu(t) = -Gy(t) - Hr(t)$$

with $F = B\mathbf{E} + QC$, $H = -CR$ and $\mathbf{E}$, G selected according to

$$PC = A\mathbf{E} + z^{-k}G$$

Show that the closed loop system poles are given by the roots of

$$FA + z^{-k}BG = 0$$

and that this can be reexpressed as the roots of

$$BP + AQ = 0$$

or

$$PF - z^{-k}QG = 0$$

Problem 8.3

Consider a GMV controller with $P = 1$, $R = 1$ and $Q = q_0$ in the usual notation, applied to the system

$$y(t) = \frac{z^{-2}(1 + 0.6z^{-1})}{1 - 0.9z^{-1}} u(t) + \frac{1-1.1z^{-1}+0.3z^{-2}}{1-0.9z^{-1}} e(t)$$

Find the value of q_0 which assigns the closed loop poles to the zeroes of $1 - 0.5z^{-1}$ and hence determine the GMV controller.

Problem 8.4

A system is described by the model:

$$Ay(t) = z^{-k}Bu(t) + Ce(t)$$

in the usual notation.
    Show that the model can be put in the alternative form:

$$y(t+k) = \tilde{A}y(t) + \tilde{B}u(t) + \tilde{C}e(t+k)$$

Hence or otherwise write down the MV controller for the system

$$(1+a_1z^{-1})y(t) = z^{-2}(b_0+b_1z^{-1})u(t)+e(t)$$

Problem 8.5

The pressure deviation $p(t)$ in a boiler from the normal operating pressure $\bar{p}$ is described by the equation

$$p(t) = \frac{(0.4-0.2z^{-1})}{(1-0.9z^{-1})(1-0.6z^{-1})} u(t-2) + \frac{(1-0.7z^{-1})}{(1-0.9z^{-1})(1-0.6z^{-1})} e(t)$$

where $e(t)$ is a zero mean gaussian white noise disturbance with variance 0.5.

(a) Calculate the MV control law for $p(t)$.

(b) Calculate the corresponding variance of $p(t)$ under MV control.

(c) Describe the implicit self-tuning MV regulator for $p(t)$.

    If the maximum permitted pressure is p_{max}, how close to p_{max} can $\bar{p}$ be set with 95% confidence that the maximum limit will not be exceeded?

Problem 8.6

In Example 8.4, use first order P,Q polynomials satisfying $P(0)=1$, $Q(1)=0$ to generate a GMV controller assigning all poles to the origin. Show that the closed-loop output behaviour is the same as for $P=1=Q$. Why?

Problem 8.7

For a CARIMA model, derive the incremental GMV controller that minimizes the variance of the pseudo-output $\phi(t)$ defined by

$$\phi(t+k) = Py(t+k) + Q\Delta u(t) - Rr(t)$$

Problem 8.8

Show that the pole assignment identity (8.67) for P,Q:

$$PB + QA = \mathbf{T}$$

where $P(0)=1$, has a unique solution if A,B are coprime and the polynomial degrees satisfy the conditions

$$n_p = n_a, \qquad n_q = n_b, \qquad n_t \leqslant n_a + n_b$$

Further, show that the corresponding controller polynomials have degrees

$$n_g = \max(n_a - 1, n_a + n_c - k)$$
$$n_f = \max(n_b + k - 1, n_b + n_c)$$
$$n_h = n_r + n_c$$

Finally, show that the identity (8.68):

$$PF - z^{-k}GQ = \mathbf{TE}$$

can be solved for P, Q and $\mathbf{E}$ if n_p, n_q are suitable chosen.

Problem 8.9

Show that, for minimum phase systems, Peterka's algorithm reduces to the standard MV regulator.

Problem 8.10

Consider the system

$$Ay(t) = z^{-k}Bu(t) + Ce(t),$$

in the usual notation. A robust (but suboptimal) modification of the MV regulator due to Astrom leads to the structure

$$u(t) = -\frac{G}{B^+F} y(t)$$

where

$$C = AF + z^{-k}GB^-$$

$$B = B^-B^+$$

and B^- contains both the nonminimum phase zeroes of B and those stable zeroes close to the unit circle. Show that, if $n_c \leqslant n_a$, then the polynomial identity can be solved uniquely for F,G if

$$n_f = k-1+n_{b-}$$

$$n_g = n_a-1$$

Show that the closed loop output is given by

$$y(t) = Fe(t)$$

For the system

$$y(t) = 2y(t-1)+u(t-1)+bu(t-2)+e(t)$$

where $b=0.9$, a controller is designed based on the estimate $\hat{b}$. Derive the MV regulator and Astrom's regulator for the estimated model. In each case, plot the variance of the closed-loop system output as a function of $\hat{b}$ in the interval $0.7 \leqslant \hat{b} \leqslant 1.1$.

Problem 8.11

If G,H are stable discrete-time transfer functions and $d(t) = (-1)^t$, show that

$$\bar{E}\{[G(z^{-1})d(t)][H(z^{-1})d(t)]\} = G(-1)H(-1)$$

Data is generated by the unknown system

$$y(t) = u(t-1)+e(t)+d(t)$$

in the usual notation, where $\{e(t)\}$ has unit variance.

It is desired to minimize the output variance and a MV self-tuning controller is constructed based on RLS estimation of the parameter α in the assumed model

$$y(t)+\alpha y(t-1) = u(t-1)+\hat{e}(t)$$

Show that there are two stationary points of the algorithm but that only one is locally stable, yielding

$$\bar{E}y^2(t) = 1 + \frac{1}{2^{1/2}}$$

What variance value could be achieved if the system were correctly known?

Problem 8.12

The discrete-time system

$$y(t) + ay(t-1) = u(t-1) + e(t) + d$$

has unknown constant parameters a,d and $\{e(t)\}$ is zero mean unit variance white noise.

It is desired to minimize the output variance and a MV self-tuning controller is constructed based on the on-line least squares estimation of the parameter α in the model

$$y(t) + \alpha y(t-1) = u(t-1) + \hat{e}(t)$$

If the parameter estimate converges to $a - \gamma_+$, show that γ_+ is the positive root of the quadratic equation

$$\gamma^2 + \gamma(1+d^2) - d^2 = 0$$

Show also that the asymptotic closed-loop behaviour of the system is characterized by a mean output offset of $d/(1+\gamma_+)$ and output variance $(1-\gamma_+^2)^{-1}$.

Problem 8.13

The discrete-time system

$$y(t) = bu(t-1) + e(t) + ce(t-1)$$

has unknown constant parameters b,c and $\{e(t)\}$ is a zero mean white noise process of variance σ_e^2.

It is desired to minimize the cost function

$$E[y(t) - r(t)]^2$$

where $\{r(t)\}$ is a deterministic reference trajectory and a self-tuning controller is constructed based on the on-line least squares estimation of the parameter β in the model

$$y(t) = \beta u(t-1) + \hat{e}(t)$$

If the parameter estimate converges, show that the convergence point is uniquely b if

$$\lim_{N\to\infty} \frac{1}{N} \sum_{t=1}^{N} r^2(t) > 0$$

Show that the solution does not yield the minimum cost.

Problem 8.14

It is required that the output y of the discrete-time system

$$y(t) = bu(t-1) + e(t)$$

in the usual notation, tracks the setpoint trajectory $\{r(t)\}$ in the sense that the control sequence $\{u(t)\}$ minimizes

$$E[y(t) - r(t)]^2$$

for each t. It is assumed that $\{r(t)\}$ is a given deterministic sequence and that the limits

$$\rho(i) = \lim_{N\to\infty} \frac{1}{N} \sum_{t=1}^{N} r(t)r(t+i) \qquad i = 0,1$$

exist and are finite.

If the parameter b is unknown and RLS estimation is used on the incorrect model

$$y(t) = \beta u(t-2) + \hat{e}(t)$$

show that the only possible finite convergence point is

$$\beta^* = b\rho(1)/\rho(0)$$

using the ODE method.

8.11 NOTES AND REFERENCES

The MV algorithm and an *ad hoc* solution of the robustness problem are discussed in:

Astrom, K.J. *Introduction to Stochastic Control*, Academic Press, 1970.

The following paper deals with MV regulation of nonminimum phase systems:

Peterka, V.
On steady state minimum variance control strategy, *Kybernetika*, **8**(3), 219–32, 1972.

The MV self-tuning regulator is described in:

Astrom, K.J. and Wittenmark, B.
On self-tuning regulators, *Automatica,* **9,** 185–99, 1973.

The detuned MV regulator was introduced as a form of pole assignment in:

Wellstead, P.E., Edmunds, J.M., Prager, D.L. and Zanker, P.M.
Self-tuning pole/zero assignment regulators, *Int. J. Control.* **30,** 1–26, 1979.

The GMV self-tuner is described in:

Clarke, D.W. and Gawthrop, P.G.
Self-tuning controller, *Proc. IEE,* **122**(9), 929–34, 1975.

The extended GMV self-tuner which incorporates pole assignment was presented in:

Allidina, A.Y. and Hughes, F.M.
Generalised self-tuning controller with pole assignment, *Proc. IEE, Part D,* **127,** 13–18, 1980.

The self-tuning active suspension application study was carried out by A.J. Truscott. Related work is described in:

Costin, M.H. and Elzinga, D.R.
Active reduction of low-frequency time impact noise using digital feedback control, *IEEE Control Systems Magazine,* 1989.

This paper provides an interesting link between minimum variance control and noise cancellation problems in signal processing (see Chapter 11). The area of suspension systems is described in traditional terms in:

Bastow, D.
Car Suspension and Handling, Pentech Press, 1987.

For an introduction to active suspension control, see for example:

Karnopp, D.C.
Active damping in road vehicle suspension systems, *Veh. Syst. Dyn.,* **12,** 291–316, 1983.

9 Multistage Predictive Control

9.1 OUTLINE AND LEARNING OBJECTIVES

The minimum variance (MV) procedures of Chapter 8 were the first to be put into a self-tuning form. The initial success of these self-tuning MV algorithms served as a great stimulus to adaptive control research. However, as noted previously, MV controllers have difficulties when the system to be controlled is discrete time nonminimum phase. The main way round this problem is to use the pole assignment procedures of Chapter 7. However, an alternative exists which retains the use of optimization criteria but avoids the problems associated with MV. The alternative is multistage predictive control in which future process variables are incorporated into the control law in such a way that the control action is less sensitive than MV to time delay mismatches and nonminimum phase systems.

In this chapter we consider a particular class of predictive control algorithms. In particular, and in keeping with the pole assignment theme of the text, we discuss a long range (multistage) predictive control algorithm which incorporates a pole placement criterion. The Generalized Pole Placement (GPP) algorithm is discussed in Section 9.2, with a self-tuning version covered in 9.3. General properties of GPP are covered in Section 9.4 and an application of the technique to robot control is given in Section 9.5.

The learning objective is to understand the basic techniques involved in multistage predictive control and to see how it relates to both standard pole assignment and MV methods.

9.2 GENERALIZED POLE PLACEMENT

In this section the algorithmic aspects of the GPP control algorithm are reviewed. In addition we indicate the role of various tuning aids which are associated with multistage predictive control algorithms.

The system models used are either of the CARIMA or CARMA forms. Recalling the discussions of Chapter 2, these forms originate with the physical system description

$$Ay(t) = Bu(t-1) + x(t) \tag{9.1}$$

where A and B are polynomials in the backward shift operator z^{-1}:

$$A = 1 + a_1 z^{-1} + \ldots + a_{n_a} z^{-n_a} \tag{9.2}$$

$$B = b_0 + b_1 z^{-1} + \ldots + b_{n_b} z^{-n_b} \tag{9.3}$$

The $k-1$ leading coefficients of B are zero if the pure transport delay of the system is k sampling intervals. In this case $y(t)$ depends only on the control signals $u(t-k-1)$, $u(t-k-2)$,. . .,.

The term $x(t)$ represents the aggregated disturbances acting on the system. If this process is nonstationary (random walk, brownian motion, etc.), then $x(t)$ can be expressed as

$$x(t) = \frac{C}{\Delta} e(t) + d \tag{9.4}$$

where d represents a nonzero mean offset (possibly a slowly varying load disturbance), $\Delta = 1 - z^{-1}$ and the increment $\Delta x(t)$ is a stationary process. Otherwise, for stationary disturbances, $x(t)$ has the following form:

$$x(t) = Ce(t) + d \tag{9.5}$$

The sequence $\{e(t), t = \ldots, -1,0,1, \ldots\}$ is zero mean uncorrelated noise with variance σ_e^2 and C is a polynomial

$$C = 1 + c_1 z^{-1} + \ldots + c_{n_c} z^{-n_c} \tag{9.6}$$

with all its zeroes within the unit circle.

Equations (9.1) and (9.4) yield the CARIMA model

$$\bar{A}y(t) = B\Delta u(t-1) + Ce(t) \tag{9.7}$$

with

$$\bar{A} = A\Delta, \qquad \Delta u(t) = u(t) - u(t-1) \tag{9.8}$$

and (9.1) and (9.5) lead to the CARMA model

$$Ay(t) = Bu(t-1) + Ce(t) + d \tag{9.9}$$

9.2.1   The GPP cost function

The essential feature of a multistage predictive control algorithm, compared with MV, is the use of a set of future signals in the cost function. Various cost functions can be constructed and one form of the so called Generalized Pole Placement (GPP) predictive controller is obtained as a result of minimizing the following multistage cost function:

$$J(N1,N) = E_t\left[\sum_{i=N1}^{N} \Phi_i^2(t) + \sum_{i=1}^{N} \lambda \Delta u^2(t+i-1) \right] \tag{9.10}$$

where

$$\Phi_i(t) = Py(t+i) - Rr(t+i) + Q\Delta u(t-1) + Sy(t) \tag{9.11}$$

$E_t[\cdot]$ denotes expectation conditioned on data up to time t and P, Q, R, S are finite polynomials in z^{-1}. The structure of these polynomials together with the way in which they are selected is discussed later. The nonnegative scalar λ is used in this connection to directly penalize control effort.

Without loss of generality, it may be assumed that N1 is unity. Note, however, that the signal $\Phi_i(t)$ depends only on those input signals $u(s)$ for $s \leqslant t+i-k-1$ and will therefore only be dependent on the signals $u(t)$, $u(t+1)$, ..., which must be chosen at time t, if $i \geqslant k+1$. This implies that we should set $N1 = k+1$ if the time delay is known.

The *output horizon* $N(>k)$ specifies the number of future outputs in the cost function. If S, λ are set to zero and only a single stage is considered ($N = 1$), then the cost function (9.10) can be compared with the GMV cost function when there is no pure time delay (See Chapter 8).

A GPP algorithm can be derived for both CARIMA and CARMA system models. A derivation of the CARIMA algorithm is presented here for $k=0$ and the modifications required for the CARMA model are then indicated.

9.2.2   A GPP controller for CARIMA models

The technical manipulations needed to derive the controller involve two polynomial partitions. The first partition plays the same role as in the derivations of both the MV and GMV controllers in that it allows $\Phi_i(t)$ in (9.11) to be expressed as the sum of two uncorrelated terms – its predicted value at time t and a prediction error. Thus:

$$PC = \bar{A}F_i + z^{-i}G_i \tag{9.12}$$

where F_i, G_i are polynomials in z^{-1} with degrees

$$\deg(F_i) = i-1; \qquad \deg(G_i) = \max(n_a, \deg(P) + n_c - i) \qquad (9.13)$$

The partition (9.12) must be performed for $i=1,2,\ldots,N$ so that, for $N>1$, two sequences of polynomials $F_1, F_2, \ldots, F_N$ and $G_1, G_2, \ldots, G_N$ are generated. These calculations can be organized into a computationally efficient algorithm (see Problem 9.1).

It follows from (9.12), (9.11) and (9.7) that the pseudo-output $\Phi_i(t)$ can be written as

$$\Phi_i(t) = F_i e(t+i) + \frac{BF_i}{C} \Delta u(t+i-1) + Q\Delta u(t-1) + \left[\frac{G_i}{C} + S\right] y(t) - Rr(t+i) \quad (9.14)$$

Again, this can be compared with the partition of the pseudo-output $\Phi(t)$ as performed in the GMV algorithm.

The second polynomial partition separates the optimization variables $\Delta u(t+N-1), \ldots, \Delta u(t)$ from data $\Delta u(t-1), \Delta u(t-2)\ldots$, which are associated with control signals already applied to the system up to time $t-1$. Provided that $b_0 \neq 0$, the identity associated with this partition is

$$BF_i = CE_i + z^{-i}\Gamma_i \qquad (9.15)$$

where

$$\deg(E_i) = i-1; \qquad \deg(\Gamma_i) = \max(n_b-1, n_c-1) \qquad (9.16)$$

and Γ_i is zero if both n_b and n_c are zero. Again, for $N>1$, the solution of (9.15) for $i=1,2,\ldots,N$ can be carried out efficiently (see Problem 9.1).

Using (9.15) in equation (9.14) yields

$$\Phi_i(t) = F_i e(t+i) + E_i \Delta u(t+i-1) + C^{-1}[(\Gamma_i+CQ)\Delta u(t-1) + (G_i+CS)y(t) - CRr(t+i)]$$

Stacking these equations for $i=1,2,\ldots,N$ into vector form yields

$$\mathbf{\Phi} = \mathbf{f} + \mathbf{E}\Delta\mathbf{u} + \mathbf{\Gamma}\Delta u(t-1) + \mathbf{G}y(t) - \mathbf{r} \qquad (9.17)$$

where

$$\mathbf{\Phi} = [\Phi_1(t), \Phi_2(t), \ldots, \Phi_N(t)]^T$$

$$\Delta\mathbf{u} = [\Delta u(t), \Delta u(t+1), \ldots, \Delta u(t+N-1)]^T$$

$$\mathbf{E} = \begin{bmatrix} e_0 & 0 & \ldots & 0 \\ e_1 & e_0 & \ldots & 0 \\ \cdot & \cdot & & \cdot \\ \cdot & \cdot & & \cdot \\ \cdot & \cdot & & \cdot \\ e_{N-1} & e_{N-2} & \ldots & e_0 \end{bmatrix}, \qquad E_i(z) = \sum_{j=0}^{i-1} e_j z^j \qquad (9.18a)$$

and $\mathbf{f}$, $\boldsymbol{\Gamma}$, $\mathbf{G}$, $\mathbf{r}$ are given by

$$\mathbf{f}^T = [F_1 e(t+1), \ldots, F_N e(t+N)] \qquad (9.18b)$$

$$\mathbf{r}^T = [Rr(t+1), \ldots, Rr(t+N)] \qquad (9.18c)$$

$$\mathbf{G}^T = C^{-1}[G_1 + CS, \ldots, G_N + CS] \qquad (9.18d)$$

$$\boldsymbol{\Gamma}^T = C^{-1}[\Gamma_1 + CQ, \ldots, \Gamma_N + CQ] \qquad (9.18e)$$

In this notation the cost function (9.10) can be put in the form

$$J(1,N) = E_t(\boldsymbol{\Phi}^T\boldsymbol{\Phi} + \lambda\Delta\mathbf{u}^T\Delta\mathbf{u}) \qquad (9.19)$$

Taking derivatives with respect to the vector of control increments $\Delta\mathbf{u}$, the cost function is minimized by the control law

$$\Delta\mathbf{u} = (\mathbf{E}^T\mathbf{E} + \lambda\mathbf{I})^{-1}\mathbf{E}^T[\mathbf{r} - \mathbf{G}y(t) - \boldsymbol{\Gamma}\Delta u(t-1)] \qquad (9.20)$$

Equation (9.20) defines a vector of future controls covering time intervals t to $t+N-1$. However, a standard feature of predictive control algorithms is that, although the controls $\Delta u(t), \ldots, \Delta u(t+N-1)$ are calculated at each sample time, *only the first* of these is applied to the system and the remaining elements discarded to be recalculated at the next sample time (see Problems 9.2 and 9.5).

The first element of $\Delta\mathbf{u}$ is obtained from equation (9.20) by writing the first row of $(\mathbf{E}^T\mathbf{E} + \lambda\mathbf{I})^{-1}\mathbf{E}^T$ as

$$\mathbf{k}^T = [k_1, \ldots, k_N] \qquad (9.21)$$

This allows the first row of equation (9.20) to be written as the *rolling horizon* control law

$$\mathscr{F}\Delta u(t) + \mathscr{G}y(t) - \mathscr{N}r(t+N) = 0 \qquad (9.22a)$$

where

$$\mathcal{F} = C + z^{-1}(\mathbf{k}^{\mathrm{T}}\mathbf{\Gamma}') + z^{-1}\lambda_N CQ \qquad (9.22\text{b})$$

$$\mathcal{G} = (\mathbf{k}^{\mathrm{T}}\mathbf{G}') + \lambda_N CS \qquad (9.22\text{c})$$

$$\mathcal{N} = RCK \qquad (9.22\text{d})$$

and

$$\mathbf{\Gamma}' = [\Gamma_1, \Gamma_2, \ldots, \Gamma_N]^{\mathrm{T}} \qquad (9.22\text{e})$$

$$\mathbf{G}' = [G_1, G_2, \ldots, G_N]^{\mathrm{T}} \qquad (9.22\text{f})$$

$$K = k_N + k_{N-1}z^{-1} + \ldots + k_1 z^{-N+1} \qquad (9.22\text{g})$$

$$\lambda_N = \sum_{i=1}^{N} k_i \qquad (9.22\text{h})$$

Before concentrating on possible choices for the weighting polynomials P,Q, R,S, it should be noted that, independent of such choices, the control law (9.22a) includes *integral action*. This is a consequence of the CARIMA assumption (9.7) coupled with the choice of cost function (9.10), (9.11). The benefits of integral action, particularly within a self-tuning framework, have already been noted in Chapters 7 and 8 in connection with servo tracking and it is therefore tempting to select the CARIMA form whatever the true system structure. We merely note that this may lead to poor regulation performance.

9.2.3 Closed loop criteria

The derivation given in the previous section is similar to that which is used in all forms of predictive controllers. The problem with this framework is that it gives only limited guidance into how to select the weighting polynomials P, Q, R, S in the cost function. The GPP approach partially resolves this problem by selecting polynomials which give desired closed-loop pole positions and hence guarantees closed loop stability. In this sense it is a multistage extension of Allidina's algorithm (Section 8.6).

The procedure operates as follows. The closed loop system equation is obtained from (9.7) and (9.22):

$$\mathcal{H}y(t) = z^{-1}B\mathcal{N}r(t+N) + \mathcal{F}Ce(t) \qquad (9.23\text{a})$$

$$H = \bar{A}\mathcal{F} + z^{-1}B\mathcal{G} \qquad (9.23\text{b})$$

The polynomials P, Q, R, S are selected so as to eliminate the steady-state output tracking error (on average) and to arbitrarily assign the closed loop poles to desired locations.

The tracking error criterion is met by selecting R so that, for a constant setpoint $r(t)$, the output and setpoint are equal. This is achieved by setting $z = 1$ (corresponding to zero frequency) in equation (9.23) and neglecting the term $\mathcal{F}Ce(t)$ associated with zero mean regulation error. This leads to the requirement that R should be chosen so that

$$\mathcal{G}(1) = \mathcal{N}(1) \tag{9.24}$$

By using (9.12), (9.22c) and (9.23b), it can be shown that this implies

$$R(1) = P(1) + S(1) \tag{9.25}$$

The closed-loop pole-selection procedure is to choose Q and S so that the characteristic polynomial $\mathcal{H}$ in equation (9.23b) corresponds with a user-specified polynomial which defines the desired closed-loop pole set. Specifically, we select Q and S so that

$$\mathcal{H} = C\mathbf{T} \tag{9.26}$$

where the zeroes of the polynomial $\mathbf{T}$ are the required closed loop poles. The polynomial C is included on the right hand side of (9.26) to cancel C in the noise transfer function numerator of equation 9.23, yielding the closed loop system equation

$$y(t) = \frac{B\mathcal{N}}{C\mathbf{T}}r(t+N-1) + \frac{\mathcal{F}}{\mathbf{T}}e(t) \tag{9.27}$$

In order to connect the selection of Q, S with the desired closed-loop pole set, it is first necessary to note that (9.22), (9.23b) and (9.26) yield

$$z^{-1}\lambda_N C(\bar{A}Q + BS) = C(\mathbf{T} - \bar{A}) - z^{-1}\mathbf{k}^T(\bar{A}\mathbf{\Gamma}' + B\mathbf{G}') \tag{9.28}$$

and this allows (9.25) to be cast in the alternative form

$$R(1) = \mathbf{T}(1)/\lambda_N B(1) \tag{9.29}$$

This expression should be compared with the corresponding equation for the selection of the precompensator R used in standard pole assignment (see Chapter 7).

Using the partitions (9.12) and (9.15), the term $(\bar{A}\Gamma' + B\mathbf{G}')$ in (9.28) is expressible as $C\mathbf{X}$. Finally, writing $\mathbf{T} - \bar{A}$ as $z^{-1}D$, (9.28) is reduced to the form

$$\lambda_N(\bar{A}Q + BS) = D - (\mathbf{k}^T\mathbf{X}) \tag{9.30}$$

This is a polynomial identity similar in form to that encountered in Chapters 7 and 8. It may be solved for the Q and S polynomials using the procedures outlined in Appendix 7.1. A unique solution exists for Q,S if A,B are coprime and the degrees of the system and design polynomials satisfy the conditions

$$\deg(Q) = \max(0, n_b - 1) \tag{9.31a}$$

$$\deg(S) = n_a \tag{9.31b}$$

$$n_t \leqslant n_a + n_b + 1 \tag{9.31c}$$

$$\deg(P) \leqslant n_a + N \tag{9.31d}$$

This completes the description of the basic GPP algorithm for CARIMA models.

A potential problem associated with the GPP algorithm and related predictive algorithms is that calculation of the control $\Delta\mathbf{u}$ requires a matrix inversion. In particular, the solution (9.20) for the future control inputs involves the inversion of the $N{\times}N$ matrix $\mathbf{E}^T\mathbf{E} + \lambda\mathbf{I}$. This matrix may be singular if λ is zero and the leading coefficients of B are also zero (that is, the transport delay k is greater than unity). In addition, the computational burden for large output horizons will be particularly heavy in a self-tuning context in which the inversion must be carried out repeatedly.

These problems can be avoided by assuming a control 'settling time' or *control horizon* $NU(<N)$ beyond which the control signal is assumed not to change, so that

$$\Delta u(t+i) = 0 \qquad i = NU, NU+1, \ldots, N-1 \tag{9.32}$$

This assumption reduces the number of optimization variables to NU and the solution is given by (9.20) where $\mathbf{E}$ is now the $N{\times}NU$ matrix

$$\mathbf{E} = \begin{bmatrix} e_0 & 0 & \ldots & 0 \\ e_1 & e_0 & \ldots & 0 \\ \cdot & \cdot & & \cdot \\ \cdot & \cdot & & \cdot \\ \cdot & \cdot & & \cdot \\ e_{N-1} & e_{N-2} & \ldots & e_{N-NU} \end{bmatrix} \tag{9.33}$$

The possible rank deficiency when λ is zero does not occur if NU is chosen to satisfy $NU \leq N-k+1$.

9.2.4   A simple example

Example 9.1

We illustrate the CARIMA GPP algorithm using the following CARIMA system:

$$y(t) = 2y(t-1)+u(t-1)+2u(t-2)+x(t) \tag{9.34a}$$

$$\Delta x(t) = e(t) \tag{9.34b}$$

where $e(t)$ is zero mean, unit variance white noise.

For simplicity, we choose

$$P = 1, \qquad R = \text{constant}, \qquad \lambda = 0 \tag{9.35}$$

and set the horizons $N = 3$, $NU = 1$.

Noting that (9.13) implies that G_i has degree one, the partition (9.12) takes the form

$$1 = (1-z^{-1})\,(1-2z^{-1})\,f_0 + z^{-1}(g_0 + g_1 z^{-1})$$

for $i = 1$, yielding

$$F_1 = 1, \qquad G_1 = 3-2z^{-1} \tag{9.36a}$$

similarly

$$F_2 = 1+3z^{-1}, \qquad G_2 = 7-6z^{-1} \tag{9.36b}$$

$$F_3 = 1+3z^{-1}+7z^{-2}, \qquad G_3 = 15-14z^{-1} \tag{9.36c}$$

To carry out the second partition (9.15), note that (9.16) tells us that Γ_i is constant. For $i = 1$, using (9.36a),

$$1+2z^{-1} = e_0 + z^{-1}\Gamma_1$$

so that

$$e_0 = 1, \qquad \Gamma_1 = 2 \tag{9.37a}$$

Similarly

$$e_1 = 5, \qquad \Gamma_2 = 6 \tag{9.37b}$$

$$e_2 = 13, \qquad \Gamma_3 = 14 \tag{9.37c}$$

For $NU = 1$, the array $\mathbf{E}$ takes the form (9.33):

$$\mathbf{E} = [1,5,13]^\mathrm{T} \tag{9.38a}$$

so that

$$\mathbf{E}^\mathrm{T}\mathbf{E} + \lambda\mathbf{I} = 195 \tag{9.38b}$$

$$\mathbf{k} = [1,5,13]^\mathrm{T}/195 \tag{9.38c}$$

$$\lambda_N = 19/195 \tag{9.38d}$$

and

$$\mathbf{\Gamma}' = [2,6,14]^\mathrm{T} \tag{9.39a}$$

$$\mathbf{G}' = [3-2z^{-1}, 7-6z^{-1}, 15-14z^{-1}] \tag{9.39b}$$

As before, we specify the closed-loop pole set by selecting

$$\mathbf{T} = 1 - 0.5z^{-1} \tag{9.40}$$

Then (9.38d) and (9.29) give

$$R = 65/38 \tag{9.41}$$

In order to solve the identity (9.30) for Q,S, note that $Q = q_0$, $S = s_0 + s_1 z^{-1}$ from (9.31). Also

$$z^{-1}D = \mathbf{T} - \bar{A} = z^{-1}(2.5 - 2z^{-1}) \tag{9.42a}$$

$$\mathbf{k}^\mathrm{T}\mathbf{\Gamma}' = 43/39 \tag{9.42b}$$

$$\mathbf{k}^\mathrm{T}\mathbf{G}' = (233 - 214z^{-1})/195 \tag{9.42c}$$

so that

$$\mathbf{k}^\mathrm{T}(\bar{A}\mathbf{\Gamma}' + B\mathbf{G}') = (448 - 393z^{-1} + 2z^{-2})/195 \tag{9.42d}$$

Substituting (9.38d), (9.42a) and (9.42d) into the identity (9.30) and solving yields

$$q_0 = \frac{25}{38}, \qquad s_0 = \frac{27}{19}, \qquad s_1 = -\frac{27}{38} \qquad (9.43)$$

Note that (9.25) implies that

$$R = 1 + s_0 + s_1 = 65/38$$

which checks with (9.41)

Finally, the CARIMA GPP controller (9.22) can be synthesized from the polynomials

$$\mathscr{F} = 1 + \frac{7}{6}z^{-1} \qquad (9.44a)$$

using (9.38d), (9.42b) and (9.43),

$$\mathscr{G} = (104 - 91z^{-1})/78 \qquad (9.44b)$$

using (9.38d), (9.42c), (9.43), and

$$\mathscr{N} = (13 + 5z^{-1} + z^{-2})/114 \qquad (9.44c)$$

using (9.38c) and (9.41).

Looking at the closed-loop system equation (9.27), we see that $\mathscr{N}$ has introduced two stable but oscillatory zeroes. If these give an unacceptable step response, then the dynamics must be further modified by a more complex choice of R.

The regulation properties of the algorithm can be measured by the output variance:

$$\mathrm{var}\, y(t) = E\left[\frac{\mathscr{F}}{\mathbf{T}}e(t)\right]^2 = \frac{127}{27} \qquad (9.45)$$

It is of interest to note that, if the CARIMA controller is applied to the system whose true representation is of the CARMA form (9.34 with Δ removed), then the closed loop output $y^*(t)$ will differ from (9.27) only in the noise term numerator where the factor Δ will appear. This may or may not degrade the regulation performance of the controller. Concentrating on the noise terms, we can write

$$y^*(t) = y(t) - y(t-1) \tag{9.46}$$

and, therefore, assuming zero mean stationary processes,

$$\text{var } y^*(t) = 2 \text{ var } y(t) - 2E[y(t)y(t-1)] \tag{9.47}$$

from which it follows that

$$0 \leqslant \text{var } y^*(t) \leqslant 4 \text{ var } y(t) \tag{9.48}$$

In our example, it turns out that

$$\text{var } y^*(t) = 64/27$$

so that the output variance (9.47) is decreased.

Figure 9.1 compares the input/output performances for these cases. Figure 9.1(a) shows the input/output behaviour for the CARIMA GPP algorithm applied to the CARMA system of equation (9.34) with the Δ removed, i.e.

$$y(t) = 2y(t-1) + u(t-1) + 2u(t-2) + e(t)$$

where $e(t)$ is zero mean white noise with variance 0.1. Figure 9.1(b) shows the input/output behaviour for the CARIMA GPP algorithm applied to the CARIMA system of equation (9.34). In both cases, the setpoint is a three-level wave form of amplitude 4,0,−4 and period 160 time steps. A visual comparison of Figures 9.1(a) and (b) confirms that the CARIMA GPP applied to the CARMA system gives rise to an output variance less than that for the CARIMA system.   □

9.2.5 The CARMA GPP controller

The controller derivation in Section 9.2.2 assumes that the underlying system takes the CARIMA form (equation (9.7)). The corresponding GPP algorithm for a CARMA system representation is now discussed.

Minimizing the cost function $J(1,N)$ subject to the CARMA model (9.9) involves technical manipulations exactly paralleling those above but with modifications to the two polynomial partitions (9.12) and (9.15). These partitions are replaced by the partitions

$$PC = AF_i + z^{-i}G_i \tag{9.49}$$

$$BF_i = C\Delta E_i + z^{-i}\Gamma_i \tag{9.50}$$

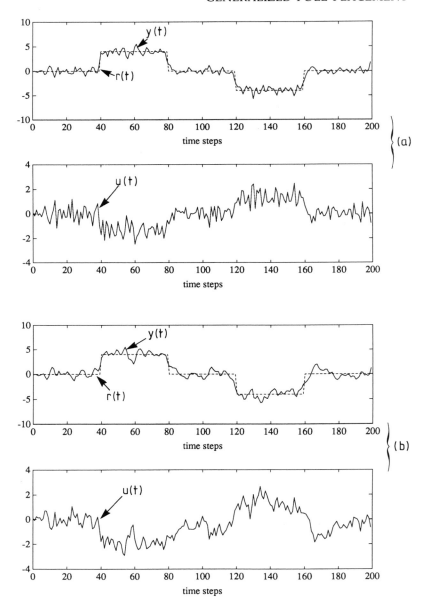

Figure 9.1 Input/output performance using (a) CARIMA GPP controller with CARMA system (Example 9.1) and (b) CARIMA GPP controller with CARIMA system (equation (9.34) in text)).

respectively, so that

$$\Phi_i(t) = F_i e(t+i) + E_i \Delta u(t+i-1) + C^{-1}[(\Gamma_i + CQ\Delta)u(t-1)$$
$$+ (G_i + CS)y(t) - CRr(t+i) + F_i d] \qquad (9.51)$$

The solution for the $\Delta \mathbf{u}$ minimizing $J(1,N)$ then follows as in (9.20).

The control law now has an offset term and this can be chosen to cancel any steady-state output tracking error due to constant load disturbance d. Also the imposition of a control horizon NU can be carried out as for CARIMA models. The reader is urged to develop this form of the algorithm as an exercise.

The key advantage in using a CARMA model is that stationary output disturbances are correctly represented. If a CARIMA model is used in such a case then significant deterioration in regulation performance is possible.

The practical disadvantage of the CARMA format occurs because of the lack of integral action in the feedback loop. Specifically, both the controller offset term and the zero frequency gain of the servo precompensator polynomial $\mathcal{N}$ must be set to ensure steady state correspondence between the reference input $r(t)$ and output $y(t)$. Such requirements are generally held to be disadvantageous because, if either the system parameters change or there is a load variation, the controller will not correctly compensate for such changes unless they are *continually* monitored and information fed to the controller synthesis procedure. This implies that, if the CARMA GPP controller is implemented in a self-tuning mode, it is inadvisable to switch off the estimator at any time. Compensation for load changes is of course automatic in control loops that incorporate integral action and the CARIMA GPP controller is no exception. System changes, however, will lead to tracking error unless the estimator is providing up-to-date parameter values to the GPP algorithm. The reason for this can be traced back to the condition on R imposed by equation (9.25). If the cost polynomials P,S are chosen to be time invariant (subject to the resulting controller stabilizing the system), then R is fixed throughout and system changes lead to no tracking error even if the estimator is switched off. The danger here (as with standard GMV) is that the chosen polynomials may cease to give satisfactory closed-loop performance (or even fail to stabilize the system) if the changes exceed certain limits. This cannot happen if S (and hence R) is tuned on-line to give a stable set of poles, but then R is intimately linked with the parameter estimates and this link must be continuously maintained if tracking error is to be avoided.

9.3   SELF-TUNING GPP CONTROL

The implementation of self-tuning versions of GPP follows the same philosophy adopted in Chapters 7 and 8 whereby the actual system parameters A, B, C are replaced by corresponding estimates $\hat{A}$, $\hat{B}$, $\hat{C}$. The following section describes an explicit self-tuning GPP algorithm for CARIMA representations. Implicit GPP self-tuning algorithms are not available.

9.3.1 An explicit GPP CARIMA self-tuning controller

Employing the certainty equivalence principle, the unknown A and B CARIMA model polynomials are replaced by the estimated values $\hat{A}$ and $\hat{B}$. Estimation of $\hat{A}$ and $\hat{B}$ (and $\hat{C}$ if necessary) is performed using the model in incremental form as follows:

$$\hat{A}\Delta y(t) = \hat{B}\Delta u(t-1) + \hat{C}\hat{e}(t) \qquad (9.52)$$

An explicit GPP self-tuning algorithm for CARIMA system models is now outlined.

As in all control algorithms (adaptive or otherwise), embodying accurate structural information in the system model increases the chances of good control performance. Knowledge of n_a, n_b, n_c is therefore important here, particularly as the controller synthesis procedure is closely linked to these structural parameters (see equations (9.30) and (9.31)). In the event of uncertainty concerning the system order, the usual tactic is to play safe by erring on the side of overparametrization. Provided that the key polynomial identity (9.30) is solved using a robust technique (see Appendix 7.1) this need not be a problem.

In addition, other parameters are required in order to set the algorithm running. These are the GPP 'tuning knobs' referred to earlier in the text and comprise:

(i) the polynomials P, R in the pseudo-output (9.11);

(ii) the output horizon N in the cost function (9.10) and the control horizon NU;

(iii) the control weighting factor λ in the cost function; and

(iv) the desired closed loop pole set as determined by the polynomial **T**.

The use and implications of the 'tuning knobs' will be discussed later in the text. The main algorithmic steps at each sample interval are:

Algorithm: Explicit CARIMA GPP Self-tuning Controller

Step (i) Estimate $\hat{A}$ and $\hat{B}$ and $\hat{C}$ in the model (9.52). Compute $\hat{\tilde{A}}$ using (9.8).
Step (ii) Solve the polynomial equation (9.30) for the polynomials Q and S.
Step (iii) Derive $\mathcal{F}$, $\mathcal{G}$ and $\mathcal{N}$ from equations (9.22) and (9.29).
Step (iv) Compute the control increment $\Delta u(t)$ from equation

$$\mathcal{F}\Delta u(t) + \mathcal{G}y(t) - \mathcal{N}r(t+N) = 0$$

Step (v) Apply the input $u(t) = u(t - 1) + \Delta u(t)$ to the system, and return to Step 1. □

In practical applications the plant may be subjected to disturbances of different types which may not be easily modelled through the C polynomial. In such cases it may be sufficient to replace C by unity or some low-order filtering polynomial which rejects high-frequency disturbance signals and results in a smoother system repsonse.

For CARMA models, the algorithm follows the same step sequence with the following differences:

(a)   An extra parameter (d) is estimated in Step (i).

(b)   The control law yields $u(t)$ not $\Delta u(t)$ in Step (iv).

(c)   A control offset term is calculated in addition to $\mathcal{F}$, $\mathcal{G}$, $\mathcal{N}$.

9.3.2   A simple example (Example 9.1 continued)

We illustrate the self-tuning CARIMA GPP algorithm by returning to Example 9.1 and casting the known parameter algorithm discussed in detail in Section 9.2.4 in an explicit self-tuning form. Figure 9.2 shows the performance of the algorithm (cf. Figures 9.1) when the system is of the CARMA form. Figures 9.2(a) and (b) show the input/output signals and the controller output. These may be compared with the equivalent fixed coefficient controller performance shown in Figure 9.1(a). Figure 9.2(c) shows the estimated system parameters obtained during self-tuning. From equation (9.34a), the correct system parameter values are $a_1 = -2$, $b_0 = 1$, $b_1 = 2$. Figure 9.2(d) shows the estimated controller parameters obtained during self-tuning. Recalling that integral action is present, the true $\mathcal{F}$ polynomial is (from (9.44a))

$$\mathcal{F} = (1 - z^{-1})(1 + \frac{7}{6}z^{-1})$$

i.e. $f_1 = 1/6$, $f_2 = -7/6$.

Likewise, from equations (9.44b and c) the true $\mathcal{N}, \mathcal{G}$ parameters are

$$g_0 = 104/78, \qquad g_1 = -91/78$$

$$n_0 = 13/114, \qquad n_1 = 5/114, \qquad n_2 = 1/114.$$

The self-tuning estimator was initialized with $\mathbf{P}(0) = 5\mathbf{I}_3$ and a constant forgetting factor $\lambda = 0.998$ was applied.   □

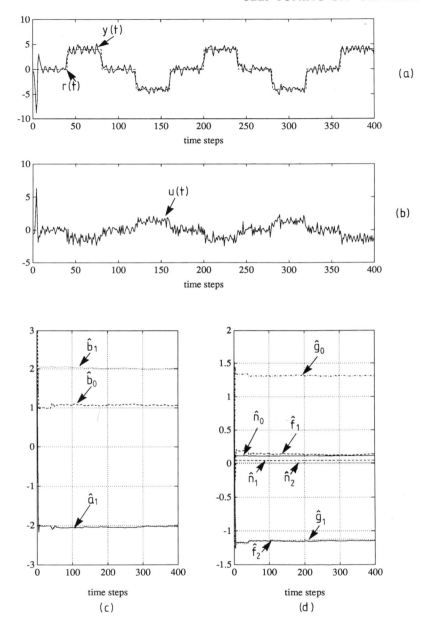

Figure 9.2
Performance of self-tuning CARIMA GPP controller with CARMA system (Example 9.1), showing (a) input/output behaviour, (b) control signal, (c) system parameter estimates and (d) controller parameters.

9.3.3 Convergence issues

Our comments concerning the convergence properties of the explicit GPP self-tuners can be brief. When the estimation model is correctly structured, then the discussion in Chapter 6 suggests that the correct input/output behaviour will result. Further, if the control input is persistently exciting (or is augmented by dither to be so), then convergence to the 'true' system parameters will occur. In those cases where there is a system/model mismatch (for example, a CARIMA model but a CARMA system), convergence problems may arise and each case must be analyzed separately (see Section 9.7 Problems).

9.4 PROPERTIES OF GPP CONTROLLERS

9.4.1 Programmed control

In the cost function (9.10) the future values of setpoint signals $r(t+i)$, $i = 1,2,$ ..., N appear. In general these future values will not be known. However, if they *are* known, the minimization of the cost function gives an optimal control law which exploits this prior knowledge. The advantages of using future setpoint values in this way have been indicated by Peterka (see Section 9.8 Notes and references) who describes the technique as *programmed control*. With reference to Figure 9.3, a programmed setpoint has its future values determined beforehand. The setpoints on a batch chemical reactor are an example of reference inputs which are preplanned to follow certain contours over the duration of a batch reaction. By contrast, many setpoint signals cannot be preplanned. Such signals are termed servo setpoints. An example of a servo setpoint would be the command signal for a given aiming signal, whereby the gun aiming system derives the setpoint from a target. It is clear that the target will try to be unpredictable.

Figure 9.3 Programmed and servo setpoint signals: (a) programmed setpoint (future values known); (b) servo setpoint (future values unknown).

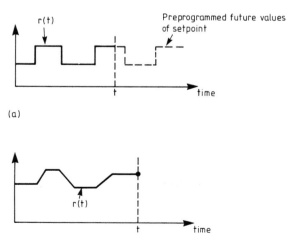

The anticipatory actuation associated with programmed control can be helpful in improving the input/output response of sluggish systems, including those with large transport delays. When applied generally, a small degree of setpoint preprogramming (say one or two time steps) can significantly reduce the magnitude of control excursions during rapid setpoint transitions. The influence of programmed control in this respect can be seen from Figure 9.4. This illustrates the control signal and system output for a system under GPP control with and without setpoint programming. Apart from the use of a programmed setpoint all other aspects of the system are identical. Note the reduced actuation signal $u(t)$ in the case of programmed control. In any situation which involves anticipatory control action (such as programmed control), it is tempting to draw analogies with phase advance compensator action. While this is valid at an intuitive level, it should be noted that (unlike phase advance compensation) programmed control does not influence the closed loop stability margins. In fact, as noted in Sections 9.2.2 and 9.2.3, the loop stability margins are set by the designer's selection of the polynomial **T** which fixes the closed loop pole set.

In situations where the future setpoint is unknown the values $r(t+i)$, $i = 1$, . . ., N must be replaced by predicted values. A common practice in such cases is to use the current value $r(t)$ for such predictions. This corresponds to the situation of regulating the output to a setpoint that is either constant or modelled as a random walk. When the setpoint varies in a more complex manner then the control achieved in this way is suboptimal and corresponds to the servo-control situation discussed earlier.

A problem which does involve prior knowledge of setpoints is the control of robotic manipulators. Here the operating task is usually a prearranged cycle of activities which is repetitively executed by the robot. Thus programmed control is particularly suited to this case, offering the possibility of significant reduction in drive motor peak currents and drive gearbox wear. The application of GPP to robot control is discussed in Section 9.5.

9.4.2   Open loop zeroes and setpoint transient performance

One reason for the use of pole placement self-tuners is that they specifically avoid the cancellation of possibly inverse unstable system zeroes. A consequence of this is that the open loop zeroes appear in the closed response characteristic in such a manner as to cause possibly poor transient response during setpoint transitions. In the context of conventional self-tuning, such problems are handled by prefiltering the setpoint in order to shape the setpoint transient response. Figure 9.5 illustrates this concept which is actually a basic foundation of feedback controller design.

Figure 9.4 (a) GPP control with a servo setpoint and (b) a GPP control with a programmed setpoint.

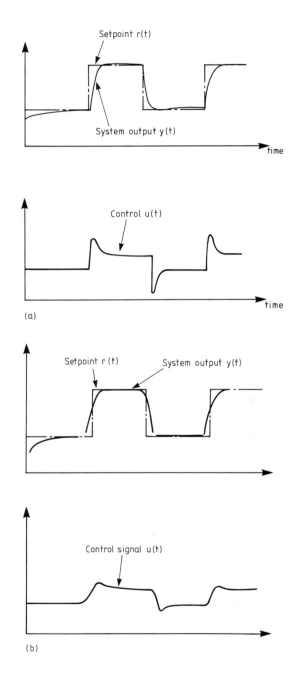

Figure 9.5
Distinguishing
the roles of
the setpoint
compensator
and feedback
compensators.

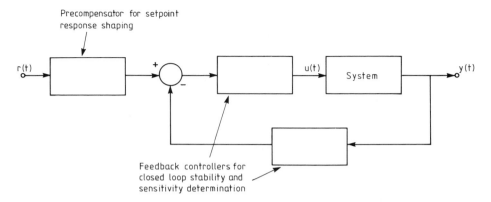

The multistage optimization approach holds out the alternative possibility of embedding the transient response shaping within self-tuning feedback action. In particular, an important feature of the GPP form of self-tuning pole assignment is the existence of various 'tuning knobs' associated with the algorithm. Amongst other things, these knobs allow the influence of open loop zeroes upon the closed loop response to be manipulated. The nature of the manipulation is difficult to quantify. Using the GPP algorithm in a commissioning mode, however, the transient response can be shaped as required by manipulation of the built-in tuning knobs *without* changing the closed loop poles. Moreover, this is done without explicit recourse to setpoint prefiltering. It is a topic of debate and personal philosophy as to whether transient shaping via GPP tuning parameters is equivalent or superior to the explicit setpoint filtering as conventionally performed.

9.4.3   Tuning parameters

The cost function which the GPP control law minimizes (equation 9.10) incorporates a number of objects which must be selected by the designer. Two of these (the polynomials Q and S) are determined so as to select a desired set of closed loop poles. The remaining objects, however, can be interpreted as tuning parameters which can be adjusted (usually in self-tuning mode) to achieve a better overall system response. The tuning parameters are:

(i) tailoring polynomials P and R;

(ii) output horizon N;

(iii) control horizon NU;

(iv) weighting coefficient λ; and

(v) closed-loop pole-set polynomial **T**.

The following paragraphs discuss the practical uses of these design factors.

Role of the P polynomial

The P polynomial is primarily useful as a means of penalizing output excursions during transient setpoint changes and loop disturbances. Since setpoint excursions $r(t)$ are directly associated with the R polynomial (see equation (9.11)), the P polynomial can be viewed as primarily penalizing the output response to loop disturbances. In particular, if it is known from engineering considerations that extraneous load disturbances occur within a specific band of frequencies, P can be selected to weight this frequency band and hence penalize corresponding output excursions. At a more formal level P can be used to obtain model-following behaviour in the case $NU = N$. In either interpretation the P object can be generalized from the polynomial form assumed above to a rational transfer function with consequent modifications to the derivation of the optimal cotroller.

Having made these remarks, however, we note that in practical demonstrations and in simulation tests P was set to unity with satisfactory results. A general recommendation would therefore be to use $P = 1$ initially and introduce filtering action as required by the problem in hand.

Role of the R polynomial

The R polynomial is primarily used to prefilter the setpoint $r(t)$ and as such contributes directly to the closed loop system zeroes (see equation 9.22(d)). In this context, R serves the setpoint precompensator role illustrated in Figure 9.5. From an engineering viewpoint this R will usually be used to bandlimit the setpoint and perform elementary signal management functions such as rate-limiting.

Role of the output horizon N

The output horizon N controls the number of future values of $y(t)$ which are weighted in the controller cost function (9.10). In practical terms it affects the control signal's ability to react before setpoint changes take place. When used in conjunction with programmed control, increasing N leads to 'smoother' system outputs and consequently less active control signals. In general it should have a value greater than the system order. Note, however, that if programmed control is applied with large values of N, then the output $y(t)$ will start to change well before the corresponding setpoint transition. This is clearly undesirable in certain setpoint profile following applications and in practice it is necessary to relate the selection of N to the requirement of the application. In a self-tuning GPP this can be done by experimenting with various values of N.

Role of the control horizon NU

The primary use of the control horizon integer *NU* is to reduce the computational effort required to invert the matrix $[\mathbf{E}^T\mathbf{E} + \lambda\mathbf{I}]$ when computing the control (e.g. equation (9.20)). It is, however, apparent from simulation results that the control horizon has a role which goes beyond computational simplification. In particular, reducing the control horizon generally produces a more sluggish output response which can have advantages for processes which are difficult to control (e.g. highly nonminimum phase). On the other hand, for systems which are relatively conventional in their dynamics, a larger value of *NU* will provide a more responsive control with better tracking action.

Role of the weighting factor λ

The control weighting factor λ plays a similar role in long range predictive control to that which it performs in the so-called lamda controller version of GMV (see Section 8.8). In particular it serves as a simple scalar weighting of excursions in the control signal. In the GPP context, therefore, it has a similar action to the control horizon. This indicates a possible redundancy in the 'tuning knob' set, since it is possible that the control horizon alone could be used to moderate excessive control actuation.

9.4.4   Relative merits of CARMA/CARIMA GPP controllers

As indicated previously, the key practical difference between CARMA GPP and CARIMA GPP self-tuning controllers is that the latter builds integral action into the feedback loop whereas the former does not. Thus the CARIMA GPP controller has an intrinsic capability of dealing with slow changes in the system output load and output nonstationary disturbances. In the same vein, setpoint tracking is automatically achieved. In the CARMA case, however, the polynomial *R* has to be recomputed at every sampling interval in order to guarantee zero steady-state tracking error. In addition, the CARMA model includes an extra parameter (the load term *d*) that must be estimated.

Despite these remarks, the CARMA GPP Controller has some computational advantages. Specifically, since the CARMA algorithm employs the polynomial *A* instead of *Ā* throughout, the polynomial *S* is of lower degree than in the CARIMA case. In turn, this implies that, the polynomials $\mathcal{F}$, $\mathcal{G}$, $\mathcal{N}$ are of lower degree and this usually gives rise to a lower output variance when noise is present. This last point is a reiteration of a previous remark concerning the problems associated with using the CARIMA model when the true process noise is stationary. It is anticipated that this problem will be particularly important in servo-mechanism self-tuning, where the predominant disturbance is wide-band sensor noise.

9.5 A GPP APPLICATION TO ROBOT CONTROL

A demonstration of the way in which GPP self-tuning controllers can be expected to operate may be gained from the following application study concerning the angular position control of a robotic manipulator. The robot manipulator arm under consideration is illustrated in Figure 9.6. It is a small five-axes manipulator built for research purposes in the Control Systems Centre (CSC), UMIST. It is a revolute joint manipulator driven by 'Harmonic Drive' servomotors which belong to a class of high torque direct current servomotors with harmonic gears for providing low backlash and high ratio single-stage speed reduction. This form of robot is quite popular in applications, such as component pick-and-place which require a small, relatively fast and accurate device.

Each robot joint has an internal velocity feedback loop in which an integral tachogenerator provides the speed signal. The motor, gear box and tachogenerator are built into a single compact unit. Information about angular position and the direction of rotation of a joint is provided by a high resolution

Figure 9.6 The CSC research robot manipulator.

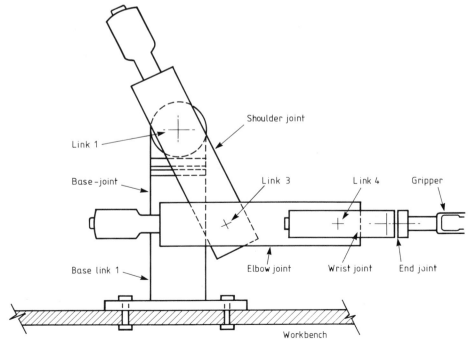

incremental encoder which is attached to the motor shaft. The servomotors are driven with commercially available saturation mode amplifiers. In addition to the usual external reference input voltages, the amplifiers also cater for local velocity feedback.

The CSC research robot was designed as a vehicle for practical research into the adaptive control of high-performance medium-load assembly manipulators. To this end, certain of its design characteristics are specific to small assembly robots. In particular, the links between each consecutive joint have been made sufficiently rigid to prevent twisting during arm operation. The centre of gravity of each joint is located at the axis of rotation of the previous joint. This reduces interaction between joints and therefore significantly simplifies the manipulator arm control problem.

The aim of this discussion is to illustrate GPP self-tuning of the CSC manipulator using a linearized discrete-time model of the joint dynamics. In particular, it is assumed that each joint of the CSC manipulator can be represented by

$$(1 + a_1 z^{-1} + a_2 z^{-2})y(t) = (b_0 z^{-1} + b_1 z^{-2})u(t) + d \tag{9.53}$$

where $u(t)$ is the actuation signal to the motor control unit, $y(t)$ is the measured joint angle and d is a (time-varying) scalar coefficient which represents the influence of gravitational loading.

Note that, because of the design of the CSC manipulator, the various joints are only lightly coupled dynamically. Thus all the system coefficients (a_1, a_2, b_0, b_1) in equation (9.53) can be assumed to be nominally constant. The only allowance for coupling between joints is the scalar d so that varying loads due to angle changes are modelled as an unknown but variable offset.

In the following series of experiments it is usually assumed that only one joint is moving at a time and the setpoint $r(t)$ is a square wave signal. When several loops are under simultaneous control each joint is modelled by an independent equation of the form (9.53).

For the tests shown the GPP self-tuner is configured in an explicit CARMA form as described in Section 9.3. The polynomial P is set at unity and the value of R tuned on-line for setpoint matching (see equation 9.25).

The desired closed-loop pole set in all experiments is given by

$$\mathbf{T} = 1 - 0.5z^{-1}$$

Unless otherwise stated the initial covariance matrix is set at $5\mathbf{I}_5$ and a conventional constant forgetting factor 0.98 used in conjunction with RLS estimation of the parameters (a_1, a_2, b_0, b_1, d).

9.5.1 Programmed control and servo control

The qualitative differences between programmed control and servo control are illustrated in Figure 9.7. This shows the shoulder joint (Figure 9.6) under GPP self-tuning control in both modes. The following remarks reiterate and clarify the points made in Section 9.4.

   (i) Comparing Figures 9.7(a) and (b), programmed control leads to a reduced actuation signal with no perceptible degradation in output response shape. The advantages of this in robot control are reduced component wear and less frequent saturation of the drive amplifier.

  (ii) Under programmed control the output $y(t)$ responds before the transition in $r(t)$. In a robot application, knowledge of this anticipatory action would have to be embedded into the robot work cycle which defines $r(t)$.

Figure 9.7
Illustrating the use of programmed and servo control of the shoulder joint. The tuning parameter settings for this experiment are $N = 4$, $NU = 2$, $\lambda = 0.5$; (a) CARMA model, programmed control and (b) CARMA model, servo control.

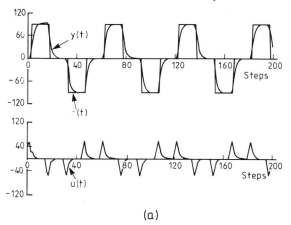

(a)

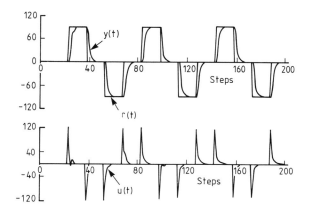

(b)

9.5.2   Use of the tuning parameters

In this section the comparative influences of the 'tuning parameters' N, NU and λ are examined. It is important to note that the examination is specific to the robot system in hand and that other applications would yield different specific results. The qualitative aspects of the discussion, however, are quite general.

The output horizon N

When operating in programmed control mode, the output horizon defines in advance at what time the system output $y(t)$ will react to setpoint changes. Accordingly, an increase in N is normally associated with an increasingly smoother system response and reduction in control signal variation. Figure 9.8 illustrates this effect for the base joint of the CSC robot using a CARMA model. In particular, as the value of N is increased from 2 to 4 the responses become increasingly sedate. Note, however, that the increase in N implies a corresponding increase in computational effort and that most of the benefit in terms of reduced actuation signal is obtained when N is increased from 2 to 3. In addition, it should be noted that large values of N under programmed control causes the output to respond significantly before a setpoint change. This behaviour is undesirable and can cause serious difficulties. In the robot example the use of programmed control with large N might cause the robot to move from a specific position before the work cycle at that position is completed.

When servo control is used the increase in N does not influence the control signal level.

The control horizon NU

As indicated earlier, the control horizon can be used to indirectly influence setpoint tracking and general transient performance. The use of NU in this way is illustrated in Figure 9.9 for the robot shoulder joint. Note the general performance improvement when NU is increased from 1 to 2 (Figures 9.9(a) and (b)). However, for increases beyond $NU = 2$, there is no perceptible change in the system performance.

The control weighting factor λ

As mentioned earlier, the control weighting factor λ is broadly equivalent in its influence to the lamda coefficient in the lamda controller (see Section 8.8). In the robotic servo mechanism application, the influence of λ is clearly

Figure 9.8
Illustrating the
use of the
output horizon
N with
programmed
control and
CARMA model
of the base joint.
Other tuning
parameter
settings for this
experiment are
$NU = 2$, $\lambda = 0.5$;
(a) $N = 2$; (b)
$N = 4$; and (c)
$N = 6$.

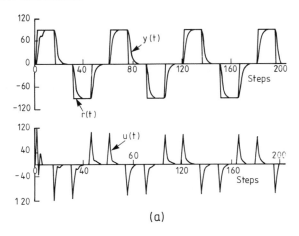

(a)

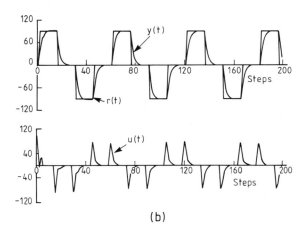

(b)

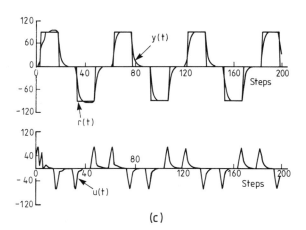

(c)

Figure 9.9
Illustrating the
use of the
control horizon
NU with
programmed
control and
CARMA model
of the shoulder
joint. Other
tuning knob
settings for this
experiment are
$N = 6$, $\lambda = 0.5$.
(a) $NU = 1$; (b)
$NU = 2$.

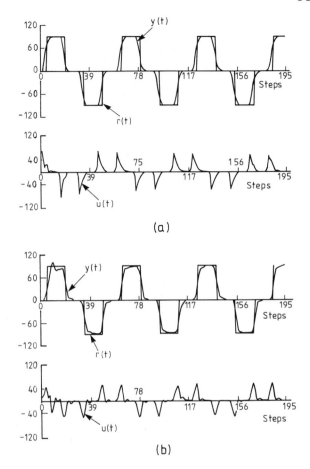

indicated in Figure 9.10. This shows the results of programmed control of the base joint with $\lambda = 0$ and $\lambda = 0.1$. Note the reduction in the control actuation and improved output response when λ is increased. Corresponding experiments with servo control displayed no significant improvement in the response as λ is increased from zero to 0.1. In fact, relatively large values of λ are required before the input/output behaviour is perceptibly altered in servo GPP control.

9.6 SUMMARY

The chapter has covered the following points:
- The use of multistage optimization to provide a predictive control algorithm
- The incorporation of pole placement within a multistage predictive control framework: the Generalized Pole Placement (GPP) algorithm

Figure 9.10
Illustrating the
use of the
control
weighting
function λ with
programmed
control and
CARMA model
of the base joint.
Other tuning
parameter
settings for this
experiment are
$N = 4$, $NU = 2$.
(a) $\lambda = 0$; (b)
$\lambda = 0.1$.

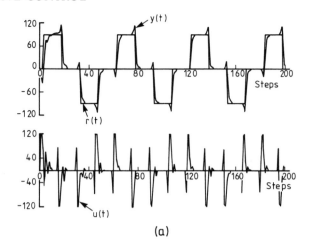

(a)

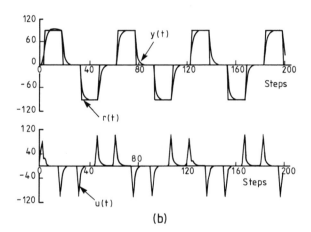

(b)

- The derivation of the GPP controller and its self-tuning realization within a CARMA or CARIMA framework
- The application of GPP to a robot control problem.

9.7   PROBLEMS

Problem 9.1

The polynomial equation

$$X(z^{-1}) = F_i(z^{-1})Y(z^{-1}) + z^{-i}G_i(z^{-1})$$

is to be solved for F_i, G_i given X,Y where

$$\deg X = n = \deg Y; \qquad X(0) = 1 = Y(0)$$
$$\deg F_i = i-1; \qquad \deg G_i = n-1$$

Show that the following recursive algorithm solves the equation for $i = 1,2,\ldots$:

(i)   Set $i = 1$
     $F_i(z^{-1}) = 1, G_i(z^{-1}) = z[X(z^{-1})-Y(z^{-1})]$
(ii) $f_i = G_i(0)$
     $F_{i+1}(z^{-1}) = F_i(z^{-1})+f_iz^{-i}$
     $G_{i+1}(z^{-1}) = z[G_i(z^{-1})-f_iY(z^{-1})]$
(iii) Set $i = i+1$; go to (ii)

Problem 9.2

The abrupt imposition of a control horizon (expressed by equation (9.32)) can be softened, while retaining the reduction in the number of optimization variables. If (9.32) is replaced by

$$\Delta u(t+i) = \alpha\Delta u(t+i-1) \qquad i = NU, NU+1,\ldots,N-1$$

for some α satisfying $|\alpha|<1$, derive the revised $N \times NU$ matrix $\mathbf{E}$ replacing (9.23) and hence the optimal control replacing (9.20).

Problem 9.3

In predictive control algorithms (such as GPP), future setpoint values are often assumed known. If the setpoint value $r(t+i)$ is not available at time t, show that it should be replaced by $\hat{r}(t+i|t)$, its best estimate at time t.

Problem 9.4

For linear stochastic models of the CARMA or CARIMA type, show that multistage cost functions of the form

$$J(N) = E_t \sum_{i=1}^{N} [y(t+i)-r(t+i)]^2$$

can be minimized term by term to yield sequentially $u(t)$, $u(t+1),\ldots, u(t+N-1)$. Is $J(N)$ the most general cost function for which this occurs?

Problem 9.5

In multistage predictive control, the optimal control for time t is calculated, not only at time t, but also at a number of previous times depending on the control horizon.

We may denote these values by $\hat{u}(t|t)$, $\hat{u}(t|t-1)$,..., where $\hat{u}(t|s)$ is the optimal $u(t)$ as calculated at time s. When we employ the rolling horizon approach, we set

$$u(t) = \hat{u}(t|t)$$

and discard all the other calculated signals. It has been suggested that a more robust approach is to set

$$u(t) = \sum_{k=1}^{NU} \gamma(k)\hat{u}(t|t-k+1)/\sum_{k=1}^{NU} \gamma(k)$$

for some $\gamma(1),\gamma(2),...,\gamma(NU)$.

Calculate this controller for the system

$$y(t) = a\,y(t-1)+u(t-1)+e(t)$$

in the usual notation, using the cost function in Problem 9.4 with $N=NU=2$. Show that the resulting closed loop system is unstable unless

$$\left|\frac{\gamma(2)a}{\gamma(1)+\gamma(2)}\right| < 1$$

Problem 9.6

Consider the CARIMA GPP example based on the model structure (9.34a). If the true system disturbance is not given by (9.34b) but by

$$\text{(i)} \quad x(t) = e(t)+ce(t-1)$$

or

$$\text{(ii)} \quad \Delta x(t) = e(t)+ce(t-1)$$

investigate the effect on the closed loop output variance, by assessing its sensitivity to c.

Problem 9.7

Develop the details of the CARMA GPP controller discussed briefly in Section 9.2.5.

Problem 9.8

Show that an alternative form of the GPP algorithm is possible based on (9.10) and (9.11) where S is set to zero and Q is replaced by the transfer function Q_n/Q_d. Show that the crucial pole assignment identity now involves the choice of the two polynomials Q_n and Q_d.

Problem 9.9

Consider the discrete-time system

$$y(t) = y(t-1) + u(t-1) + e(t)$$

in the usual notation. Derive the minimum mean square error linear output predictors $\hat{y}(t+1|t)$ and $\hat{y}(t+2|t)$ for a given input sequence.

It is required to regulate the system output to zero. Derive the lamda controller and hence compute the closed-loop output variance when λ is unity.

An alternative two-stage controller is proposed. At time t, the control signals $u(t)$, $u(t+1)$ are computed to minimize the cost function

$$J(2) = E_t \sum_{i=1}^{2} [y^2(t+i) + \lambda u^2(t+i-1)]$$

Show that

$$u(t) = -\frac{1+2\lambda}{1+3\lambda+\lambda^2} y(t) \qquad (*)$$

and derive $u(t+1)$.

If the algorithm is operated in a 'rolling horizon' mode, the controller (*) is used at each time t, so that $u(t+1)$ is calculated but not implemented. If λ is unity, show that the output variance achieved in this mode is lower than for the standard lamda controller.

9.8   NOTES AND REFERENCES

The GPP method is described fully in:

Lelic, M.A. and Zarrop, M.B.
  A General Pole Placement Self-Tuning Controller, UMIST Control Systems Centre Report No. 652, 1986.
Lelic, M.A. and Zarrop, M.B.
  A generalized pole-placement self-tuning controller: part 1 basic algorithm, *Int. J. Control*, **46**(2), 547–68, 1987.

The application of GPP to robotics is more fully discussed in:

Lelic, M.A. and Wellstead, P.E.
  Generalized pole placement self-tuning controller: part 2 application to robot manipulator control, *Int. J. Control*, **46**(2), 569–607, 1987.

The related GPC technique is discussed in:

Clarke, D.W., Mohtadi, C. and Tuffs, P.S.
  Generalized predictive control: part 1 The basic algorithm, *Automatica*, **23**(2), 137–48, 1987.

Clarke, D.W., Mohtadi, C. and Tuffs, P.S.
Generalized predictive control: part 2 extensions and interpretations, *Automatica*, **23**(2), 149–60, 1987.

The control horizon concept is drawn from the Dynamic Matrix Control Method in:

Cutler, C.R. and Ramaker, B.L.
Dynamic matrix control — a computer control algorithm, *Proc. JACC, San Francisco*, paper WP5-B, 1980.

Other multistep predictive self-tuners are given in:

DeKeyser, R.M.C., and Van Cauwenberghe, A.R.
A self-tuning multi-step predictor application, *Automatica*, **17**(1), 167–74, 1979.
Greco, C., Menga, G., Mosca, E. and Zappa, G.
Performance improvements of self-tuning controllers by multistep horizons: the MUSMAR approach, *Automatica,* **20**(5), 681–99, 1984.
Peterka, V.
Predictor based self-tuning control, *Automatica,* **20**(1), 39–50, 1984.
Ydstie, B.E.
Extended horizon adaptive control, *Proc. IFAC Ninth World Congress*, Budapest, Hungary, 1984.

Part 3

Self-Tuning Signal Processing

This part concerns the area of adaptive signal processing. Three basic aspects are addressed: filtering, prediction and smoothing. Signal processing is an important area of self-tuning theory and application. The discussion given here is meant to serve as an introduction to the area and to set a wider perspective for the material of Parts 2 and 4.

10 Prediction

10.1 OUTLINE AND LEARNING OBJECTIVES

In many engineering and economic situations it is frequently the case that future knowledge of a signal will greatly improve the performance of a control or signal processing system. When the signal is deterministic, for example a sine wave, then it is easy to predict exactly what value it will take at any time in the future. When the signal has a random component, however, it is no longer possible to make an exact forecast. In such circumstances it is usual to construct a predictor based on some criterion of 'goodness'. In this connection the optimization ideas associated with minimum variance (MV) regulation (Chapter 8) can be applied to the task of predicting future values of a random signal. This chapter explains how this can be done for stationary random processes, such that, given past and current measurements of a signal, it is possible to estimate its value at an arbitrary future time. Figure 10.1 illustrates this concept.

In this chapter, Section 10.2 describes the basic theory of MV prediction. Section 10.3 explains some extensions and modifications to the basic algorithm. Section 10.4 outlines the self-tuning version of the predictor and 10.5 describes an application of self-tuning prediction to a complex industrial process.

Figure 10.1
Illustrating the *k*-step ahead prediction problem.

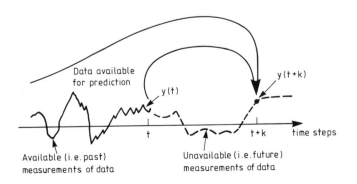

The learning objectives are to understand the methods of stochastic prediction based upon transfer function models of random signals, and to see how they can be implemented in either fixed or self-tuning form.

10.2 MINIMUM VARIANCE PREDICTION

Consider a discrete time random signal $y(t)$ described by

$$y(t) = \frac{C}{A} e(t) \qquad (10.1)$$

where, as usual,

$$C = 1 + c_1 z^{-1} + \ldots + c_{n_c} z^{-n_c}$$

$$A = 1 + a_1 z^{-1} + \ldots + a_{n_a} z^{-n_a}$$

and $\{e(t)\}$ is a discrete white noise sequence with variance σ_e^2.

If we wish to make a prediction of $y(t)$ at k time steps in the future then we can begin by writing

$$y(t+k) = \frac{C}{A} e(t+k) \qquad (10.2)$$

and then demand a 'good' estimate (prediction) of $y(t+k)$ based on data available at time t. Denoting this predictor by $\hat{y}(t+k|t)$, a common criterion of 'goodness' is to choose it to minimize the cost function

$$V_k = E\langle \tilde{y}^2(t+k) \rangle \qquad (10.3)$$

where $\tilde{y}(t+k)$ is the output prediction error

$$\tilde{y}(t+k) = y(t+k) - \hat{y}(t+k|t) \qquad (10.4)$$

In general, the resulting minimum mean square error (MMSE) predictor is given by the conditional expectation

$$\hat{y}(t+k|t) = E[y(t+k)|t] \qquad (10.5)$$

(see Appendix) but this predictor may not be simple to construct and may be nonlinear in the data.

From equation (10.2)

$$E[y(t+k)|t] = E\left[\left\{\frac{C}{A}e(t+k)\right\}|t\right]$$ (10.6)

so that we require estimates of past, current and future noise signals based on data available at time t.

If the noise is *independent* (gaussian, say) rather than merely uncorrelated then

$$E[e(t+i)|t] = E[e(t+i)] = 0 \quad \text{if } i>0$$ (10.7)

Using the same partition of C/A as in Chapter 8, we can write

$$\frac{C}{A} = F + z^{-k}\frac{G}{A}$$

i.e.

$$C = FA + z^{-k}G$$ (10.8)

where

$$F = 1 + f_1 z^{-1} + \ldots + f_{k-1} z^{-k+1}$$
$$G = g_0 + g_1 z^{-1} + \ldots + g_{n_a-1} z^{-n_a+1}$$

Then, using (10.7) in (10.6),

$$E[y(t+k)|t] = E\left[\left\{\frac{G}{A}e(t)\right\}|t\right]$$ (10.9)

If, in addition, it is assumed that the data available at time t extends to the infinite past, then, inverting the model (10.1),

$$E[e(t+i)|t] = \frac{A}{C}y(t+i) \quad \text{if } i \leqslant 0$$ (10.10)

Of course, the infinite past is never available but the stability of A/C ensures that this 'reconstruction' of $e(t+i)$ will still be accurate provided that t is sufficiently large.

From (10.9) and (10.10)

$$E[y(t+k)|t] = \frac{G}{C} y(t) \tag{10.11}$$

Even if the noise is not independent, it is easy to see that (10.11) is the *linear* predictor that is 'best' in the sense of minimizing the cost function V_k. If any predictor $\hat{y}(t+k|t)$ is linear in the data (and therefore linear in $e(t)$, $e(t-1)$, . . .), then, using (10.8),

$$V_k = E\left[\frac{G}{A} e(t) - \hat{y}(t+k|t)\right]^2 + (1+f_1^2 + \ldots + f_{k-1}^2)\sigma_e^2$$

$$= E\left[\frac{G}{C} y(t) - \hat{y}(t+k|t)\right]^2 + \text{constant}$$

and the minimizing predictor is

$$\hat{y}(t+k|t) = \frac{G}{C} y(t) \tag{10.12}$$

Note that

$$y(t+k) = \hat{y}(t+k|t) + Fe(t+k) \tag{10.13}$$

so that

$$\tilde{y}(t) = Fe(t) \tag{10.14}$$

and

$$E[\tilde{y}(t)] = E[y(t) - \hat{y}(t|t-k)] = 0 \tag{10.15}$$

so that $\hat{y}(t|t-k)$ is an *unbiased* estimate of $y(t)$.

For such predictors V_k is the predictor variance and therefore we can also term (10.12) a 'minimum variance' predictor.

From (10.1), (10.12) and (10.14) the predictor $\hat{y}(t+k|t)$ can be written as

$$\hat{y}(t+k|t) = \frac{G}{AF} \tilde{y}(t) \tag{10.16}$$

This is a useful form for self-tuning prediction. It can also be used to interpret the prediction algorithm in terms of MV control. Specifically, if the prediction error $\tilde{y}(t)$ is interpreted as the regulation error in a feedback regulator, then,

Figure 10.2
Interpreting the
k-step ahead
predictor as a
MV regulation
problem.

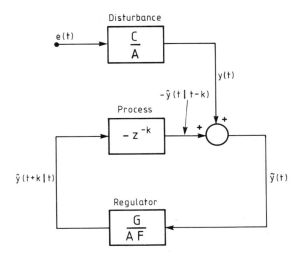

the predictor equation (10.16) can be interpreted as a regulator equation. Figure 10.2 illustrates this idea, whereby the 'process' is a pure delay $-z^{-k}$.

Example 10.1

Consider the ARMA(1,1) process

$$y(t) = ay(t-1) + e(t) + ce(t-1) \tag{10.17}$$

so that $A = 1-az^{-1}$, $C = 1+cz^{-1}$.

For the one-step-ahead predictor $(k=1)$, $F=1$ and therefore the partition (10.8) is always trivial, yielding

$$z^{-1}G = C-A = (c+a)z^{-1}$$

i.e. $G(z) = c+a$ so that

$$\hat{y}(t+1|t) = \left[\frac{c+a}{1+cz^{-1}}\right] y(t)$$

or, recursively expressed,

$$\hat{y}(t+1|t) = (c+a)y(t) - c\hat{y}(t|t-1) \tag{10.18}$$

For $k = 2$, we have to solve the identity,

$$1 + cz^{-1} = (1 - az^{-1})(1 + f_1 z^{-1}) + z^{-2} g_0$$

yielding

$$f_1 = c + a, \quad g_0 = a(c + a)$$

so that

$$\hat{y}(t+2|t) = a(c+a)y(t) - c\hat{y}(t+1|t-1) \tag{10.19}$$

The variances of the prediction errors in the two cases are σ_e^2 and $[1+(c+a)^2]\sigma_e^2$ respectively and these can be used as measures of confidence in the predictions. For example, under a gaussian assumption, we expect the actual value of $y(t+1)$ to lie in the interval $\hat{y}(t+1|t) \pm 2\sigma_e$ with 95% probability.

Figures 10.3(a) and (b) compare $y(t)$ with $\hat{y}(t|t-1)$ and $\hat{y}(t|t-2)$ respectively when

$$a = 0.5, \quad c = 0.2, \quad \sigma_e = 0.1$$

Note that the two step ahead prediction $\hat{y}(t|t-2)$ is a rather poor estimate of the actual process output. However, when the upper and lower 95% confidence intervals for $\hat{y}(t+|t-2)$ are also plotted (Figure 10.3(c)), it can be seen that the interval $\hat{y}(t|t-2) \pm 2\sigma_e(1+(c+a)^2)^{1/2}$ spans the process $y(t)$ in an appropriate way. In this particular example, the quality of prediction deteriorates rapidly as k is increased.   □

The prediction error variance for a k step ahead predictor is given by

$$V_k = (1 + f_1^2 + \ldots + f_{k-1}^2)\sigma_e^2$$

This increases monotonically with the prediction horizon k but

$$\lim_{k \to \infty} E\{\tilde{y}(t)^2\} = (1 + f_1^2 + \ldots)\sigma_e^2$$

$$= E\left[\frac{C}{A}e(t)\right]^2 = E\{y^2(t)\}$$

which is finite for a stationary process.

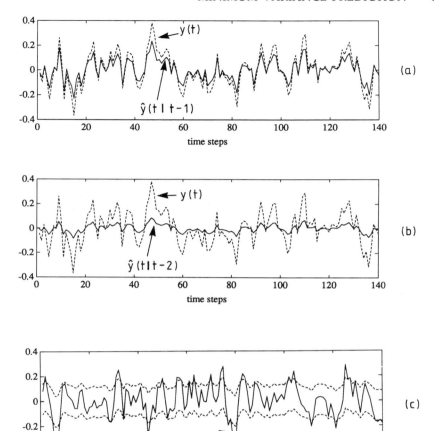

Figure 10.3
Comparison of $y(t)$ with (a) $\hat{y}(t|t-1)$ and (b) $\hat{y}(t|t-2)$ for the system (Example 10.1). Plot (c) compares $y(t)$ with the 95% confidence intervals for the prediction $\hat{y}(t|t-2)$.

In an off-line algorithm the predictor relies upon a knowledge of the transfer function C/A. In practice this knowledge might arise from physical insight into the noise generating mechanism or estimates of A and C obtained using a maximum likelihood type algorithm (see Chapter 3). In an on-line situation the estimation of the process model (equation (10.1)) can be combined with the predictor solution to give a self-tuning predictor. Self-tuning predictor algorithms are discussed in Section 10.4.

The following section describes some useful extensions to the MV prediction theory.

10.3   INCREMENTAL PREDICTION AND MEASURABLE DISTURBANCES

The theoretical development in the previous section assumed that the signal to be predicted is generated by a simple ARMA model (equation (10.1)). Recalling the discussion in Chapter 2, however, a signal source is frequently more complex than can be adequately described by an ARMA representation. In particular, in prediction applications it is frequently the case that the signal has a nonzero mean, possibly a drift component and a deterministic component. As an example consider Figure 10.4 which shows the quarterly number of passengers carried by air and sea in the United Kingdom for the years 1961 to 1978. The time interval between each data point is three months. For transportation planners it is very interesting to be able to predict the number of passengers to be carried in future years. This involves predicting several time steps ahead from the three monthly data as in Figure 10.3. Thus, for example, a four step ahead predictor would provide a forecast of passenger movements one year ahead.

The reason for showing this data is to illustrate a typical record of real data which is used for prediction and yet which contains a number of signal components which do not fit into the ARMA model proposed in the previous section. In the case of the passenger data, these signal components are comprised as follows:

(i) A disturbance component $\mathscr{D}(t)$ with both a constant and trend term i.e.
$$\mathscr{D}(t) = d_0 + d_1 t$$
This represents the nonzero mean level of the data and the apparently linearly increasing mean value of the passenger movements. This latter

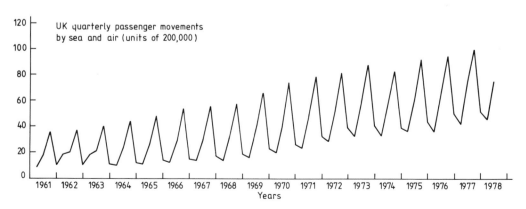

Figure 10.4 Plot of the quarterly passenger movements by sea and air for the United Kingdom, in units of 200 000.

component represents the generally increasing trend in the number of people travelling.

(ii) An approximately sinusoidal component of known period (i.e. one year). This can be modelled (as suggested in Section 2.3.1) by a sine wave with unknown phase and amplitude. This sinusoidal component represents the cyclical variation in travel, with large numbers in summer and smaller numbers in winter.

The presence of such disturbances in a signal demands some modifications to the basic prediction algorithm. The following subsections outline the most widely used modifications and indicate the areas of application.

10.3.1 Incremental predictor

In some cases the signal $y(t)$ will appear to have a random drift component and may exhibit a nonzero mean. This form of behaviour indicates that the signal is most appropriately modelled by an ARIMA model:

$$y(t) = \frac{C}{(1-z^{-1})A} e(t) \tag{10.20}$$

In this case we can use the standard procedure for prediction detailed in Section 10.2 by either

(i) replacing A in the polynomial partition (10.8) by the polynomial $(1-z^{-1})A$; or

(ii) applying the unmodified algorithm to the incremental process $\Delta y(t)$, defined by

$$\Delta y(t) = \frac{C}{A} e(t) \tag{10.21}$$

These alternatives give predictor equations of different forms (for $k>1$) but yield the same predicted values of $y(t)$ (see Problem 10.2).

Note that, if the signal described by equation (10.20) also contains a constant nonzero mean, the incremental form (10.21) removes it from the prediction data description.

Example 10.1 (continued)

Consider the incremental form of equation (10.17):

$$\Delta y(t) = a\Delta y(t-1)+e(t)+ce(t-1)$$

Modifying the predictor equation (10.18) yields

$$(\Delta\hat{y})(t+1|t) = (c+a)\Delta y(t)-c(\Delta\hat{y})(t|t-1)$$

where the notation emphasizes that we are constructing an estimate of the first difference and *not* taking the first difference of the estimate. This implies the expansion

$$\hat{y}(t+1|t)-y(t) = (c+a)[y(t)-y(t-1)]-c[\hat{y}(t|t-1)-y(t-1)]$$

which simplifies to

$$\hat{y}(t+1|t) = (c+a+1)y(t)-ay(t-1)-c\hat{y}(t|t-1)$$

This can be checked by using (10.12) and noting that

$$z^{-1}G =C-(1-z^{-1})A$$

so that

$$G = (c+a+1)-az^{-1}$$

Similarly, equation (10.19) is replaced by

$$(\Delta\hat{y})(t+2|t) = a(c+a)\Delta y(t)-c(\Delta\hat{y})(t+1|t-1)$$

leading to the predictor equation

$$\hat{y}(t+2|t) = \hat{y}(t+1|t) +a(c+a) [y(t)-y(t-1)]-c[\hat{y}(t+1|t-1)-\hat{y}(t|t-1)]$$

Note that this involves the one-step-ahead predictor. Alternatively, using ΔA in the standard partition for $k=2$, this coupling is avoided and the predictor equation is

$$\hat{y}(t+2|t) = [a^2+(a+1)(c+1)]y(t)-a(a+c+1)y(t-1)-c\hat{y}(t+1|t-1)$$

Figure 10.5 compares the one and two step ahead predictions for the incremental form of Example 10.1. Note that the apparent high quality of the predictions (compared with Figure 10.3) is due to the incremental form of the model (10.20) and the scaling of the graphs.   □

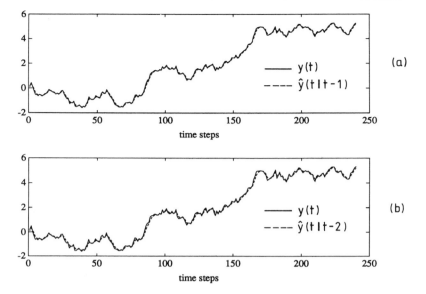

Figure 10.5
Comparison of
$y(t)$ (full line)
with (a) $\hat{y}(t|t-1)$
(broken line)
and (b) $\hat{y}(t|t-2)$
(broken line) for
the incremental
version of
Example 10.1.

10.3.2   Prediction with deterministic signal components

If the $y(t)$ process is described by an ARMA process plus a disturbance process
$\mathcal{D}(t)$ and possibly a contribution from a measurable source $v(t)$, then it can be
modelled as described in Chapter 2 (Section 2.3.3) by the expression

$$Ay(t) = Ce(t) + \mathcal{D}(t) + Dv(t) \tag{10.22}$$

The correct procedure here is to consider the deterministic component as a
known input term and treat it for prediction purposes in exactly the same
way as the input term $Bu(t-k)$ of the CARMA model in the development of
the MV controller (see Section 8.2.4). This immediately leads to the predictor

$$\hat{y}(t+k|t) = \frac{1}{C}\left[Gy(t) + DFv(t+k) + F\mathcal{D}(t+k) \right] \tag{10.23}$$

(cf. equation (8.26)). Equation (10.14) for the prediction error $\tilde{y}$ remains valid.

10.4   SELF-TUNING PREDICTION

10.4.1   Basic predictor

The predictor (10.12) can be put into self-tuning form in one of two ways.
The first is via an explicit algorithm in which a recursive estimator is used to

form estimates of A and C which are then used at each time interval to compute the values of F and G via the identity (10.8). These in turn are used to compute the prediction using either equation (10.12) or (10.16).

In practice, however, it is more usual to use an implicit self-tuning algorithm in which the predictor coefficients are directly estimated using RLS. The implicit predictor uses equation (10.16) in the form

$$\hat{y}(t|t-k) = - \sum_{i=1}^{n_q} q_i \hat{y}(t-i|t-i-k) + \sum_{i=0}^{n_g} g_i \tilde{y}(t-k-i) \qquad (10.24)$$

where

$$Q = AF = 1 + q_1 z^{-1} + \ldots + q_{n_q} z^{-n_q}$$

and

$$n_q = n_a + k - 1$$
$$n_g = n_a - 1$$

Within a self-tuning framework, the unknown signal $\tilde{y}$ is replaced by the prediction error ϵ to yield the estimation model:

$$y(t) = - \sum_{i=1}^{n_q} q_i \hat{y}(t-i|t-i-k) + \sum_{i=0}^{n_g} g_i \epsilon(t-k-i) + \hat{e}(t) \qquad (10.25)$$

This equation can be used as a basis for estimation in an implicit self-tuning predictor as follows:

Algorithm: Implicit Self-tuning Predictor

At time step t:

Step (i)   Use RLS to estimate the parameters $\hat{q}_i$, $\hat{g}_i$ in the least squares model

$$y(t) = - \sum_{i=1}^{n_q} \hat{q}_i \hat{y}(t-i|t-i-k) + \sum_{i=0}^{n_g} \hat{g}_i \epsilon(t-k-i) \qquad (10.26)$$

Step (ii)   Use the estimated parameters to compute the k step ahead prediction as

$$\hat{y}(t+k|t) = -\sum_{i=1}^{n_q} \hat{q}_i \hat{y}(t+k-i|t-i) + \sum_{i=0}^{n_g} \hat{g}_i \epsilon(t-i) \tag{10.27}$$

Step (iii)   Wait for interval t to end, then put $t \rightarrow t+1$ and loop back to Step (i).   □

Note that the number of parameters to be estimated differs between the explicit and implicit algorithms. In the former, n_a+n_c coefficients of the A, C polynomials appear. For the implicit algorithm, the number is

$$n_g+n_q+1 = \max(n_a+n_c, 2n_a+k-1)$$

which is never less than n_a+n_c and is greater if

$$k > n_c - n_a + 1$$

The advantage of the implicit algorithm is that it does not require the partition process (equation (10.8)) to be performed. Thus, at the expense of an increased computational load for the recursive estimator, we have avoided the solution of equation (10.8). For this reason the implicit form of self-tuning prediction is often used.

Example 10.1 *(continued)*

Figures 10.6 and 10.7 compare the estimated predictor parameters for both the explicit and implicit self-tuning predictors when $a = 0.5$, $c = 0.2$, $\sigma_e = 0.1$ and $k = 2$. In particular, Figure 10.6 shows the results of explicit self-tuning prediction, with Figure 10.6(a) showing the parameter estimates and Figure 10.6(b) the corresponding F and G polynomial coefficients. In this case $f_1 = 0.7$, $g_0 = 0.35$. The explicit self-tuning was performed at time step t by

(i) updating AML (RML$_2$) estimates using $y(t)$;

(ii) using the estimated A,C objects to synthesize F,G (equation (10.8));

(iii) computing the prediction $\hat{y}(t+k|t)$ using equation (10.12).

The quality of the explicit self-tuning predictor can be assessed from Figure 10.6(c). This compares the theoretical accumulated loss with the actual accumulated loss using self-tuning. The theoretical accumulated loss at time step t is, in this case,

$$V(t) = tV_k = t\sigma_e^2(1+f_1^2)$$

Figure 10.6
Illustrating the
explicit self-
tuning two-step
ahead predictor
estimates for
Example 10.1.
Plot (c) compares
the theoretical
and actual
accumulated loss
functions.

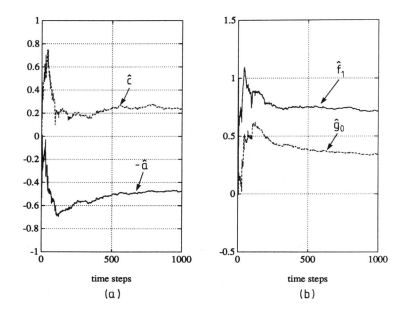

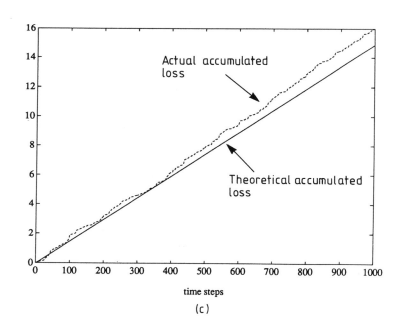

The actual accumulated loss is computed using

$$V_a(t) = \sum_{i=1}^{t} \epsilon^2(i)$$

The implicit self-tuning predictor results are shown in Figure 10.7. The estimated predictor parameters are shown in Figure 10.7(a). In this case $q_1 = 0.2$, $q_2 = -0.35$ and $g_0 = 0.35$. The actual and theoretical accumulated losses are compared in Figure 10.7(b). Note that, after an initial divergence during the tuning-in phase, the two traces run parallel. This indicates that the self-tuning predictor is producing good output predictions despite the fact that the individual predictor estimates $\hat{g}_0$, $\hat{q}_1$, $\hat{q}_2$ are some way from their theoretical values.

As a final qualitative indication of the implicit self-tuning predictor's performance, Figure 10.8 compares $y(t)$ with its two-step ahead prediction under self-tuning. This should be compared with Figure 10.3(b) for the fixed coefficient predictor.

Throughout the above self-tuning simulations, an AML(RML$_2$) estimator was used with $\mathbf{P}(0) = 100\mathbf{I}_2$ and a forgetting factor $\lambda = 0.998$. Zero initial conditions were assumed on all parameter estimates. $\qquad\square$

10.4.2   Predictor with auxiliary variables

Suppose that it is required to predict a process using an implicit self-tuning predictor when auxiliary variables are present. In practice these auxiliary variables may be associated with drift disturbances or deterministic components as described in Section 10.3.2. Assume for convenience that the deterministic components can all be presented by a single polynomial D operating on a known signal $v(t)$. The signal description is

$$Ay(t) = Ce(t) + Dv(t) \tag{10.28}$$

Using equations (10.23) and (10.28), the predictor is given by

$$\hat{y}(t+k|t) = \frac{G}{AF}\tilde{y}(t) + \frac{D}{A}v(t+k) \tag{10.29}$$

This equation may be compared with equation (10.16) for the basic predictor. In addition, the same idea as used in equation (10.21) may be employed to write the implicit predictor for the process (equation (10.28)) as

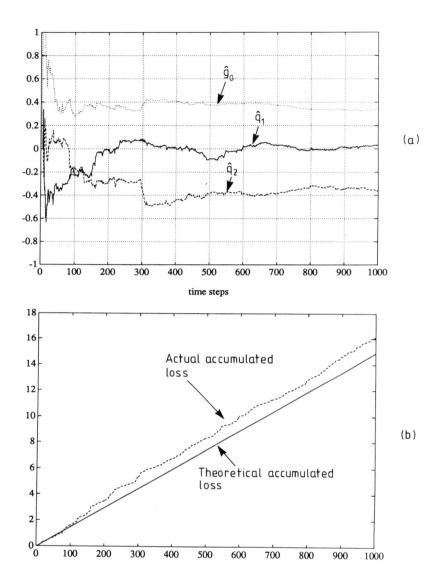

$$\hat{y}(t|t-k) = -\sum_{i=1}^{n_q} q_i \hat{y}(t-i|t-i-k) + \sum_{i=0}^{n_g} g_i \tilde{y}(t-k-i) + \sum_{i=0}^{n_p} p_j v(t-i) \quad (10.30)$$

where $P = DF$

$$= p_0 + p_1 z^{-1} + \ldots + p_{n_p} z^{-n_p}$$

and $n_p = n_d + n_f$

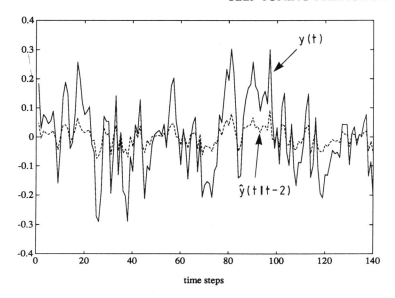

Figure 10.8 Comparing the implicit self-tuned predictor output $\hat{y}(t|t-2)$ with $y(t)$.

By applying the same procedure with equation (10.30) as was applied with equation (10.23), the following implicit self-tuning predictor with auxiliary variables can be obtained:

Algorithm: Implicit Self-tuning Predictor with Auxiliary Variables

At time step t:

(i) Use RLS to estimate the parameters $\hat{q}_i$, $\hat{g}_i$ and $\hat{p}_i$ in the model

$$y(t) = -\sum_{i=1}^{n_q} \hat{q}_i \hat{y}(t-i) + \sum_{i=0}^{n_g} \hat{g}_i \epsilon(t-k-i) + \sum_{i=0}^{n_p} \hat{p}_i v(t-i) \qquad (10.31)$$

(ii) Use the estimated parameters to compute the k-step ahead prediction as

$$\hat{y}(t+k|t) = -\sum_{i=1}^{n_q} \hat{q}_i \hat{y}(t+k-i|t-i) + \sum_{i=0}^{n_q} \hat{g}_i \epsilon(t-i) + \sum_{i=0}^{n_p} \hat{p}_i v(t+k-i) \qquad (10.32)$$

(iii) Wait for interval t to end, set $t \rightarrow t+1$ and loop back to Step (i).   □

Note that the above example has assumed that all additional signals can be modelled by a single auxiliary variable $v(t)$ and polynomial D. In fact, drift terms and contributions from additional sources can easily be added using a similar procedure to that employed here.

10.5   APPLICATION — SELF-TUNING PREDICTION OF HOT METAL QUALITY IN A BLAST FURNACE

Statistical predictors have a wide range of practical applications in almost all areas of endeavour. Wherever an historical record of variables exist there is a wish to predict the future values of the variables. In engineering and technical applications the main use of predictors is as operator/management aids. For example, in a complex industrial process it is usually desirable to employ skilled human supervisors to oversee the plant behaviour and make major operational decisions. Their decisions can be supported by prediction algorithms which forecast the future behaviour of key process variables. Thus in a hypothetical situation (Figure 10.9), a supervisor, confronted with a complex display panel, can be assisted in his/her decision making by a set of predictions. As an extension of this, the predictions can be linked to controller algorithms (operated open loop) to advise on suitable control settings. The supervisor can either use these advisory settings or ignore them if his/her experience suggests other actions are appropriate.

Iron and steel making complexes are typical of industrial processes which are operated in this open loop supervisory mode. In the following paragraph

Figure 10.9 Illustrating the possible use of predictions of process variables (PV) in assisting process supervisors.

we describe how the prediction methods described in this chapter may be used as advisory tools in one aspect of steelmaking: blast furnace operation.

One of the key concerns of the steelmaking industry in recent years has been to understand more fully the inner workings of the blast furnace. The primary outcome of this attention has been a number of furnace models based upon an analysis of the physical and chemical nature of the ironmaking process. These, together with cumulative plant experience, have deepened the ironmaker's understanding of his plant. However, recent Japanese experience during the excavation of a quenched furnace has created some doubts concerning the mechanisms at work inside the furnace, to the extent that many research workers are reappraising the theory behind the ironmaking process. Added to this is the fact that when existing furnace models are put to work they often yield disappointing results. To be specific, the implementation of blast furnace control schemes usually involves a furnace model which can be used to predict information concerning the next cast. Unfortunately, due to the fundamental complexity of the blast furnace, these predictions are often poor. Evidently, analytical models are less than satisfactory as on-line blast furnace prediction tools because of the fundamental difficulty involved in matching a detailed and intricate mathematical model to a vastly complicated and inherently time-varying physical process. However, if blast furnace automation is to proceed, then some means must be found to formulate a model which can reliably and consistently predict the quality of the hot metal output of the furnace. A possible way of doing this is to use prediction methods to construct a predictive model of furnace behaviour.

It is known, however, that fixed coefficient predictor models give poor results when applied to blast furnace predictions. The main reason for this is that the dynamical characteristics of a blast furnace change significantly over time. Some of these variations are due to changes in the nature and composition of the material which is input to a furnace. However, a further source are the changes within the furnace which take place during the ironmaking process.

The above reasoning suggests that a self-tuning predictor may be an appropriate technique for forecasting the behaviour of a blast furnace. The specific aim is to produce an operator's aid which will predict the quality of hot metal produced from the furnace. The rationale for using a self-tuning predictor is that the time-varying parameter tracking features of a recursive estimator can follow changes in the furnace dynamics in a manner which physiochemical models cannot do. A statistical predictor is also appropriate because our *only* requirement is to forecast the future values of furnace hot metal product. There is no requirement to reproduce or represent the exact dynamic behaviour of the ironmaking process.

To understand the predictor application to a blast furnace it is useful to review the basic operation of a blast furnace seen as a system for making

Figure 10.10
System
representation of
a blast furnace.

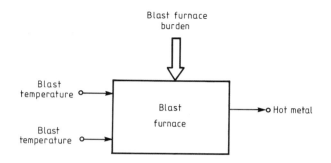

iron. To this end, Figure 10.10 shows a block diagram representation of a blast furnace. The primary input is the burden, comprising a mixture of various minerals with the main components being iron oxide, limestone and other material required in the chemical reduction of iron oxide to iron. The iron oxide is converted by chemical reactions in the furnace with the aid of additional control inputs from the temperature and humidity of the fuel blown into the base of the furnace. During the operation of the furnace, hot iron accumulates at the base of the furnace. At periodic intervals the hot iron is drawn off or 'cast'. These cast intervals (usually of the order of two hours) fix the sampling interval of the process, since it is only at these times that the hot metal quality can be measured. A number of quality criteria are applied, but all hinge upon the content of hot metal. In particular the following items are important:

(i) the percentage iron content;

(ii) the percentage silicon content; and

(iii) the percentage carbon content.

The requirement, therefore, is for a self-tuning predictor which is capable of predicting these three output variables on the basis of a predetermined two-hour sampling interval. As indicated above, the blast furnace is not susceptible to simple analytical modelling and hence it is difficult to select a model order for the predictor on the basis of physical considerations. The procedure adopted for model order detection was therefore to apply the predictor with increasing model orders until the resulting prediction error obtained no longer decreased significantly when the predictor order was further increased. In the experiment described here, a model order $n_a = 3$, $n_c = 3$ was found adequate.

   In fact, the experiment described here is a feasibility study in which the self-tuning predictor was applied to a complex blast furnace computer simulation. The aim was to demonstrate that a self-tuning predictor was a suitable technique to be considered for furnace forecasting. The results shown here are a small sample of the tests actually conducted during the feasibility study. The first result illustrates the tuning-in properties of the self-tuning predictor when predicting the two output variables: the percentages of carbon and silicon. The hot metal from a cast is about 95% iron, with additional small percentages of carbon and silicon. The percentage silicon in particular is used as an indicator of hot metal quality. Figure 10.11 shows typical tuning runs for %C and %Si predictions one cast ahead. The fast convergence of these predictions (within 10 casts) is representative of the performance of self-tuning predictors.

The second result illustrates the tracking behaviour of the self-tuning predictor when the blast temperature is increased by 30% at time step 400. Figure 10.12 shows the corresponding %C and %Si predictions for this situation. Note that in this case the prediction period has been reduced to eight minutes. This is possible on a computer-simulated blast furnace and was necessary in this case in order to demonstrate the behaviour of the self-tuner when more frequent hot metal data was supplied. The recursive estimator in this case was given an adaptive capability by employing a random walk (see Chapter 4) which was applied to the covariance matrix when the step in blast temperature occurred.

As noted above, the results described in this section are just a small subset of results from a feasibility study which was undertaken in order to demonstrate to steel industry managers that self-tuning predictors could perform well as forecasting tools. In particular, the results shown here are predictions obtained by treating the %C and %Si separately as the outputs of univariate ARMA processes. A stage of refinement beyond this is to treat the hot metal composition as a multivariable ARMA process. In particular, the process would have %Fe, %C and %Si as the three output variables. Beyond this it would also be reasonable to include information on blast temperature and humidity as measurable auxiliary variables, as described in Section 10.4.2.

10.6 SUMMARY

This chapter has described the basic techniques for self-tuning prediction. Specific points covered are:

- Outline of MV prediction theory
- Extensions to include additional disturbances and measurements
- Self-tuning prediction algorithms
- A typical application involving the prediction of the output of a complex industrial process plant.

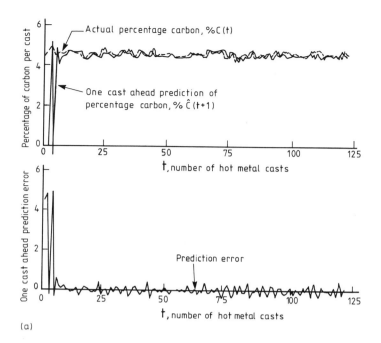

(a)

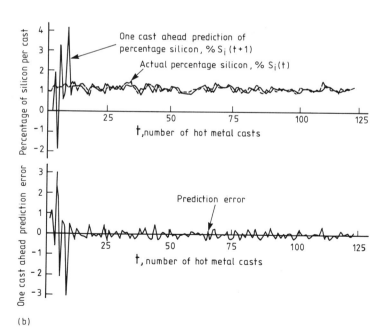

(b)

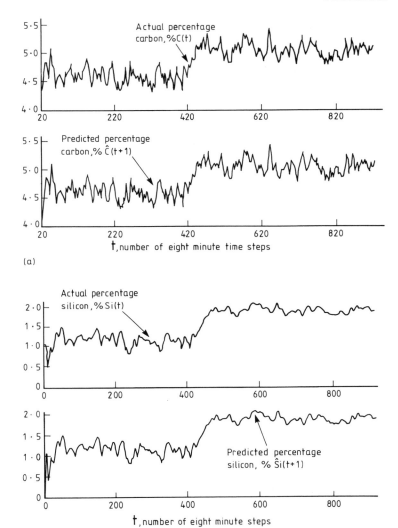

Figure 10.12 Illustrating the retuning performance of an implicit predictor of (a) %C and (b) %Si in response to a step change in blast temperature (the blast temperature has been raised by 30% at the 400th cast).

10.7   PROBLEMS

Problem 10.1

Consider the discrete time ARMA(1,1) process

$$Ay(t) = Ce(t)$$

in the usual notation.

(a) Show that the optimal k-step predictor $\hat{y}(t+k|t)$ can be expressed in terms of the prediction error $\tilde{y}(t)=y(t)-\hat{y}(t|t-k)$ filtered by a rational discrete time transfer function. Hence write down the implicit self-tuning form of the predictor.

(b) Show that the optimal prediction problem can be interpreted as a MV regulation problem.

(c) Determine the two-step predictor for the process

$$y(t) = \frac{1-0.4z^{-1}}{1-0.9z^{-1}} e(t)$$

Problem 10.2

Given the polynomials A,C, denote by (F_k,G_k) and (F'_k,G'_k) the polynomial pairs satisfying respectively the equations

$$C = AF+z^{-k}G$$

$$C = \Delta AF'+z^{-k}G'$$

where $\deg F = k-1 = \deg F'$.
  Show that

$$\Delta G_k = (G'_k - G'_{k-1})$$

and hence show that the alternative approaches to incremental prediction discussed in Section 10.3.1 are equivalent.

Problem 10.3

An industrial sensor is designed to measure temperature on the surface of an object by repeatedly scanning across the surface. The measurements are made at equally spaced intervals during the scan such that the variable $y(m,n)$ indicates the temperature at the mth sampling interval on the nth scan.
  The data can be modelled by the difference equation

$$y(m,n)+a_iy(m-1,n) = e(m,n) + c_1e(m-1,n) + c_2e(m,n-1)$$

where $e(m,n)$ is a bivariate white noise sequence such that

$$E\{e(m,n)e(m-i,n-j)\} \quad \begin{array}{ll} = \sigma_e^2 & j = i = 0 \\ = 0 & \text{otherwise} \end{array}$$

$$Ee(m,n) = 0$$

Obtain the one step ahead predictor $\hat{y}(m+1,n)$ for $y(m,n)$.

Assume that data up to position m is available on the current (nth) scan and that data for all m is available on previous scans (e.g. scan $n-1$, $n-2$ etc.).

Hint: Cast the data model into operator form using the shift operator notation

$$w^{-1}y(m,n) = y(m-1,n)$$

$$z^{-1}y(m,n) = y(m,n-1)$$

and obtain a model in operator form

$$A(w^{-1})y(m,n) = C(z^{-1},w^{-1})e(m,n)$$

Then solve an identity of the form

$$C(z^{-1},w^{-1}) = A(w^{-1}) + w^{-1}G(z^{-1},w^{-1})$$

to obtain the predictor.

Alternative Hint: See Part 4 which contains a whole chapter on two-dimensional predictors.

Problem 10.4

A batch chemical reactor has a product yield $y(t)$ which increases linearly with time as the reaction proceeds but is also subject to an additional stochastic fluctuation. Suppose that the product yield can be modelled as:

$$y(t) = \frac{(1+c_1 z^{-1})}{1+a_1 z^{-1}+a_2 z^{-2}}\, e(t) + \frac{d_0+d_1 t}{1+a_1 z^{-1}+a_2 z^{-2}}$$

where the first term on the right-hand side denotes the stochastic fluctuations in $y(t)$ and the second term denotes the linearly increasing component of $y(t)$.

In order to give the process operators a forecast of the process behaviour, a two step ahead predictor of $y(t)$ is required. This will be displayed on the operator's console as a 'preview sensor' which can be used to judge when the process is to be stopped.

(a) Show how you would construct a two-step ahead predictor for $y(t)$.

(b) Show how the predictor would be implemented as an explicit self-tuning algorithm.

(c) Additional important operator information is the average rate of increase of yield during a run and the variance of the yield about the linearly increasing value. Show how the self-tuning predictor could be used to supply this data.

Problem 10.5

It is required to predict the output $y(t+r)$ of the CARMA model

$$Ay(t) = Bu(t-k)+Ce(t)$$

in the usual notation, based on data $y(t)$, $y(t-1)$, ..., $u(t)$, $u(t-1)$, ..., . Derive the MMSE linear predictor in the cases $r>k$ and $r\leqslant k$ when the input is given by:

(a)   a deterministic function of time

(b) $F_1u(t)+G_1y(t) = 0$ (linear regulator)

(c) $A_1u(t) = C_1w(t)$ (ARMA model)

where $E[w(i)e(j)] = 0$, $i, j = 1, 2, ..., $.

Problem 10.6

An output predictor $\hat{y}(t+r|t)$ for the standard ARMA model uses a data window, thus

$$\hat{y}(t+r|t) = \lambda_0 y(t)+\lambda_1 y(t-1) + ... + \lambda_N y(t-N)$$

Show that the MMSE predictor of this form selects $\lambda_0, ..., \lambda_N$ to minimize the variance of

$$\frac{(\wedge C-G)}{A} e(t)$$

in the usual notation, where $\wedge = \lambda_0+\lambda_1 z^{-1} + ... + \lambda_N z^{-N}$.
    Calculate the predictor and prediction error variance in the cases:

(a) $A = 1, r > n_c$

(b) $C = 1, A = 1-az^{-1}$

Problem 10.7

This is not really a problem but an assignment in applying a self-tuning predictor to some real economic data. If your predictions are accurate you will have realized the financier's dream – a reliable mechanism for predicting stock market performance! The data tabulated is the monthly Financial Times Index of ordinary industrial shares for the years 1931 to 1970. The tables are in six blocks, each covering five years.
    The twelve numbers in each column are the share prices for the months of that year. The first number in a column is the price in January, the next is for February and so on down the column.
    When using this data it will be interesting for you to recall some historical events which affected stock market performance. These were:

(i) the Great Depression of the 1930s;

(ii) the Second World War;

(iii) the devaluations of sterling in September 1949 and November 1967.

Table P10.7

1931	1932	1933	1934	1935
70.44	60.4	67.6	84.95	89.72
67.6	59.57	67.57	84.6	94.8
68.22	62.6	66.47	88.26	92.64
66.04	58.72	66.85	83.95	93.15
57.47	54.77	69.5	83.65	97.88
57.75	52.47	73.22	86.74	100
62.32	58.74	70.32	86.77	99.1
57.55	61.82	79.55	86.9	100.8
56.57	64.95	81.14	88.95	99.3
65.28	65.08	83.75	90.15	98.5
67.37	65.82	82.7	92.58	103.6
60.02	65.4	82.2	93.07	105.4

1936	1937	1938	1939	1940
108.5	123.8	95.3	77.7	75.5
112.4	121.5	90.8	81.7	76.9
111.5	118.3	86.1	84.2	78.9
112.6	115.6	84.1	79.2	76.7
110.5	114.4	85.3	81.2	70.1
110.0	111.3	82.7	80.9	57.3
113.3	110.4	87.1	78.8	57.8
116.3	113.2	84.2	78.5	61.4
118.7	108.3	80.4	70.6	63.5
122.2	103.8	82.2	71.4	65.8
123.3	99.8	83.5	74.5	69.5
121.9	97.0	79.3	74.3	69.8

1941	1942	1943	1944	1945
71.9	80.7	86.2	103.6	113.4
69.8	78.1	96.7	103.9	113.6
67.8	75.9	97.1	103.2	115.4
67.2	78.2	98.1	105.3	116.9
68.3	77.6	99.3	108.4	114.4
71.3	79.5	98.8	112.1	113.4
74.3	80.2	100.9	114.3	116.0
77.4	82.4	104.6	114.2	111.3
78.8	85.1	105.3	109.6	113.6
78.4	88.6	104.5	109.8	114.9
80.6	83.1	100.7	112.2	116.2
80.1	92.8	102.3	112.0	113.2

Table P10.7 (*cont.*)

1946	1947	1948	1949	1950
115.4	136.5	125.3	122.2	104.6
115.7	129.2	115.7	120.7	105.8
114.6	129.5	113.5	113.8	104.6
117.4	131.5	119.0	114.8	106.6
124.7	135.4	120.7	114.1	108.7
127.5	135.6	115.4	103.6	114.0
126.3	130.7	111.1	103.6	111.3
127.0	116.7	113.5	102.7	113.0
124.6	112.4	115.7	106.2	115.8
120.3	113.6	118.0	103.4	117.0
128.5	116.8	121.0	101.7	117.2
133.6	122.8	120.4	105.1	114.7

1951	1952	1953	1954	1955
118.6	117.2	117.5	133.0	191.1
122.1	112.9	122.3	137.4	190.2
120.7	109.0	124.3	136.6	180.7
127.0	112.3	121.3	144.2	185.2
136.1	109.6	116.4	149.0	189.7
138.0	104.3	117.4	152.3	211.8
134.2	103.7	119.9	158.8	219.1
131.9	113.6	123.3	166.2	201.9
134.4	115.7	124.8	171.2	194.7
136.4	114.5	127.9	176.3	187.3
128.4	114.6	130.0	179.3	189.6
122.5	155.9	128.5	179.8	197.2

1956	1957	1958	1959	1960
193.1	183.2	164.0	219.1	331.1
180.3	185.9	159.3	217.0	322.5
175.8	187.1	161.5	218.7	317.8
187.7	197.7	168.9	224.7	315.4
188.1	204.5	158.0	233.4	308.9
177.7	203.3	173.6	237.0	318.0
179.4	204.5	176.0	236.7	313.1
182.7	199.2	184.5	250.4	325.7
181.2	190.2	193.0	255.1	331.1
178.1	170.6	205.3	285.0	325.9
168.9	166.6	211.8	302.1	312.1
174.0	166.7	216.0	318.5	300.6

Table P10.7 (*cont.*)

1961	1962	1963	1964	1965
312.2	305.4	288.4	336.7	338.5
321.3	302.5	294.7	330.8	348.0
337.7	285.7	304.0	337.8	335.3
353.6	301.1	308.5	346.5	339.5
360.2	292.2	311.5	343.2	350.0
334.0	264.2	308.0	343.5	331.3
311.3	264.5	313.1	354.6	317.6
312.7	276.0	321.4	361.0	322.6
307.3	274.6	326.4	367.2	329.7
290.8	274.6	338.7	361.8	345.1
298.6	269.4	342.4	345.0	350.3
298.0	286.7	344.9	332.6	341.5

1966	1967	1968	1969	1970
345.4	317.4	401.2	509.9	414.0
355.0	315.9	412.9	485.4	402.7
347.9	322.1	413.3	470.5	393.2
345.5	338.0	452.7	465.6	384.6
363.3	343.2	464.2	431.9	343.1
363.4	347.8	466.4	398.9	355.5
346.4	352.7	481.3	378.1	340.7
308.0	355.7	491.4	376.3	340.2
305.6	373.1	509.1	384.6	347.5
300.4	393.1	488.2	376.2	369.0
293.6	407.9	483.7	379.6	335.4
305.6	392.5	493.0	392.3	331.4

Problem 10.8

Determine the two-step predictor for the process

$$y(t) = \frac{1 - 0.4z^{-1}}{1 - 0.9z^{-1}} e(t)$$

Write down the implicit self-tuning one-step prediction algorithm for the process detailing for each time t the data to be stored and the sequence of steps to be taken.

The predictor parameters are estimated using RLS estimation and local convergence analysis of the algorithm is carried out using the ODE approach. If $\boldsymbol{\theta}$ is a convergence point of the estimated parameter vector, derive the rational transfer function H as a function of θ so that

$$\epsilon(t) = Hy(t)$$

Hence show that the first-order stationarity conditions at the convergence point are a finite set of equations, each of the form

$$E[Pe(t) . Qe(t)] = 0$$

where P,Q are rational transfer functions dependent on θ.

10.8   NOTES AND REFERENCES

The MV predictor is fully discussed in:

Astrom, K.J.
  Introduction to Stochastic Control Theory, Academic Press, 1970.

The idea of a self-tuning predictor was introduced by Bjorn Wittenmark in:

Wittenmark, B.
  A self-tuning predictor, *IEEE Trans.* **AC-19**(6), 848–51, 1974.

The self-tuning predictor idea can be extended to the multivariate case. See for example:

Tanttu, J.T.
  A self-tuning predictor for a class of multivariable stochastic processes, *Int. J. Control*, **32**(2), 359–70, 1980.

The analysis and modelling of blast furnaces is a very difficult problem, with many specialists devoting their entire lives to its study. The model used in the application reviewed here is that described in:

Fielden, C.J. and Wood, B.I.
  A dynamical digital simulation of the blast furnace, *J. Iron and Steel Inst.*, July, 650–8, 1968.

11 Self-tuning Filters

11.1 OUTLINE AND LEARNING OBJECTIVES

In this chapter we consider the problems and techniques associated with self-tuning filtering. The problem of signal filtering occurs in many (if not most) areas of information engineering. The filtering algorithms considered here cover those which have a useful and widely applied self-tuning form. For example, filters are frequently used to remove noise from measured data to provide a smoothed measurement. Smoothing algorithms of this form are discussed in Section 11.3. In addition, self-tuning filters are widely used for channel equalization in telecommunications systems. Section 11.4 deals with this form of filter. A wide range of signal processing applications call for the removal of measurable but unwanted signal components. This requirement is met by self-tuning interference cancelling filters. These are described in Section 11.5.

The learning objective is to gain an appreciation of the algorithms used in filtering applications and to understand their self-tuning implementation and use.

11.2 BACKGROUND

Almost all data recorded from a real process or transmitted through a communication system is subject to distortion. The distortion can take a number of forms, the following two being the most common:

(a) *Addition of an unwanted signal component*

This situation is illustrated in Figure 11.1, in which the signal measurement or transmission mechanism allows an unwanted signal component (noise $n(t)$) to be added to the true signal $s(t)$. Self-tuning filters are designed to process the measured signal $y(t)$ and remove the influence of the noise signal component. Two types of filter are applicable here depending upon the nature of the unwanted signal. When the unwanted signal is an unknown noise

Figure 11.1
A signal
measurement or
transmission
mechanism in
which a signal
source $s(t)$ is
corrupted by an
unwanted signal
$n(t)$.

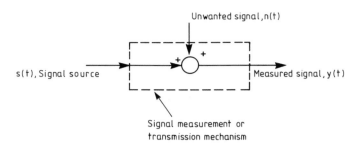

source, then a filter or smoothing algorithm is applicable. This type of algorithm is discussed in Section 11.3. On the other hand, Section 11.5 discusses self-tuning filters which perform this function when the unwanted signal is from a *known* source. This procedure is known as *interference cancellation filtering*.

(b)   *Signal distortion by an unwanted system transfer function*

This situation is illustrated in Figure 11.2, in which the signal measurement/transmission process is nonideal. In particular, the signal source is modified by the action of a transfer function H_c which characterizes the measurement/signal transmission system. The aim of the self-tuning signal processor in this case is to remove the effect of the transfer function H_c. Section 11.4 discusses this problem under the heading of *channel equalization*.

In order to motivate the discussion of self-tuning filters it is useful to understand the relationship between corresponding control and signal processing strategies. Consider first the interference cancelling filter mentioned above and described in Section 11.5. The aim is to remove from a signal an unwanted signal (Figure 11.1). This task is analogous to the problem of feedforward disturbance rejection in control engineering. Indeed, the methods used for these tasks are rather similar in algorithmic terms.

In the same vein, the channel equalizing filter mentioned above and described in Section 11.4 is analogous to the procedure used in control systems whereby the system transfer function is cancelled by an inverse model and

Figure 11.2
A signal
measurement or
transmission
mechanism in
which a signal is
distorted by the
channel transfer
function H_c.

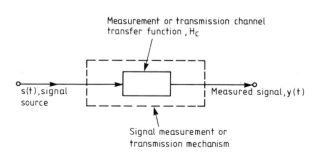

replaced by a 'desired transfer function'. Again, many of the algorithms used are the same in the channel equalization and control situations. At a deeper technical level, the smoothing algorithms of Section 11.3 were originally developed from the same basis as the prediction algorithms discussed in Chapter 10. Moreover, as has been indicated in Chapter 10, the ideas of signal prediction and process regulation are intimately linked. Finally, and at a very practical level, filters of the kind discussed in Section 11.3 are routinely used to condition the output signals from control system sensors prior to using those signals in a feedback system.

11.3 SMOOTHING AND FILTERING

11.3.1 Problem description

Consider the situation where we need to obtain filtered or smoothed estimates of a signal which is corrupted by unknown noise. In particular, we suppose that the measured data is $y(t)$ given by

$$y(t) = s(t) + v(t) \tag{11.1}$$

where, as shown in Figure 11.3, $s(t)$ is the desired signal and $v(t)$ is an independent unmeasurable noise source. The noise source may be caused in a control system by sensor or measurement noise, while in a signal processing environment a common source is noise associated with a signal transmission channel. In order to extract $s(t)$ from $y(t)$ we need in general to know the covariance structure of the corrupting noise. In this particular case we will assume that $v(t)$ is white noise.

The terms *filtering* and *smoothing* may need some explanation. In the current context a fixed k-lag smoother is an algorithm which uses the data set $y(t')$ for $-\infty < t' \leq t+k$ (Figure 11.4(a)) in order to construct an estimate of $s(t)$. Likewise a filter algorithm uses the data set $y(t')$ for $-\infty < t' \leq t$ in order to estimate $s(t)$ (Figure 11.4(b)). A filter can in these terms be interpreted as a

Figure 11.3 A measurable signal $y(t)$, represented as the sum of a desired signal $s(t)$ and an unknown noise process $v(t)$.

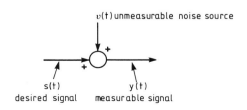

Figure 11.4
Illustrating the
distinction
between (a)
smoothing and
(b) filtering.

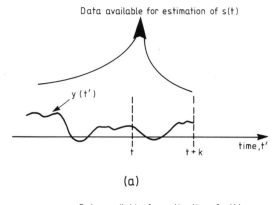

(a)

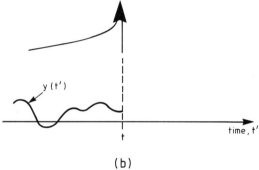

(b)

zero-lag smoother. In addition, a k-lag smoother has a complementary relationship with a k-step predictor.

Section 11.3.2 describes fixed coefficient smoother design for the case of a fixed lag k, a white noise disturbance $v(t)$ and a signal generating mechanism defined by

$$s(t) = \frac{C}{A} e(t) \tag{11.2}$$

where C, A are polynomials of degree n in the backward shift operator and $e(t)$ is zero mean white noise.

The self-tuning versions of these algorithms are given in Section 11.3.3.

11.3.2 Fixed-lag smoothing

The k-lag smoothed estimate of $s(t)$ that is 'best' in the minimum mean square sense is given by

$$\hat{s}(t|t+k) = E\{s(t)|y(0), \ldots, y(t+k)\} \tag{11.3}$$

(see Appendix and Section 10.2).

Now suppose that an ARMA representation of equation (11.1) exists and takes the form

$$y(t) = \frac{D}{A}\,\tilde{e}(t) \tag{11.4}$$

where D is a polynomial of degree n in the shift operator, $\tilde{e}(t)$ is a white noise process with variance $\sigma_{\tilde{e}}^2$ and A, C, D are related by the power spectral relation:

$$\sigma_{\tilde{e}}^2 DD^* = \sigma_v^2 AA^* + \sigma_e^2 CC^* \tag{11.5}$$

The innovations are uncorrelated and equation (11.3) can be written as

$$\hat{s}(t|t+k) = E\{y(t) - v(t)|\tilde{e}(0), \ldots, \tilde{e}(t+k)\}$$

$$= y(t) - \sum_{i=0}^{t+k} E\{v(t)|\tilde{e}(i)\} \tag{11.6}$$

Using

$$E\{v(t)|\tilde{e}(i)\} = E[v(t)\tilde{e}(i)]\,\tilde{e}(i)/\sigma_{\tilde{e}}^2 \tag{11.7}$$

it follows that

$$\hat{s}(t|t+k) = y(t) - \sum_{i=0}^{t+k} E\{v(t)\tilde{e}(i)\}\tilde{e}(i)/\sigma_{\tilde{e}}^2 \tag{11.8}$$

From equations (11.4), (11.2) and (11.1), the noise sources are related by

$$\tilde{e}(t) = \frac{A}{D}\,v(t) + \frac{C}{D}\,e(t) \tag{11.9}$$

so that

$$E\{v(t)\tilde{e}(i)\} = h_{i-t}\,\sigma_v^2 \qquad i \geq t \tag{11.10}$$
$$= 0 \qquad i < t$$

where

$$\frac{A}{D} = H = \sum_{j=0}^{\infty} h_j z^{-j} \qquad (11.11)$$

and $\{h_j\}$ is the impulse response sequence of the inverse ARMA model. Note that A/D is stable and can therefore be expanded in this way.

The *k-lag smoother* is, therefore, from equation (11.8)

$$\hat{s}(t|t+k) = y(t) - (\sigma_v^2/\sigma_{\tilde{e}}^2) \sum_{i=0}^{k} h_i\, \tilde{e}(t+i) \qquad (11.12)$$

Equation (11.12) is the smoother for $s(t)$ based upon a fixed lag of k steps. As is clear from the equation it requires a knowledge of $\{h_j\}$ from the inverse ARMA model, together with the ratio $\sigma_v^2/\sigma_{\tilde{e}}^2$. From the factorization (11.5) we have

$$\sigma_{\tilde{e}}^2 (1 + d_1 z + \ldots + d_n z^n)(z^n + d_1 z^{n-1} + \ldots + d_n)$$

$$= \sigma_{\tilde{e}}^2 (c_1 z + \ldots + c_n z^n)(c_1 z^{n-1} + \ldots + c_n)$$

$$+ \sigma_v^2 (1 + a_1 z + \ldots + a_n z^n)(z^n + a_1 z^{n-1} + \ldots + a_n) \qquad (11.13)$$

and setting z equal to zero this yields

$$\frac{\sigma_v^2}{\sigma_{\tilde{e}}^2} = \frac{d_n}{a_n} \qquad (11.14)$$

Note that $C(0)$ is assumed zero in (11.13) so that the signal $s(t)$ is assumed to have no pure white noise component. This is without loss of generality as any such component can be included in $v(t)$.

From (11.1) and (11.11), the smoothing error is given by

$$s(t) - \hat{s}(t|t+k) = \frac{\sigma_v^2}{\sigma_{\tilde{e}}^2} \sum_{i=0}^{k} h_i\, \tilde{e}(t+i) - v(t) \qquad (11.15)$$

Using (11.10), the error variance is

$$E[s(t) - \hat{s}(t|t+k)]^2 = \sigma_v^2 \left[1 - \left(\frac{\sigma_v^2}{\sigma_{\tilde{e}}^2}\right) \sum_{i=0}^{k} h_i^2 \right] \qquad (11.16)$$

For $k=0$, we have the *filtered estimate* $\hat{s}(t|t)$ with variance

$$E[s(t) - \hat{s}(t|t)]^2 = \sigma_v^2 (1 - \sigma_v^2/\sigma_{\tilde{e}}^2) \tag{11.17}$$

As k increases, the variance (11.16) decreases monotonically, asymptotically achieving the value

$$\lim_{k \to \infty} E[s(t) - \hat{s}(t|t+k)]^2 = \sigma_v^2 \left[1 - E\left\{ \frac{A}{D} v(t) \right\}^2 \bigg/ \sigma_{\tilde{e}}^2 \right] \tag{11.18}$$

Example 11.1

Consider the process

$$s(t) = \frac{1}{1 - 0.95z^{-1}} e(t-1), \qquad e(t) \sim N(0, \sigma_e^2)$$

$$y(t) = s(t) + v(t), \qquad v(t) \sim N(0, \sigma_v^2)$$

where $\sigma_e^2 = 1$, $\sigma_v^2 = 10$.

The ARMA model takes the form

$$y(t) = \frac{1 + dz^{-1}}{1 - 0.95z^{-1}} \tilde{e}(t)$$

where d, $\sigma_{\tilde{e}}$ are found from (11.13) or, equivalently, by equating spectral density functions:

$$10(1.9025 - 1.9000 \cos\omega) + 1 = \sigma_{\tilde{e}}^2 (1 - d^2 + 2d\cos\omega)$$

For $\omega = 0$, π, respectively

$$\sigma_{\tilde{e}}^2 (1 + d)^2 = 1.025$$

$$\sigma_{\tilde{e}}^2 (1 - d)^2 = 39.025$$

Dividing to eliminate $\sigma_{\tilde{e}}$ and taking square roots leads to the reciprocal values

$$d = -0.721 \quad or \quad -1.387$$

Taking the value for which $|d| < 1$ (so that H is stable) leads to

$$d = -0.721, \quad \sigma_{\tilde{e}} = 3.629$$

so that $d/a = 0.759$ and

$$h_i = -0.229(0.721)^{i-1} \quad (i \geqslant 1), h_o = 1$$

From (11.16)

$$E[s(t) - \hat{s}(t|t+k)]^2 = 1.577 + 0.829(0.721)^{2k} \quad (k \geqslant 0) \tag{11.19}$$

□

11.3.3 Self-tuning smoothing

The self-tuning implementation of fixed-lag smoothing algorithms hinges upon the fact that a recursive estimator applied to a process formed using (11.1) and (11.2) will estimate the coefficients of the corresponding composite ARMA model. These estimates can then be used to determine the smoother coefficients. Accordingly, the self-tuning smoother is an explicit algorithm which uses an RELS or AML method (see Chapter 3) to determine estimates of A and D and then directly synthesize (h_i) from the defining relation (11.11). The k-lag self-tuning smoother takes the form:

Algorithm: Explicit Self-tuning Smoother

At time step t:

(i) Estimate the polynomials $\hat{A}$, $\hat{D}$ using $y(t)$ and an RELS or AML recursion (see Chapter 3).

(ii) Determine $\{\hat{h}_i\}$ from

$$\hat{H}\hat{D} = \hat{A}$$

(iii) Calculate the smoothed signal as

$$\hat{s}(t-k|t) = y(t-k) - \frac{\hat{d}_n}{\hat{a}_n} \sum_{i=0}^{k} \hat{h}_i \tilde{e}(t-k+i) \tag{11.20}$$

(iv) Set $t \rightarrow t+1$ and return to step (1).   □

Note that the prediction error $\epsilon(t)$ is used as an estimate of the signal $\tilde{e}(t)$ and is supplied by the recursive estimator.

Example 11.1 (continued)

Figure 11.5(a) shows the behaviour of the estimates $\hat{a}$, $\hat{d}$ when RELS is used to self-tune in the filter case $k=0$. The initial conditions for this estimation experiment were $\mathbf{P}(0) = 200\mathbf{I}_2$, $\hat{\boldsymbol{\theta}}(0)=0$. No forgetting or covariance management were applied.

Figure 11.5 Self-tuning smoothing, showing (a) the parameter estimates $\hat{a}$, $\hat{d}$ for Example 11.1 and (b) the corresponding estimated smoother parameters.

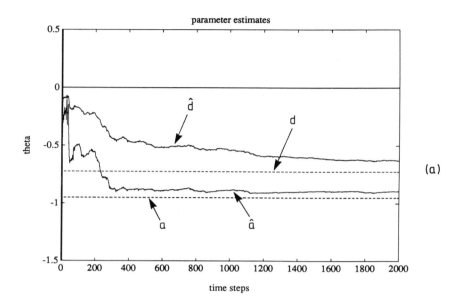

(a)

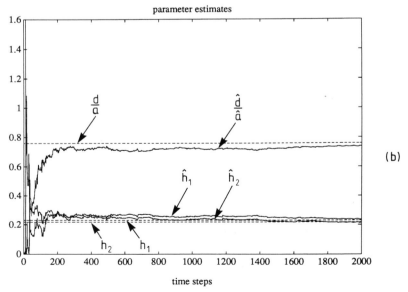

(b)

Table 11.1 Average loss per step ($N = 1000$)

k	Optimal	Self-tuning
0	2.406	2.452
1	2.008	2.105
2	1.801	1.795
∞	1.577	—

Table 11.1 gives the average loss per step

$$V_N = \frac{1}{N} \sum_{t=1}^{N} [s(t) - \hat{s}(t|t+k)]^2$$

incurred by the self-tuning smoother compared with the exact values calculated from equation (11.19).

Note that the estimates $\hat{a}$, $\hat{d}$ in Figure 11.5(a) converge slowly. They should be compared with the corresponding estimates of $\hat{d}/\hat{a}$, $\hat{h}_1$ and $\hat{h}_2$ shown in Figure 11.5(b). As in other self-tuning algorithms, these key quantities converge rather quickly to the neighbourhood of the correct values, despite the slow convergence of the individual system parameter estimates.   □

11.4 CHANNEL EQUALIZATION

In many industrial measurement systems, and especially in telecommunication channels, the transmitted signal ($s(t)$ in Figure 11.2) is significantly modified by the measurement or communications channel transfer function H_c. This is a particular problem if the channel is dispersive. A dispersive system is one in which the delay through the system is a function of frequency. In such cases transmitted data samples may overlap at the receiving end of the transmission channel, causing a distortion known as 'intersymbol interference'.

A solution to this problem is to filter the received signal $y(t)$ by the inverse of the channel transfer function as shown in Figure 11.6. However, changes in the channel transfer function characteristic make it desirable to use a self-adaptive filter. Figure 11.7 shows a possible form of channel equalizer. The scheme is based upon a direct estimation of the channel characteristics, whereby a recursive estimator is used to estimate the coefficients of the transfer function H_c using the model

$$y(t) = H_c s(t) \tag{11.21}$$

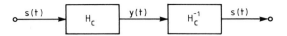

Figure 11.6 Illustrating the idea of channel equalization by using the inverse of the channel transfer function as a filter.

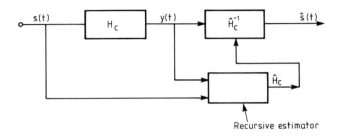

Figure 11.7 Self-tuning channel equalization based upon recursive estimation of H_c.

The form used for the channel transfer function may be autoregressive (AR), moving average (MA), or autoregressive, moving average (ARMA), depending upon the application. In signal processing applications an AR system is sometimes called an 'all-pole' transfer function. Likewise an MA system may be called a finite impulse response (FIR) transfer function. Other alternative descriptions include infinite impulse response (IIR), which may be used to describe AR or ARMA systems.

If the channel contains a significant known time delay k, then this can be incorporated into the estimation model thus:

$$y(t) = H_c s(t-k) \tag{11.22}$$

In this way the estimated transfer function will not include the signal propagation time delay. The estimated model $\hat{H}_c$ is then used in inverse form (see Figure 11.7) to produce the compensated signal $\hat{s}(t)$:

$$\hat{s}(t) = \hat{H}_c^{-1} y(t) \tag{11.23}$$

If the inverse model is an exact estimate of the channel transfer function, and the channel is noise-free then $\hat{s}(t)$ will be an exact replica of the signal. If a delay is incorporated into the model, as in equation (11.22), then $\hat{s}(t)$ will be an exact replica of the delayed signal $s(t-k)$.

Example 11.2

Consider a system in which a signal transmission channel is given by

$$y(t) = -a_1 y(t-1) + b_0 s(t-1) + e(t)$$

where $a_1 = -0.7$, $b_0 = 0.7$ and $e(t)$ is a zero mean white noise signal with variance 0.25. In order to obtain a channel equalizing filter, a square wave dither signal is applied to the input $s(t)$. The dither signal has unit amplitude and period 40 time steps. RLS is used to estimate the parameters a_1, b_0 and the compensated signal $\hat{s}(t)$ is obtained using

$$\hat{s}(t) = \hat{b}_0^{-1} (y(t) + \hat{a}_1 y(t-1))$$

Figure 11.8 shows the results of estimation and the corresponding signals. Note that the compensated signal $\hat{s}(t)$ converges after about 40 time steps. Actually, this is much slower than would be expected in practice. The estimator convergence rate has been artificially slowed down by using an initial covariance matrix of $\mathbf{P}(0) = 0.1\mathbf{I}_2$.   □

Difficulties arise with this inverse filtering approach when the transfer function H_c is nonminimum phase. The difficulty is exactly the same as that encountered in MV control applications in which the inverse system is unstable.

Figure 11.8 Self-tuning channel equalization for a minimum phase system (Example 11.2), showing (a) the system parameter estimates and (b) the channel input $s(t)$, the channel output $y(t)$ and the compensated output $\hat{s}(t)$.

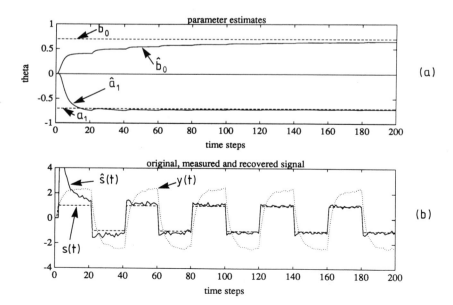

In channel equalization, the nonminimum phase problem manifests itself as follows. If H_c is described by an ARMA model

$$y(t) = \frac{B}{A} s(t) \qquad (11.24)$$

then the equalizing filter is given by

$$\hat{s}(t) = \frac{A}{B} y(t) \qquad (11.25)$$

However, if B is nonminimum phase, then the equalizing filter will be unstable. When the equalizing filter poles exactly match the channel zeroes this instability in H_c^{-1} does not matter since the unstable poles of H_c^{-1} are cancelled by nonminimum phase zeroes of H_c. In a practical situation, however, exact pole/zero cancellation will not occur. A further difficulty arises if *any* additional signal enters the system of Figure 11.7 and corrupts $y(t)$. Then the output $\hat{s}(t)$ will contain unstable components of output despite exact cancellation between $\hat{H}_c^{-1}$ and H_c.

Example 11.3

Consider a channel equalization problem in which

$$H_c = \frac{(b_0 + b_1 z^{-1})}{1 + a_1 z^{-1}} z^{-1}$$

and the parameter values are $a_1 = 0.5$, $b_0 = 1$, $b_1 = 1.2$. Note the numerator of H_c has a nonminimum phase zero at $z = -1.2$.

RLS is used to estimate the parameters of H_c and the equalizing filter is implemented as

$$\hat{s}(t) = \hat{b}_0^{-1} (y(t) + \hat{a}_1 y(t-1) - \hat{b}_1 \hat{s}(t-1))$$

Figure 11.9 shows the parameter estimates and signals associated with this experiment. A unit square wave of period 40 time steps was used as a dither signal $s(t)$ and a zero mean white noise of variance 0.3 corrupted the measured channel output $y(t)$. The choice $P(0) = 0.1 I_3$ was used in order to slow down the convergence of the parameter estimates. The unstable nature of the equalizing filter is clearly demonstrated as $\hat{b}_1$ becomes greater than unity at the 120th time step. From this point on, the oscillatory behaviour of $\hat{s}(t)$ increases such that at the 400th time step (not shown) the algorithm failed. □

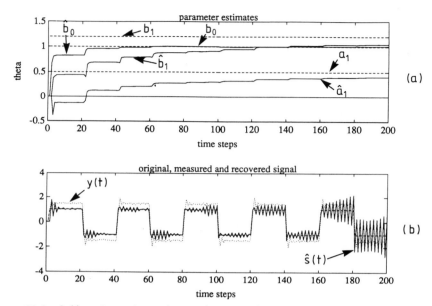

Figure 11.9 Self-tuning channel equalization for a nonminimum phase system (Example 11.3), showing (a) the system parameter estimates and (b) associated signals. Note that the recovered input signal $\hat{s}(t)$ increases in an unstable manner once the estimated $\hat{b}_1$ coefficient enters the nonminimum phase region.

As a consequence of the frequent instability of H_c, algorithms for channel equalization are much concerned with avoiding the problem of inverse unstable channel transfer functions. Several possibilities exist. These are:

(a) to estimate an approximate AR model of the system (11.24), e.g.

$$\hat{A}y(t) = \hat{b}_0 \, s(t-1) \tag{11.26}$$

The inverse of this is a MA filter and will be stable.

Example 11.3 (continued)

A RLS algorithm was used to estimate the parameters of an AR approximation (11.26) to the nonminimum phase H_c of Example 11.3. The order of $\hat{A}$ was set at 4 and the equalized output $\hat{s}(t)$ calculated according to

$$\hat{s}(t-1) = \hat{b}_0^{-1} \left(y(t) + \sum_{i=1}^{4} \hat{a}_i y(t-i) \right)$$

Figure 11.10(a) shows the corresponding parameter estimates and Figure 11.10(b) the associated signals. Note that the equalized output $\hat{s}(t)$ remains

Figure 11.10
Self-tuning
channel
equalization for
a nonminimum
phase system
(Example 11.3)
using an
approximate AR
model of order
$n_a = 4$. Note that
the recovered
signal $\hat{s}(t)$ is
stable.

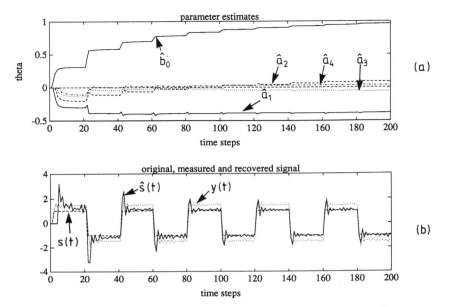

stable but, as expected, is not a faithful replica of $s(t)$. For the purpose of demonstration, the convergence of the RLS algorithm was slowed down in this example by setting $\mathbf{P}(0) = 0.1\mathbf{I}_5$. □

The disadvantage of using a MA approximation in channel equalization is that the model polynomial must often be of high order to adequately characterize the system transfer function. In communications systems, however, a high order AR model is often acceptable. In a self-tuning equalizer, the recursive estimation of the coefficients of the AR model would be performed using a 'fast' algorithm (see Chapter 5). Indeed, one of the reasons for the development of 'fast' estimators was the need for computationally efficient algorithms for recursive estimation of high order MA or AR models. High order in such applications may mean 50 or more coefficients.

(b) The second option is to estimate H_c in ARMA form:

$$y(t) = H_c s(t) = \frac{B}{A} s(t) \tag{11.27}$$

and to factor the B polynomial into the form

$$B = B^+ B^- \tag{11.28}$$

where B^+ has all its zeroes inside the unit disk and B^- has all its zeroes outside the unit disk.

The inverse filter to be applied is then formed from A and the inverse stable factors of B, e.g.

$$\hat{s}(t) = \frac{A}{B^+ B^-(1)} \, y(t) \tag{11.29}$$

and it is accepted that after convergence the equalized channel output $\hat{s}(t)$ will contain the nonminimum phase component of B, i.e.

$$\hat{s}(t) = \frac{B^-}{B^-(1)} \, s(t) \tag{11.30}$$

The term $B^-(1)$ in equations (11.29) and (11.30) denotes the value of B^- evaluated at $z=1$. It is used in order to ensure that the DC (zero frequency) gain of the equalizing filter is correct.

Example 11.3 (continued)

Figure 11.11 shows the signal behaviour when the equalizing filter for the system of Example 11.3 is synthesized using equation (11.29). In this case the equalized channel output is given by

$$\hat{s}(t) = (\hat{b}_0 + \hat{b}_1)^{-1}(y(t) + \hat{a}_1 y(t-1))$$

Note that the equalized signal $\hat{s}(t)$ in Figure 11.11 is not a faithful replica of $s(t)$. However, it remains stable even when the estimated B object becomes nonminimum phase (at time step $t=40$ in this test).   □

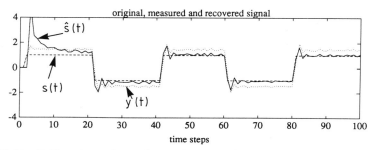

Figure 11.11 Self-tuning channel equalization for a nonminimum phase system (Example 11.3) using the method of removing the nonminimum phase factor.

(c) A further engineering modification is to use a delay k in the estimation equation (11.22) which is much larger than the channel delay time. An ARMA model is then estimated using the large delay. The resulting estimated ARMA model can be minimum phase even though the actual channel is nonminimum phase. The equalizing filter is then obtained by inverting the estimated (minimum phase) ARMA model. The use of delays larger than the true channel delay is explained further elsewhere (see the note on Widrow in Section 11.9 Notes and references). Overestimation of the process time delay has also been proposed in MV control to overcome nonminimum phase behaviour.

11.5 INTERFERENCE CANCELLING FILTERS

11.5.1 Basic problem

The basic idea in interference cancelling filters can be explained with reference to Figure 11.1. In this figure the signal $s(t)$ is corrupted by an unwanted signal $n(t)$. When the unwanted signal is an unknown noise source then the smoothing algorithms of Section 11.3 are appropriate. In this section we consider the situation in which the unwanted signal is from a *known* source. It is further assumed that it is possible to measure a signal $n_r(t)$ which is related to the noise signal $n(t)$. This situation arises frequently in signal processing but is particularly common in communications and acoustic applications. A possible situation drawn from acoustic noise signal processing is depicted in Figure 11.12. Here the signal $s(t)$ is a speaker's voice which we wish to transmit in uncorrupted form. Unfortunately, an acoustic noise source

Figure 11.12 Illustrating a possible acoustic application of the interference cancelling filter. The filter H_n is selected to remove the contribution $n(t)$ to the signal $y(t)$.

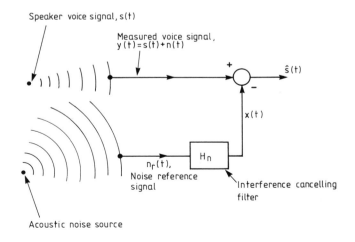

is present such that the microphone detects a combination of the speaker's voice and the acoustic noise.

Hypothetically, the speaker might be an aircraft pilot talking to his ground controller via a microphone in his headseat. The acoustic noise source $n(t)$ in such a case would be from the aircraft engine as detected by the pilot's headset microphone. The noise reference signal $n_r(t)$ could in this situation be provided by a further cockpit microphone situated in such a position that it detects the aircraft noise, but not the pilot's voice. The function of an interference cancelling filter in this case would be to allow the ground controller to hear the pilot's voice, uncorrupted by cockpit noise.

In cancelling the noise $n(t)$ the assumption is made that the interference $n(t)$ is related to the noise reference $n_r(t)$ by a causal transfer function H_n. Thus

$$n(t) = H_n n_r(t) \tag{11.31}$$

The basic procedure is to take the noise reference signal (in the aircraft case a measurement of the acoustic noise source alone), filter it with the transfer function H_n and subtract it from the measured signal $y(t)$.

Usually, the relationship between $n(t)$ and $n_r(t)$ is unknown so that the problem must be recast as follows. Assume that the signal $s(t)$ and the noise $n(t)$ are uncorrelated, the noise $n(t)$ and $n_r(t)$ are correlated and all signals are zero mean. The mean square value of the signal $\hat{s}(t)$ is, with reference to Figure 11.12, given by

$$E\{\hat{s}^2(t)\} = E\{(s(t)+n(t)-x(t))^2\} \tag{11.32}$$

Because the desired signal and the noise source are uncorrelated and zero mean, this can be rewritten as

$$E\{\hat{s}^2(t)\} = E\{s^2(t)\} + E\{(n(t)-x(t))^2\} \tag{11.33}$$

In attempting to cancel the interference noise source it is necessary to minimize the second term on the right-hand side of equation (11.33). Since the signal and noise are independent, this is equivalent to minimizing $E\{\hat{s}^2(t)\}$ itself. The signal $x(t)$ is the only adjustable object on the right-hand side of equation (11.33). Thus the mean square value of $\hat{s}(t)$ is minimized by selecting

$$x(t) - n(t) = 0$$

i.e.

$$H_n n_r(t) - n(t) = 0 \tag{11.34}$$

However, this is also the criterion which we would use to cancel the effect of $n(t)$ upon the signal $y(t)$ in Figure 11.12. Thus the optimum interference cancelling filter H_n is that filter which minimizes the mean square value of $\hat{s}(t)$.

The form which the noise cancelling filter H_n takes will, in general, depend upon the nature of the corrupting noise source. It is possible, however, to illustrate the general form of H_n under the assumption that $s(t)$, $n(t)$ and $n_r(t)$ are stationary zero mean processes. In this case it is possible to represent the situation given in Figure 11.12 by the block diagram given in Figure 11.13. In this figure the corrupting noise $n(t)$ and the noise reference signal are assumed to be generated by a zero mean white noise source $e(t)$ operating upon stable, minimum phase filters, as follows:

$$\left. \begin{aligned} n(t) &= \frac{C_1}{A_1}\, e(t) \\ n_r(t) &= \frac{C_2}{A_2}\, e(t) \end{aligned} \right\} \tag{11.35}$$

The process $e(t)$ is assumed to be independent of $s(t)$. It follows from substitution of equations (11.35) into equation (11.34) that, for exact cancellation of $n(t)$, the noise cancelling filter should be

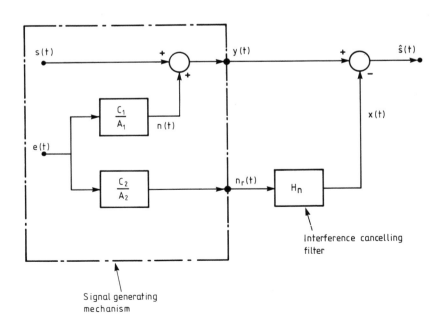

Figure 11.13
Theoretical representation of an interference cancelling filter.

$$H_n = \frac{C_1 A_2}{A_1 C_2} \tag{11.36}$$

Thus, under the stationarity assumption on the processes $s(t)$, $n(t)$, $n_r(t)$, the interference cancelling filter H_n is the stable minimum phase object given by equation (11.36).

Example 11.4

Consider the system in which

$$n(t) = -0.5n(t-1) + e(t) - 0.4e(t-1)$$

$$n_r(t) = +0.9n_r(t-1) + e(t) - 0.8e(t-1)$$

The interference cancelling filter is

$$H_n = \frac{(1-0.4z^{-1})(1-0.9z^{-1})}{(1+0.5z^{-1})(1-0.8z^{-1})} \qquad \square$$

11.5.2   Self-tuning interference cancellation

In order to formulate an interference cancelling algorithm based upon a least squares estimation procedure, it is necessary to restrict the form which H_n can take. In particular, in order to avoid biasing of the estimates caused by the correlated structure of the signal $s(t)$, it is necessary to require that H_n be a moving average (MA) process. With reference to Figure 11.13 and equation (11.36), this is achieved by requiring that $A_1 = 1$, $C_2 = 1$. The noise cancelling filter will then be moving average:

$$H_n = B \tag{11.37}$$

where $B = C_1 A_2$. The self-tuning interference cancelling filter then follows by writing the measured signal $y(t)$ in terms of the signal $n_r(t)$, thus:

$$y(t) = Bn_r(t) + \hat{s}(t) \tag{11.38}$$

However, the measured data can also be used to estimate the parameters of B by formulating the following least squares model

$$y(t) = Bn_r(t) + \hat{e}(t) \tag{11.39}$$

It follows from the assumption concerning the data and the discussion in the previous paragraph that the recursive least squares estimate of H_n, based upon the model (11.39), will lead to an error $\hat{e}(t)$ which will converge to the desired signal $s(t)$. The error $\hat{e}(t)$ in (11.39) then forms a suitable estimate $\hat{s}(t)$ of the original signal.

A noise cancelling filter B can therefore be designed by a least squares estimation procedure in which the residual provides the required estimate of $s(t)$. Thus the estimated parameters of B are in fact a by-product; the real result lies in the error $\hat{e}(t)$. The determination of $\hat{e}(t)$ can be put into a self-tuning format by applying a recursive least squares estimator to a model equation which includes a suitable structure for B. However, from equation (11.37) the structure of B is

$$B = b_0 + \ldots + b_{n_b} z^{-n_b}$$

The layout of a self-tuning noise cancelling filter is shown in Figure 11.14. The corresponding self-tuning algorithm is given below.

Algorithm: Self-tuning Interference Cancelling Filter

At the tth time step:

(i) Use a RLS estimator to update the parameter estimates in the model

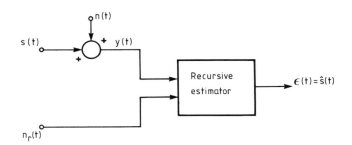

Figure 11.14
Illustrating the idea of a self-tuning interference cancelling filter.

$$y(t) = \hat{B} \, n_r(t) + \hat{e}(t) \tag{11.40}$$

(ii) Use the least square prediction error $\epsilon(t)$ as an estimate of the desired signal, thus

$$\hat{s}(t) = \epsilon(t) \tag{11.41}$$

(iii) Wait till time step t has elapsed, set t to $t+1$ and loop back to step (i)☐

This algorithm assumes that the relationship between $n_r(t)$ and $n(t)$ is a causal moving-average process. The moving average assumption is required in order to ensure that least squares can be used to determine $\epsilon(t)$. When H_n is in fact an ARMA process, then a high order moving-average process may be required to obtain an acceptable degree of approximation.

Example 11.5

Consider an interference cancelling problem in which (Figure 11.13) $s(t)$ is a square wave of unit amplitude and period 40 time steps. The signal $s(t)$ is corrupted by a noise source $n(t)$ generated from unit variance white noise $e(t)$ with $C_1 = 1-0.5z^{-1}$, $A_1 = 1$. Figures 11.15(a) and (b) show the signal $s(t)$ and the noise corrupted measured signal $y(t)$. Note that the original square wave is barely discernible in the measured $y(t)$.

The reference noise signal $n_r(t)$ is generated from $e(t)$ with $C_2 = 1$, $A_2 = 1-0.8z^{-1}$. A self-tuning interference cancelling filter was applied using the model

$$y(t) = (b_0+b_1z^{-1} + b_2z^{-2})n_r(t) + \hat{e}(t)$$

where, from the defined C_1, A_2, the true parameters are $b_0 = 1$, $b_1 = -1.3$, $b_2 = 0.4$.

Figure 11.15(c) shows the parameter estimates obtained during self-tuning and Figure 11.15(d) shows $\epsilon(t)$ which forms the estimate of the original signal $s(t)$.   ☐

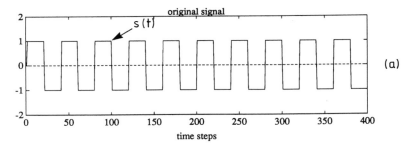

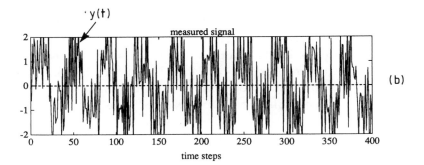

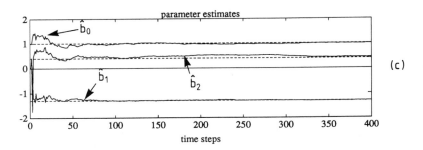

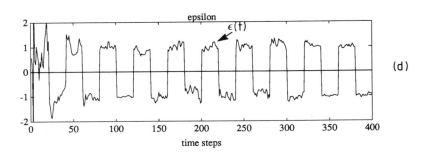

Figure 11.15 Self-tuning interference cancelling results (Example 11.5), showing (a) the original signal $s(t)$, (b) the measurable signal $y(t)$, (c) the estimated parameters of the noise cancelling filter and (d) the estimate of the original signal as embodied in the estimation error $\epsilon(t)$.

11.6   APPLICATION: SENSOR SIGNAL CONDITIONING

There are numerous applications of adaptive or self-tuning noise cancelling and equalizing filters. Many of these examples are in the telecommunications areas where the relatively high cost of adaptive signal processing hardware can be more easily absorbed than in other engineering spheres.

An excellent telecommunications example of an interference cancelling self-tuning filter is the modern form of adaptive echo cancellers used in long distance telephone circuits. The TRT CEN device described in Chapter 1 is an example of such a device which is used to suppress the echo of a speaker's voice which will be returned to him over a long haul telephone link.

The reader is referred to Chapter 1 for a description of this application (see also Section 11.9 Notes and references). In this section we consider applications of the interference cancelling filter which has been found of practical value in sensor signal processing in a control engineering environment.

Two applications are presented. The first is a simulated example of sensor drift compensation. The second describes the compensation of a sensor signal for supply interference.

11.6.1   Cancelling sensor drift

It frequently occurs that the measured output of a sensor is subject to a slow drift which is proportional to changes in measurable ambient conditions, such as temperature or pressure. Suppose we denote the true sensor output as $s(t)$, the measured sensor output $y(t)$ and the drift signal $n(t)$. If $n_r(t)$ is the measurable ambient variable, then we can write

$$n_r(t) = d\, n(t) \tag{11.42}$$

where d is the constant of proportionality. Using the ideas in Section 11.5, it is possible to compensate for the drift using a self-tuning algorithm based on the model

$$y(t) = b_0 n_r(t) + \hat{e}(t) \tag{11.43}$$

where the parameter b_0 should converge to d^{-1} and $\epsilon(t)$ to the drift free signal $s(t)$.

Example 11.6

Suppose the true sensor output $s(t)$ is a square wave of unit amplitude and period 40 time steps and the drift signal $n(t)$ is obtained from the integrated noise process

$$n(t) = \frac{e(t)}{1-z^{-1}}$$

where $e(t)$ is zero mean white noise, normally distributed with variance 0.3. The reference disturbance $n_r(t)$ is given by

$$n_r(t) = 0.5\, n(t)$$

Figure 11.16(a) shows the parameter estimate $\hat{b}_0$ for a self-tuning run in which $b_0 = 2$. Figure 11.16(b) compares the drift-corrupted signal and the recovered drift-free sensor signal. □

Figure 11.16
Self-tuning
interference
cancelling
applied to the
sensor drift
elimination
problem
(Example 11.6),
showing (a) the
estimated drift
gain, (b) the
drifting sensor
output $y(t)$ and
the drift
corrected sensor
signal $\epsilon(t)$.

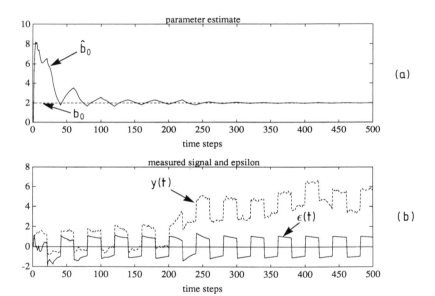

11.6.2 Cancelling supply interference

Consider the situation in which a sensor output is corrupted by 50 Hz interference from the main AC supply. This can be a particular problem in situations where the signal being sensed is very weak and high levels of amplification are required to bring it up to a reasonable voltage level. Typically, the signal picks up very small levels of supply interference which are a significant nuisance after amplification. Often such problems can be avoided by the screening of cables, circuitry or other hardware modifications. When this is not possible a signal processing algorithm is used to remove the

Figure 11.17 A
self-tuning notch
filter for the
cancellation of
50 Hz
interference.

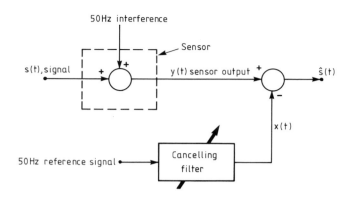

Figure 11.17 A self-tuning notch filter for the cancellation of 50 Hz interference.

unwanted 50 Hz frequency. A standard way of cancelling such single frequency interference is to use a so-called 'notch filter'. This is a filter which passes all frequencies except the unwanted frequency, in this case 50 Hz. In this section we consider a self-tuning alternative to notch filtering. Specifically, a self-tuning interference cancelling filter is proposed in which the interference signal is a 50 Hz sine-wave derived from the main AC supply. Figure 11.17 shows the layout for the cancelling filter, in which the reference signal is a 50 Hz sine wave. The task of the cancelling filter is to adapt the gain and phase of the reference sine wave to match that of the interference. This can be done by structuring the cancelling filter such that

$$x(t) = d_s \sin \omega t + d_c \cos \omega t \tag{11.44}$$

where d_s, d_c are unknown gains which must be adjusted in order to match the gain and phase of $x(t)$ to that of the interference signal.

Recall that t is the discrete time variable, so that the frequency ω in equation (11.44) is given by

$$\omega = 2\pi f / \tau_s \tag{11.45}$$

where τ_s is the sampling interval and f is 50 Hz.

The 50 Hz interference cancelling filter functions with the following model of the measured sensor output $y(t)$:

$$y(t) = d_s \sin \omega t + d_c \cos \omega t + \hat{e}(t) \tag{11.46}$$

The coefficients d_s, d_c are estimated recursively and when they converge to their correct values the prediction error $\epsilon(t)$ will converge to the signal $s(t)$.

The self-tuning 50 Hz interference cancelling filter is as follows:

Algorithm: Self-tuning 50 Hz Interference Cancelling Filter

At time step t:

(i) Use recursive least squares to update the parameter estimates of d_s and d_c in the model equation (11.46).

(ii) Use the prediction error $\epsilon(t)$ as an estimate of the desired signal $\hat{s}(t)$. Thus

$$\epsilon(t) = \hat{s}(t) \qquad (11.47)$$

(iii) Set $t \rightarrow t + 1$ and return to step (i).   □

It is useful to note that the modelling of periodic signals of known shape and frequency but unknown magnitude and phase is generally straightforward and widely used. The topic was discussed in Chapter 2 under disturbance modelling. The estimation and removal of periodic signals is also a common practice in prediction (see Chapter 10).

As a specific application of the 50 Hz interference removal filter, consider the situation in which strain measurements are being made on machines in a power station environment which involve significant 50 Hz radiation. Despite extensive cable and amplifier screening, the amplified signal contains a significant 50 Hz component (Figure 11.18(a)). A self-tuning 50 Hz noise canceller was applied, with the results shown in Figure 11.18(b). Note the rapid initial tuning-in of the filter and subsequent good interference cancellation performance. In this application a recursive least squares algorithm was used with a forgetting factor of 0.995. Lower forgetting factors can be used if greater tracking capability is required. However, reducing the forgetting factor also reduces the interference rejection properties of the filter.

11.7 SUMMARY

The chapter has discussed the topic of self-tuning filtering from three distinct viewpoints:

- Smoothing of signals corrupted by unknown noise
- Channel equalization of signals distorted by measurement or transmission circuits
- The cancellation of interference due to measurable signals

Figure 11.18
Results from a
self-tuning
50 Hz cancelling
filter applied to
a strain gauge
output signal:
(a) original
strain gauge
amplifier output;
(b) strain gauge
amplifier output
after interference
cancelling; and
(c) parameter
estimates for
cancelling filter.

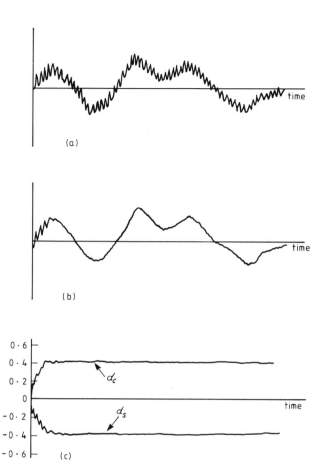

The final application illustrated the use of self-tuning interference cancellation
to implement notch filters in sensor signal processing.

11.8   PROBLEMS

Problems 11.1

Show that the stationary stochastic process

$$y(t) = \frac{C_1}{A_1} e_1(t) + \frac{C_2}{A_2} e_2(t)$$

where $e_1(t)$, $e_2(t)$ are independent white noise processes, can be expressed in the
standard form

$$Ay(t) = Ce(t)$$

Problem 11.2

The stable process

$$y(t) = \sum_{i=1}^{\infty} h_i\, u(t-i)$$

is modelled by

$$y(t) = Bu(t-k)$$

and the coefficients $(b_0, \ldots, b_{n_b})$ estimated by RLS. If the input is chosen as zero mean white noise, show that the only possible convergence point is $(h_k, h_{k+1}, \ldots, h_{k+n_b})$. Thus give an example of a nonminimum phase system leading to an inverse stable model.

Problem 11.3

What modifications must be made to the fixed-lag smoother of Section 11.3.2 in each of the following cases:

(a) $C(0) \neq 0$;

(b) $\{v(t)\}$ is not white;

(c) $v(t)$ correlated with $e(t-1)$?

Problem 11.4

Show that the filtered estimate $\hat{s}(t|t)$ satisfies a recursion of the form:

$$\hat{s}(t|t) = \sum_{i=1}^{n} [\alpha_i \hat{s}(t-i|t-1) + \beta_i y(t+1-i)]$$

Problem 11.5

Consider the restricted complexity smoother
$$\hat{s}(t|t+k) = G(z^{-1})y(t+k)$$

where G is a (possibly infinite) polynomial. Show that the variance of the smoothing error is quadratic in the coefficients of G and hence write down the set of linear equations for the coefficients that yield the minimum variance.

In Example 11.1, calculate the optimal two-lag smoother for $n_g = 0$ and the associated error variance.

11.9 NOTES AND REFERENCES

The Weiner smoothing problem is discussed in numerous books and articles. See for example:

Anderson, B.D.O. and Moore, J.
Optimal Filtering, Prentice Hall, 1979.

The innovations algorithms for smoothing are described in:

Kailath, T. and Frost, T.
An innovations approach to least squares estimation: part II – linear smoothing in additive white noise, *IEEE Trans.*, **AC-13**, 655–60, 1968.

The description given here follows that given for self-tuning smoothers in:

Hagander, P. and Wittenmark, B.
Self-tuning filters for fixed-lag smoothing, *IEEE Trans.*, **IT-23**, 377–84, 1977.

The idea of smoothing and filtering to remove sensor or channel noise can be extended to two-dimensional signals and data. Such extensions are useful in image enhancement. The techniques for two-dimensional self-tuning smoothing are dealt with in Chapter 12. The theoretical basis for two-dimensional self-tuning was first established in:

Wellstead, P.E. and Caldas-Pinto, J.R.
Self-tuning filters and predictors for two-dimensional systems, *Int. J. Control*, **42**(2), 457–78, 1985.

One of the earliest, if not the earliest, reported self-tuning filters is that reported in:

Gabor, D., Wilby, W.P.L. and Woodcock, R.
Universal non-linear filter, predictor and simulator which optimizes itself by a learning process, *Proc. IEE, part B*, **108**, 422–38, 1961.

The aim of this chapter has been to give a flavour of noise cancelling filter techniques. For more detail see one of the many excellent books on self-tuning signal processing and adaptive filters. Of particular interest is:

Widrow, B. and Stearn, S.D.
Adaptive Signal Processing, Prentice Hall, 1985.

This focuses upon the pioneering work of Widrow and his followers at Stanford University. It discusses all aspects of adaptive time domain filtering with especial emphasis upon equalization, interference cancelling and adaptive arrays.

Also:

Cowan, C.F.N. and Grant, P.M. (eds)
Adaptive Filters, Prentice Hall, 1985.
is a good account of modern adaptive filtering with a wide range of contributions from numerous distinguished authors. The book is particularly informative on the currently available technologies for self-tuning filtering. It also contains a chapter on adaptive filtering in telecommunications which gives a fairly detailed treatment of

telephone echo cancelling. This should be used to obtain background on the theory underlying the echo canceller mentioned in Section 11.6 and Chapter 1.

Other books recommended in the area include:

Bellanger, M.
  Digital Processing of Signals: Theory and Practice, Wiley, 1984.
Teichler, J.R., Johnson, C.R. and Larimore, M.G.
  Theory and Design of Adaptive Filters, Wiley, 1987.

The latter book is interesting in that it gives details of adaptive filter implementations using the Texas Instruments range of digital signal processing chips.
  The TRT product (mentioned in Section 11.6 and more fully in Chapter 1) is fully described in the reference:

Erdreich, M., Lassaux, J. and Mamann, J.M.
  30-channel digital echo canceller for telephone circuits, *Commutation and Transmission,*
  4, 53–66, 1985.

For a general introduction to echo cancelling see:

Weinstein, S.B.
  Echo cancellation in the telephone network, *IEEE Commun.* **15**, 9–15, 1977.

The interference cancelling concept, noise control and optimal regulation are very closely related. An interesting example of active noise control and minimum variance regulation is presented in:

Costin, M.H. and Elzinga, D.R.
  Active reduction of low-frequency tire impact noise using digital feedback control,
  IEEE Control Systems, August, 1989.

This approach to vehicle noise control should be compared with the active suspension regulation study described in Chapter 8. The related area of adaptive sound control is a vigorous area of development in acoustics. Some typical references are:

Ross, C.F.
  An algorithm for designing a broadband active sound control system, *J. Sound and Vibration,* **80**, 373–80, 1989.
Eriksson, L.J., Allie, M.C. and Greiner, R.A.
  The selection and application of an IIR adaptive filter for use in active sound attenuation, *IEEE Trans.* **ASSP-35**, 433–7, 1987.

Part 4

Special Topics

In this part we cover aspects of self-tuning which either do not fit directly into the control/filtering framework or are concerned with advanced techniques. Specifically:

(a) Chapter 12 deals with the advanced topics of two-dimensional self-tuning prediction and smoothing.

(b) Chapter 13 discusses the extremum control problem within a self-tuning framework.

(c) Chapter 14 presents a widely used frequency domain form of self-tuning.

12 Two-Dimensional Self-tuning Algorithms

12.1 OUTLINE AND LEARNING OBJECTIVES

This chapter is concerned with self-tuning signal processing of two-dimensional data sources. In particular, it is shown that the one-dimensional smoothing and prediction algorithms of Chapter 11 can be extended to two dimensions.

The layout of the chapter is as follows. Section 12.2 discusses the need for two-dimensional signal processing. Section 12.3 explains how two-dimensional data sources are represented for signal processing. The algorithms for prediction and smoothing in two dimensions are covered in Sections 12.4 and 12.5 respectively. The mechanisms for operating these algorithms in a self-tuning form are discussed in Section 12.6. The chapter is concluded with some applications (12.7) to image enhancement. The notes and references section (12.10) guides the reader to the source material for this chapter and to other more general references on two-dimensional signal processing.

The learning objective is to understand the need for and uses of two-dimensional signal processing algorithms and their self-tuning versions.

12.2 BACKGROUND AND MOTIVATION

The techniques and algorithms presented in the first three parts of this book concerned one-dimensional (1D) data sources. The one dimension refers to the independent variable of which the measured signals are a function. In almost all cases it has been assumed that time is the independent variable. Indeed, this is the natural state of affairs in control systems and conventional signal processing. There are, however, a range of signal processing problems with more than one independent variable. Moreover, these problems need not necessarily involve time as an independent variable. A good example of this is the restoration of photographic images. Here there are two independent variables, the x and y directions in the photographic plane (Figure 12.1). A

Figure 12.1 The independent variables x, y in the 2D signal processing of a photographic image.

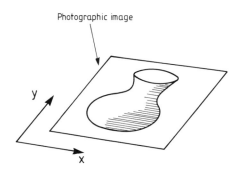

Photographic image

y

x

typical task would be to remove random noise from an image prior to the application of some feature extraction algorithm. Figure 12.2 shows a noisy image before and after filtering with a noise suppressing algorithm. In Figure 12.2(a), an image of a vase on a table has been corrupted with white noise in order to simulate imperfections in the photographic process. Figure 12.2(b) shows the noise corrupted image after a simple two-dimensional (2D) noise suppressing filter has been applied. The purpose of the filter is to restore the image by making the vase more readily discernible.

The image restoration example is an example of 2D signal processing in which there is a need for filtering action which uses information in both x and y directions. Normal 1D filtering would not yield satisfactory results in this case simply because it ignores much of the useful information in the image. (In fact it ignores all information except that on a line through the current filtering point.)

Other practical situations exist where data occurs naturally in two dimensions. For example, scanning radar systems may involve a radar transmitter/receiver head which rotates continuously (Figure 12.3). The signal received during one complete rotation can be considered as one 'line' of a 2D data field. The complete data field would then have angle (between 0° and 360°) as the independent variable on the horizontal axis and the rotation number on the vertical axis (Figure 12.3). The signal processing problem in the radar application could be filtering of the received signals, in order to separate valid echoes from clutter (i.e. independent noise). Equally, the requirement might be to predict the value of the received signal at some future sweep number and angle.

The preceding paragraphs indicate two applications in which 2D data may naturally occur. Numerous other examples exist, ranging from the study of agricultural crop treatments to the prediction of periodic time series. The kind of operations which 2D data of these kinds require are either filtering or prediction. In particular, the image restoration example requires 2D algorithms which are the equivalent of the 1D smoothers and filters considered in Chapter

Figure 12.2 An example of image restoration in which a noise corrupted photograph has been filtered using a 2D noise removal algorithm: (a) noise corrupted image of a vase on a table and (b) the filtered image of the vase.

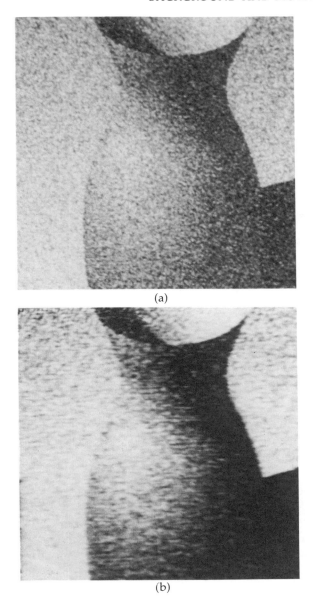

(a)

(b)

11. The radar prediction example requires 2D algorithms which are extensions of the 1D predictors given in Chapter 10.

The reason why 2D signal processing is required is hopefully clear from the examples. It is however pertinent to ask 'what makes 2D different from 1D?' Two-dimensional algorithms are different from 1D algorithms in two essential ways. First, they are computationally much more demanding and algor-

Figure 12.3 A rotating radar dish collects similar information every complete revolution. The data associated with each new revolution can be considered as an additional line in a 2D data field.

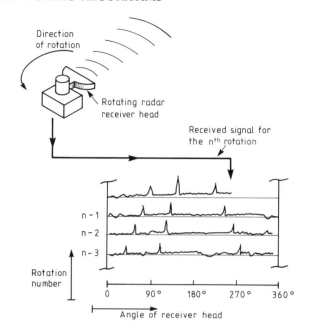

ithmically complex. Second, from a theoretical viewpoint, the 1D polynomial methods which have underpinned the algorithms of Parts 2 and 3 do not directly apply in 2D systems. More will be said of this later when 2D algorithms are developed. All that needs to be said at this point is that it is only possible to develop 2D self-tuning algorithms in certain special cases. What makes them important is that the special cases turn out to be useful in a wide range of real problems.

12.3 MODELS FOR TWO-DIMENSIONAL SYSTEMS

The system and signal models discussed in Chapter 2 were for 1D processes. The object of this section is to develop the corresponding models for 2D systems. In order to do this it is first necessary to decide the form of the 2D data field and the manner in which it will be processed by a signal processing algorithm. The most practically useful assumption on the form of 2D data is that it exists as a set of equispaced points in a 2D matrix (Figure 12.4). If the original source of the data is a photograph, for example, then the points in the data field are obtained by scanning the photograph in both the x and y directions. During the scanning operation the intensity level at equispaced intervals is digitized. Thus the continuous spatial variables x, y are replaced by the indices m and n, and the continuous 2D function which defines the

Figure 12.4
Two-
dimensional
data represented
as a matrix of
observed data
points.

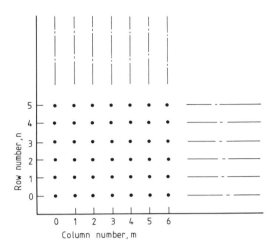

image $s(x,y)$ is replaced by a 2D matrix of points $s(m,n)$. In image processing each one of these points in the matrix is called a *pixel*. This term is frequently adopted in general 2D signal processing. If an image is digitized in this way, a discrete 2D process $s(m,n)$ is obtained, where $s(m,n)$ denotes the signal value at the pixel located at the intersection of row n and column m. Just as for 1D systems, it is convenient to use shift operators which can be used to denote relative pixel locations. We introduce the column and row shift operators w and z in two dimensions as follows:

(i) column shift (Figure 12.5(a))
$$w^{-i}s(m,n) = s(m-i,n)$$

(ii) row shift (Figure 12.5(b))
$$z^{-j}s(m,n) = s(m,n-j)$$

Thus in general we can refer to the signal $s(m-i,n-j)$ as $w^{-i}z^{-j}s(m,n)$. This notation will prove useful when defining 2D models and algorithms.

12.3.1   Data scanning and supports

There is great freedom in selecting the way in which 2D data are scanned and processed. This contrasts with the processing of 1D data where it is normally only possible to scan the data in one way, as they become available in real time. In 2D signal processing, however, the data are often predigitized and processed off-line. This means that all the data are available at any one time and can be processed in any order. Some of the most ingenious 2D signal processing algorithms take advantage of this freedom. We will assume,

Figure 12.5
Illustrating the
2D shift
operators: (a) the
column shift
operator w and
(b) the row shift
operator z.

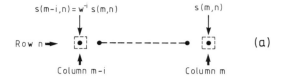

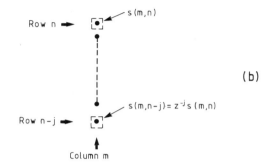

however, that a conventional raster scanning method is used. Raster scanning is the procedure whereby the data are processed one row at a time moving from left to right (Figure 12.6). It is commonly adopted as the standard scanning method in 2D signal processing because electromechanical devices often naturally collect data in this way. The raster scanning convention allows us to define the causality of 2D systems as they are considered here. Specifically, and with reference to Figure 12.7, when processing the data point $s(m,n)$, we assume that all data to the left on the current row and on all previous rows are available. This area of available data is denoted as $S(m,n)$ in Figure 12.7 and is referred to as the nonsymmetric half-plane (NSHP).

Figure 12.6 The
raster scanning
method for 2D
data. The pixels
are processed
one line at a
time working
from left to right
and moving up
one line when
the current line
is finished.

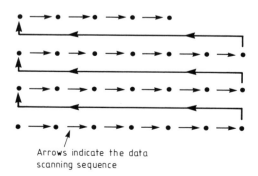

Figure 12.7 The set of natural past data for a signal $s(m,n)$ at pixel (m,n) is denoted $S(m,n)$ and consists of all data to the left on the current line and on all previous lines. The set $S(m,n)$ is termed the Nonsymmetric Half-Plane (NSHP).

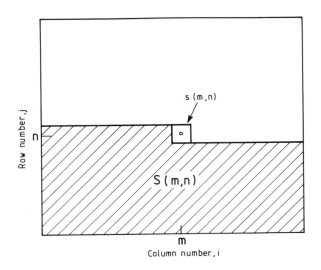

12.3.2 Two-dimensional ARMA models

A 2D equivalent of the 1D ARMA models used in Chapter 2 can be defined to represent 2D random processes. In particular, if a 2D data field is available in raster scan mode, then it is possible to represent the value of a signal at pixel (m,n) as an autoregressive moving average of the set of available data on the current line and all previous lines. Thus the NSHP $S(m,n)$ constitutes the natural past for the signal $s(m,n)$ (Figure 12.7). If we assume that the signal $s(m,n)$ is generated via a 2D white noise process $e(m,n)$ which has zero mean and variance σ_e^2, then it is possible to propose a 2D ARMA model with the form

$$\sum_{(i,j)\in\beta} a(i,j)\, s(m-i,n-j) = \sum_{(i,j)\in\alpha} c(i,j)\, e(m-i,n-j) \tag{12.1a}$$

The 2D white noise sequence has correlation structure

$$\begin{aligned} E\{e(m,n)e(m-i,n-j)\} &= \sigma_e^2 & i=j=0 \\ &= 0 & \text{otherwise} \end{aligned}$$

Using the shift operator notation, equation (12.1a) can be put in the alternative form:

$$A(w^{-1},z^{-1})s(m,n) = C(w^{-1},z^{-1})e(m,n) \tag{12.1b}$$

where the polynomials A and C are defined by

$$A(w^{-1},z^{-1}) = \sum_{i,j \in \beta} a(i,j)w^{-i}z^{-j}$$

$$C(w^{-1},z^{-1}) = \sum_{i,j \in \alpha} c(i,j)w^{-i}z^{-j}$$

$\left.\right\}$ (12.1c)

The coefficients $a(i,j)$ are associated with the autoregressive part of the model and have nonzero values over a region β in $S(m,n)$. The region β is termed the *support* of the coefficient set $\{a(i,j)\}$. In 2D terms, the support β is analogous to the degree of a 1D autoregressive (AR) polynomial. The coefficients $c(i,j)$ are associated with the moving average (MA) part of the model. The support for the set $\{c(i,j)\}$ is α.

The coefficient $a(0,0)$ in the AR part is assumed to be unity so that the signal model can be written as

$$a(m,n) = - \sum_{(i,j) \in \beta - (0,0)} s(i,j)\, s(m-i,n-j) + \sum_{(i,j) \in \alpha} c(i,j)\, e(m-i,n-j) \quad (12.2)$$

where the notation $\beta - (0,0)$ means 'the region β excluding the point $(0,0)$'.

Note that equation (12.2) resembles the well-known difference equation representation of a signal in which the current value is given as a sum of weighted values of past signal values and noise values. The size and shape of the supports β, α determine which past values of signal and noise contribute to these sums. In this context, a widely used support type is the NSHP support shown in Figure 12.8 and denoted $R(M,N)$. This form of support spans N previous rows and incorporates pixels M columns either side of the current pixel. With this form of support the polynomials $A(w^{-1},z^{-1})$ and $C(w^{-1},z^{-1})$ take the form

Figure 12.8 The NSHP support $R(M,N)$.

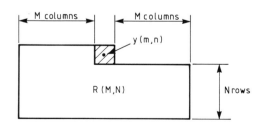

$$A(w^{-1},z^{-1}) = \sum_{i=0}^{M}\sum_{j=0}^{N} a(i,j)w^{-i}z^{-j} + \sum_{i=1}^{M}\sum_{j=1}^{N} a(-i,j)w^{i}z^{-j}$$

$$C(w^{-1},z^{-1}) = \sum_{i=0}^{M}\sum_{j=0}^{N} c(i,j)\,w^{-i}z^{-j} + \sum_{i=1}^{M}\sum_{j=1}^{N} c(-i,j)\,w^{i}z^{-j}$$

(12.3)

Example 12.1

A 2D ARMA model with the supports $\beta = \alpha = R(1,1)$ can be written out as:

$$
\begin{aligned}
s(m,n) = &- a(1,0)s(m-1,n) - a(-1,1)s(m+1,n-1) \\
&- a(0,1)s(m,n-1) - a(1,1)s(m-1,n-1) \\
&+ c(1,0)e(m-1,n) - c(-1,1)e(m+1,n-1) \\
&+ c(0,1)e(m,n-1) + c(1,1)e(m-1,n-1) \\
&+ e(m,n)
\end{aligned}
$$

Using the shift operator form, this model can be rewritten as:

$$A(w^{-1},z^{-1})s(m,n) = C(w^{-1},z^{-1})e(m,n)$$

$$A(w^{-1},z^{-1}) = 1 + a(1,0)w^{-1} + a(-1,1)wz^{-1} + a(0,1)z^{-1} + a(1,1)w^{-1}z^{-1}$$

$$C(w^{-1},z^{-1}) = 1 + c(1,0)w^{-1} + c(-1,1)wz^{-1} + c(0,1)z^{-1} + c(1,1)w^{-1}z^{-1}$$

The coefficients of $A(w^{-1},z^{-1})$, $C(w^{-1},z^{-1})$ are associated with their corresponding pixel locations in Figure 12.9.   ☐

The assumption throughout this chapter is that the 2D signal source has NSHP supports $\beta = \alpha = R(M,N)$ and that the data is available in a corresponding raster form. It will be appreciated that many alternative supports and scanning protocols may be used. The ones proposed here are the most useful in the current discussion.

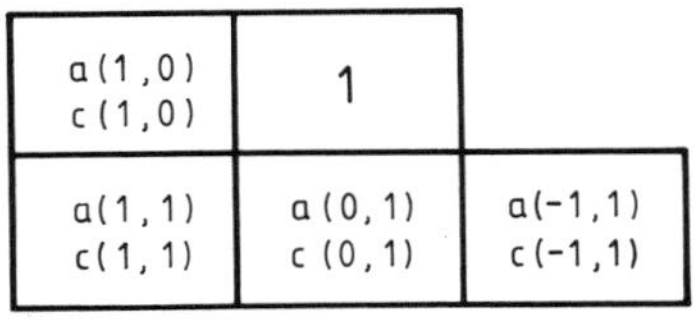

$a(1,0)$ $c(1,0)$	1	
$a(1,1)$ $c(1,1)$	$a(0,1)$ $c(0,1)$	$a(-1,1)$ $c(-1,1)$

Figure 12.9 The support $R(1,1)$ and corresponding $A(w^{-1}, z^{-1})$, $C(w^{-1}, z^{-1})$ coefficient locations for Example 12.1.

It will also be assumed that we are considering stable, minimum phase 2D random processes. This assumption imposes restrictions on the ARMA coefficients in (12.1) which translate (as in 1D) into stability requirements on certain polynomials. The stability of 2D polynomial objects is not a simple matter. For further details, the reader is referred to the references in Section 12.10. The exposition below comments on 2D polynomial properties only where necessary and the need for certain polynomial operations to be stable will become obvious.

12.3.3 Two-dimensional models for smoothing

When we wish to smooth a 2D data field the assumption is that the image signal $s(m,n)$ is created by a process of the form (12.1). The current 2D datum, however, is $y(m,n)$ given by:

$$y(m,n) = s(m,n) + v(m,n) \qquad (12.4)$$

where $v(m,n)$ is sensor or measurement noise, which is represented by a 2D gaussian white noise signal. It is assumed that $v(m,n)$ has zero mean, variance σ_v^2 and is independent of $s(m,n)$. In an image processing application, $v(m,n)$ could represent the influence of film grain, quantization or imperfections in the imaging system.

In order to obtain a smoother or filter for the signal $s(m,n)$, it will be assumed that the process $y(m,n)$ can be equivalently represented in the ARMA form:

$$y(m,n) = - \sum_{(i,j)\in\beta-(0,0)} a(i,j)\, y(m-i,n-j) + \sum_{(i,j)\in\gamma_p} d(i,j)\, \tilde{e}(m-i,n-j) \qquad (12.5a)$$

or alternatively

$$A(w^{-1},z^{-1})y(m,n) = D(w^{-1},z^{-1})\tilde{e}(m,n) \qquad (12.5b)$$

where $\tilde{e}(m,n)$ is a 2D white noise signal with zero mean and variance $\sigma_{\tilde{e}}^2$. The support for the $d(i,j)$ coefficients should consist of all data on the N rows shown in Figure 12.10. In practice it can be restricted to the support γ_p given by $R(M,N)$ augmented by p columns at either end (see again Figure 12.10). In this case $D(w^{-1},z^{-1})$ is given by:

$$D(w^{-1}z^{-1}) = \sum_{i=0}^{M+p} \sum_{j=0}^{N} d(i,j)w^{-i}z^{-j} + \sum_{i=1}^{M+p} \sum_{j=1}^{N} d(-i,j)w^{i}z^{-j}$$

Figure 12.10
The support γ_p
for the ARMA
model moving
average
coefficient set
$\{d(i, j)\}$.

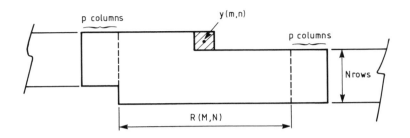

12.3.4 Two-dimensional models for prediction

For the 2D prediction problem, the 2D data field is assumed to take the form of a 2D ARMA process given by:

$$y(m,n) = - \sum_{(i,j) \in \beta - (0,0)} a(i,j)\, y(m-i,n-j) + \sum_{(i,j) \in \alpha} c(i,j)\, e(m-i,n-j) \quad (12.6a)$$

or alternatively

$$A(w^{-1},z^{-1})y(m,n) = C(w^{-1},z^{-1})e(m,n) \quad (12.6b)$$

where, as before $e(m,n)$ is a 2D gaussian white noise process and in this case $c(0,0)$ is assumed to be unity.

12.4 TWO-DIMENSIONAL PREDICTION

In one dimension (Chapter 10) a predictor uses past data on the time axis in order to predict k time steps ahead. The 2D predictor considered here uses past data on the current line and all previous lines in order to predict k pixels ahead on the current line (Figure 12.11). By comparing this figure with Figure 10.1 it is clear that, by incorporating information from previous lines, there is potential for improving the quality of prediction. There are two reasons for this: first, from a statistical viewpoint, the more relevant the data that is incorporated into an estimator, the better the estimate. Second, from a heuristic viewpoint, the information included in pixel $y(m+k,n-1)$ and its neighbouring pixels are likely to be highly relevant in predicting the values which $y(m+k,n)$ will take.

The data available for prediction consists of the current pixel $y(m,n)$ and the natural past $S(m,n)$. The combination of these two can be denoted $S(m+1,n)$, so that the predictor of $y(m+k,n)$ is defined by the conditional expectation

Figure 12.11
Two-
dimensional *k*-
pixel ahead
prediction in
which the
current pixel and
the NSHP *S(m,n)*
(consisting of all
previously
scanned lines) is
available for
prediction.

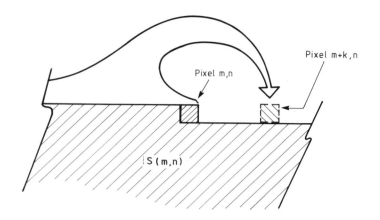

$$\hat{y}(m+k,n) = E\{y(m+k,n) \mid y(i,j) \in S(m+1,n)\} \qquad (12.7)$$

The *k*-step prediction error is

$$\tilde{y}(m+k,n) = y(m+k,n) - \hat{y}(m+k,n) \qquad (12.8)$$

and the loss function to be minimized is

$$V_k = E\{\tilde{y}^2(m+k,n)\} \qquad (12.9)$$

To minimize this loss function, recall the partitioning idea which was used in the 1D case (Chapter 10). We can use this here in order to split the required signal into a part that is reconstructible at pixel (m,n) and a part that is not. Applying this idea (which we justify in a moment) gives

$$y(m+k,n) = y'(m+k,n) + \sum_{p=0}^{k-1} f(p,0)\, e(m+k-p,n) \qquad (12.10)$$

The term $y'(m+k,n)$ is that part of $y(m+k,n)$ which is calculable at (m,n), in the sense that it can be expressed as a function of current and past values of $y(m,n)$. It can be written in the form:

$$y'(m+k,n) = -\sum_{(i,j)\in\beta-(0,0)} a(i,j)\, y'(m+k-i,n-j) + \sum_{(i,j)\in\theta} g(i,j)\, e(m-i,n-j) \qquad (12.11)$$

where the object $\{g(i,j)\}$ has the support region θ, which is shown pictorially in Figure 12.12 and described further on. The second term on the right hand

Figure 12.12
The support
region θ for the
{g(i, j)} set used
in prediction.

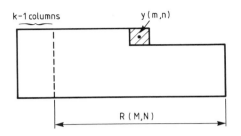

side of equation (12.10) represents a weighted sum of unknown errors, i.e. not available at the current point. The sequence {f(p,0)} represents the set of coefficients multiplying these driving noise values from 'future time'. In terms of the sequences {a(i,j), c(i,j)}, the sequence {f(p,0) : p=0,1 ..., k−1} is the first k elements of the zero row of the 2D impulse response expansion of y(m,n). Note that it is possible to express the 2D process y(m,n) of equation (12.6) in 2D impulse response form:

$$y(m,n) = \sum_{S(m,n)} f(i,j)\, e(m-i,n-j) + e(m,n) \tag{12.12}$$

where {f(i,j)} is the 2D impulse response sequence with support S(m,n), the NSHP.

The computation of {f(p,0)} and {g(i,j)} is explained later using a 2D polynomial identity. For the moment, however, we note that, in the decomposition leading to equation (12.10), the term y'(m,n) is independent of (and therefore uncorrelated with) the remaining terms. This implies that y'(m+k,n) contains all the available information concerning y(m+k,n) that is contained in the current pixel (m,n) and its past. The loss function is thus minimized by the choice of y'(m+k,n) as the predictor, i.e.

$$\hat{y}(m+k,n) = y'(m+k,n)$$

The k-step ahead predictor then becomes:

$$\hat{y}(m+k,n) = - \sum_{(i,j)\in\beta-(0,0)} a(i,j)\, \hat{y}(m+k-i,n-j) + \sum_{(i,j)\in\theta} g(i,j)\, e(m-i,n-j) \tag{12.13}$$

Consider now the nature of the decomposition in equations (12.10) and (12.11) in 2D polynomial terms, since this may be used to justify the technique for calculating the coefficients {f(p,0)} and {g(i,j)}.

By substitution of equation (12.11) in (12.10) to eliminate $y'(m+k,n)$, and noting the definition (equation (12.6)) of $y(m,n)$, it is possible to write the following identity:

$$\sum_{(i,j)\in\beta} c(i,j)w^{-i}z^{-j} = \sum_{(i,j)\in\beta} a(i,j)w^{-i}z^{-j}\left[\sum_{p=0}^{k-1} f(p,0)w^{-p}\right] + w^{-k}\sum_{(i,j)\in\theta} g(i,j)w^{-i}z^{-j}$$

(12.14)

This can be put in more compact form by using the 2D polynomials $C(w^{-1},z^{-1})$, $A(w^{-1},z^{-1})$, $F_1(w^{-1})$ and $G(w^{-1},z^{-1})$. The polynomials $C(w^{-1},z^{-1})$, $A(w^{-1},z^{-1})$ are defined as in equation (12.3), while the polynomials $F_1(w^{-1})$ and $G(w^{-1},z^{-1})$ are given by

$$\left.\begin{array}{l} F_1(w^{-1}) = F(w^{-1},1) = 1 + f(1,0)w^{-1} + \ldots + f(k-1,0)w^{-k+1} \\ G(w^{-1},z^{-1}) = \sum_{(i,j)\in\theta} g(i,j)w^{-i}z^{-j} \end{array}\right\}$$

(12.15)

Using these definitions equation (12.14) can be written as a polynomial identity, thus:

$$C(w^{-1},z^{-1}) = A(w^{-1},z^{-1})\,F_1(w^{-1}) + w^{-k}G(w^{-1},z^{-1})$$

(12.16)

This is a 2D diophantine equation and is similar in form to that used in 1D MV control (Chapter 8) and prediction (Chapter 10).

The unknown coefficients of $F_1(w^{-1})$ and $G(w^{-1},z^{-1})$ can be obtained from equation (12.14) in a straightforward manner. In particular, by equating coefficients of like powers of w^{-i},z^{-j}, a set of linear simultaneous equations can be obtained from equation (12.14). The only problems in solving the identity concern the ranges on the indices (i,j) required for a unique solution. For a general $R(M,N)$ support, the support region for $G(w^{-1},z^{-1})$ is defined as:

$$\theta = \{(i,j): 0 \leqslant i \leqslant M-1, j=0\}\ \mathbf{U}\{(i,j): -M-k \leqslant i \leqslant M-1, 1 \leqslant j \leqslant N\}.$$

The corresponding $G(w^{-1},z^{-1})$ polynomial is given by

$$G(w^{-1},z^{-1}) = \sum_{i=0}^{M-1}\sum_{j=0}^{N} g(i,j)w^{-i}z^{-j} + \sum_{i=1}^{M+k}\sum_{j=1}^{N} g(-i,j)w^{i}z^{-j}$$

(12.17a)

The product $A(w^{-1},z^{-1})F_1(w^{-1})$ is denoted by $Q(w^{-1},z^{-1})$ with the support region:

$$\gamma = \{(i,j) : 0 \leq i \leq M+k-1, j = 0\} \cup \{(i,j) : -M \leq i \leq M-1, 1 \leq j \leq N\}$$

The corresponding $Q(w^{-1},z^{-1})$ polynomial is given by

$$Q(w^{-1},z^{-1}) = \sum_{i=0}^{M+k-1}\sum_{j=0}^{N} q(i,j)w^{-i}z^{-j} + \sum_{i=1}^{M}\sum_{j=1}^{N} q(-i,j)w^{i}z^{-j} \qquad (12.17b)$$

which is readily obtained from inspection of the supports of $A(w^{-1},z^{-1})$ and $F_1(w^{-1})$. See Figures 12.12 and 12.13 for sketches of these regions.

Example 12.2

To clarify the manipulations associated with 2D prediction, the case of a two-pixel ahead predictor $\hat{y}(m+2,n)$ is derived for the process defined by:

$$y(m,n) = -\,a(1,0)y(m-1,n) - a(-1,1)y(m+1,n-1)$$
$$- a(0,1)y(m,n-1) - a(1,1)y(m-1,n-1)$$
$$+ c(1,0)e(m-1,n) + c(-1,1)e(m+1,n-1)$$
$$+ c(0,1)e(m,n-1) + c(1,1)e(m-1,n-1)$$
$$+ e(m,n)$$

In this case the polynomials $A(w^{-1},z^{-1})$ and $C(w^{-1},z^{-1})$ are given by:

Figure 12.13
The support region γ for $Q(w^{-1}, z^{-1})$.

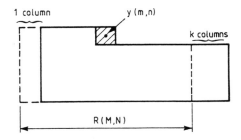

$$A(w^{-1},z^{-1}) = 1 + a(1,0)w^{-1} + a(-1,1)wz^{-1} + a(0,1)z^{-1} + a(1,1)w^{-1}z^{-1}$$

$$C(w^{-1},z^{-1}) = 1 + c(1,0)w^{-1} + c(1,1)wz^{-1} + c(0,1)z^{-1} + c(1,1)w^{-1}z^{-1}$$

Using these relations to solve the identity given by equations (12.14) and (12.16) for k equal to two, it is possible to show that:

$$F_1(w^{-1}) = 1 + f(1,0)w^{-1} = 1 + [c(1,0) - a(1,0)]w^{-1}$$

and $G(w^{-1},z^{-1})$ is given by:

$$G(w^{-1},z^{-1}) = g(0,0) + g(-3,1)w^3z^{-1} + g(-2,1)w^2z^{-1} + g(-1,1)wz^{-1} + g(0,1)z^{-1}$$

where

$$g(0,0) = -a(1,0)f(1,0)$$
$$g(0,1) = -a(1,1)f(1,0)$$
$$g(-1,1) = c(1,1) - a(1,1) - a(0,1)f(1,0)$$
$$g(-2,1) = c(0,1) - a(0,1) - a(-1,1)f(1,0)$$
$$g(-3,1) = c(-1,1) - a(-1,1)$$

Note that $G(w^{-1},z^{-1})$ above has a support which corresponds to the definition of θ and that the expressions for the coefficients of $F_1(w^{-1})$ and $G(w^{-1},z^{-1})$ uniquely satisfy the 2D diophantine equation (12.16).   □

From equation (12.13), the predictor can be written as

$$\hat{y}(m+k,n) = \frac{G(w^{-1},z^{-1})}{A(w^{-1},z^{-1})} e(m,n) \tag{12.18}$$

or, substituting for $e(m,n)$, cast in the form

$$\hat{y}(m+k,n) = \frac{G(w^{-1},z^{-1})}{C(w^{-1},z^{-1})y(m,n)} \tag{12.19}$$

The prediction error can be written in the following form:

$$\tilde{y}(m+k,n) = F_1(w^{-1})e(m+k,n) \tag{12.20}$$

Using this and equation (12.16) leads to the alternative predictor equation:

$$\hat{y}(m+k,n) = \frac{G(w^{-1},z^{-1})}{Q(w^{-1},z^{-1})}\,\tilde{y}(m,n) \tag{12.21}$$

These relations (equations (12.13), (12.19) and (12.21)) can be compared directly with the corresponding 1D prediction expressions in Chapter 10.

As in the 1D case, the predictor form given in equation (12.21) is widely used, since it expresses the prediction in terms of past predictions and prediction errors. As will be made clear later, equation (12.21) is particularly suited for an implicit self-tuning implementation of the 2D predictor. With this algorithm, it is possible to find the $\{q(i,j)\}$ and $\{g(i,j)\}$ objects by direct estimation, eliminating the need to solve the identity (12.16).

Since the prediction error (12.20) contains all the 'future' errors, the theoretical variance of the prediction estimates can be expressed as:

$$V_k = \sigma_e^2 \left[\sum_{p=0}^{k-1} f^2(p,0) \right]$$

12.5   TWO-DIMENSIONAL SMOOTHING

In one dimension (Chapter 11), the aim is to smooth or filter a 1D signal using only the past time data. In two dimensions the aim is to smooth a 2D data field $y(m,n)$ using all the information available from the current line and all the information from all previously scanned lines. Figure 12.14 depicts this situation in which all values of the measured data in the natural past

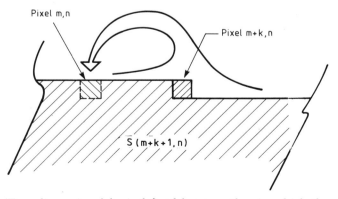

Figure 12.14   Two-dimensional k-pixel fixed lag smoother in which the current pixel and the natural past $S(m + k + 1, n)$ is available for obtaining a smoothed estimate of $s(m,n)$ from $y(m + k, n)$.

$S(m+k+1,n)$ are available in order to obtain a smoothed estimate of $s(m,n)$. The objective is therefore to remove the unknown sensor noise $v(m,n)$ from the measurement $y(m,n)$ in some way, where

$$y(m,n) = s(m,n) + v(m,n) \tag{12.4)bis}$$

Given the representation of $y(m,n)$ in ARMA form (equation 12.5) and recalling the analogous 1D procedure (Chapter 11), the k fixed lag smoothed estimate of $s(m,n)$ is

$$\hat{s}(m,n) = \sum_{(i,j)\in S(m+k+1,n)} E\{s(m,n) \mid \tilde{e}(i,j)\}$$

$$= \sum_{(i,j)\in S(m+k+1,n)} E\{y(m,n) - v(m,n) \mid \tilde{e}(i,j)\}$$

$$= y(m,n) - \sum_{(i,j)\in S_1(m,n)} E\{v(m,n) \mid \tilde{e}(i,j)\} \tag{12.22}$$

where $S_1(m,n) = \{(i,j) : m \leqslant i \leqslant m+k, j = n\}$, i.e. the k points on the current line between (m,n), the point at which smoothing takes place, and $(m+k,n)$, the current datum location.

Now note that:

$$E\{v(m,n) \mid \tilde{e}(i,j)\} = E\{v(m,n) \, \tilde{e}(i,j)\} \, \tilde{e}(i,j)/\sigma_{\tilde{e}}^2 \tag{12.23}$$

This enables equation (12.22) to be written as:

$$\hat{s}(m,n) = y(m,n) - \sum_{(i,j)\in S_1(m,n)} e\{v(m,n) \, \tilde{e}(i,j)\} \, \tilde{e}(i,j)/\sigma_{\tilde{e}}^2 \tag{12.24}$$

The summation term on the right-hand side of (12.24) can be simplified by noting that it is possible to invert the ARMA form of $y(m,n)$ (equation 12.5(b)) to yield:

$$\tilde{e}(i,j) = \frac{A(w^{-1},z^{-1})}{D(w^{-1},z^{-1})} \, y(i,j) \tag{12.25}$$

provided the transfer function is stable. The signal $\tilde{e}(m,n)$ can also be equivalently represented by its 2D impulse response sequence $\{h(p,q)\}$, giving:

$$\tilde{e}(i,j) = \sum_{(p,q)\in S(i,j)} h(p,q) \, y(i-p,j-q) \tag{12.26}$$

where again $S(i,j)$ denotes the natural past of the pixel (i,j) and encompasses that part of the current (ith) row already scanned, together with all previous rows. In addition, from the equivalence of equations (12.25) and (12.26) it follows that:

$$H(w^{-1},z^{-1}) = \frac{A(w^{-1},z^{-1})}{D(w^{-1},z^{-1})} \qquad (12.27)$$

where $H(w^{-1},z^{-1})$ is the 2D transform of the impulse response sequence $\{h(p,q)\}$.

Using the definition of $y(m,n)$ from equations (12.1), (12.4) and (12.5), $\tilde{e}(i,j)$ can be written as:

$$\tilde{e}(i,j) = \frac{C(w^{-1},z^{-1})}{D(w^{-1},z^{-1})} e(i,j) + \frac{A(w^{-1},z^{-1})}{D(w^{-1},z^{-1})} v(i,j) \qquad (12.28)$$

Noting the orthogonality of $e(i,j)$ and $v(i,j)$, the expectation embedded in equation (12.24) can be rewritten as:

$$E\{v(m,n)\,\tilde{e}(i,j)\} = E\{v(m,n) \sum_{(p,q)\in S(i,j)} h(p,q)\,y(i-p,j-q)\}$$

$$= E\{v(m,n) \sum_{(p,q)\in S(i,j)} h(p,q)\,v(i-p,j-q)\}$$

$$= \begin{cases} \sigma_v^2\,h(i-m,j-n) & i \geq m, j \geq n \\ 0 & \text{otherwise} \end{cases} \qquad (12.29)$$

Making note of the definitions of the supports $S_1(m,n)$ and $S(i,j)$, equation (12.29) becomes:

$$E\{v(m,n)\,\tilde{e}(i,j)\} = \begin{cases} \sigma_v^2 h(p,0) & 0 \leq p \leq k \\ 0 & \text{otherwise} \end{cases} \qquad (12.30)$$

Finally, using (12.30) in (12.24), the smoothed value of $s(m,n)$ based on data in $S(m+k+1,n)$ can be estimated by:

$$\hat{s}(m,n) = y(m,n) - \frac{\sigma_v^2}{\sigma_{\tilde{e}}^2} \sum_{p=0}^{k} h(p,0)\,\tilde{e}(m+p,n) \qquad (12.31)$$

where $\{h(p,0)\}$ corresponds to the first row of the impulse response expansion $\{h(i,j)\}$.

From the defining relation (12.27), the sequence $\{h(p,0)\}$ can be calculated from the equation set:

$$a(i,0) = \sum_{(p,j) \,:\, p+j=i} h(p,0)\, d(j,0) \quad (h(0,0) = 1) \tag{12.32}$$

In addition, the variance of the smoothing error can be computed as:

$$V_k = E\left\{[s(m,n) - \hat{s}(m,n)]^2\right\}$$

$$= \sigma_\nu^2 \left[1 - \frac{\sigma_\nu^2}{\sigma_{\tilde{e}}^2} \sum_{p=0}^{k} h^2(p,0)\right] \tag{12.33}$$

For systems in which $c(0,0)$ is zero in the coefficient set $\{c(i,j)\}$, the ratio of σ_ν^2 to $\sigma_{\tilde{e}}^2$ (which appears in both the smoother equation (12.31) and in the variance calculation (12.33)), can be determined from

$$\frac{\sigma_\nu^2}{\sigma_{\tilde{e}}^2} = \frac{d(M,N)}{a(M,N)} \tag{12.34}$$

This relationship is obtained by setting $w = z = 0$ in the power spectral equivalence relation which links $A(w^{-1},z^{-1})$, $C(w^{-1},z^{-1})$ and $D(w^{-1},z^{-1})$. Moreover, it applies in cases where the supports β, α are the same and equal to $R(M,N)$.

The power spectral equivalence relation mentioned above is obtained by equating the power spectra of the right- and left-hand sides of equation (12.28). This gives the relationship:

$$D(w^{-1},z^{-1})\, D(w,z)\, \sigma_{\tilde{e}}^2 = A(w^{-1},z^{-1})\, A(w,z)\, \sigma_\nu^2 + C(w^{-1},z^{-1})\, C(w,z)\, \sigma_c^2 \tag{12.35}$$

where $A(w^{-1},z^{-1})$, $C(w^{-1},z^{-1})$ and $D(w^{-1},z^{-1})$ are the 2D z-transforms of $\{a(i,j)\}$, $\{c(i,j)\}$ and $\{d(i,j)\}$ respectively.

The power spectral identity (12.35) is equivalent to the spectral factorization identity familiar in 1D signal processing (Chapter 11). The equivalence is, however, not complete, since the fundamental theorem of algebra does not apply to 2D polynomials. The practical implication of this is that, in general, $D(w^{-1},z^{-1})$ is of infinite degree. Hence, the truncation of this support (Figure 12.10) to γ_p represents an approximation. It can be shown that, for low pass processes, the approximation is rather good and the truncated D polynomial is stable. Nonetheless, it must be borne in mind that many of the relationships (like the spectral factorization theorem) do not apply in 2D.

12.6   SELF-TUNING IN TWO DIMENSIONS

The adaptation of the 2D fixed coefficient synthesis rules (given in the previous sections) to a self-tuning format follows the same path as in 1D self-tuning. Specifically, the parameters of an appropriate 2D model are estimated recursively. The estimated parameters resulting at each recursive step are then used in an appropriate synthesis rule to obtain the desired predictor or smoother parameters. These parameters are then used in the prediction/smoothing law to process the data.

The recursive estimation of 2D data models is no different from 1D data models, in that the data and unknown parameters are arranged in a vector form

$$y(m,n) = \mathbf{x}^{\mathrm{T}}(m,n)\,\hat{\boldsymbol{\theta}} + \hat{e}(m,n) \tag{12.36}$$

where $x(m,n)$ is a vector of the 2D observations and $\hat{\boldsymbol{\theta}}$ is a vector of the unknown 2D model parameters.

At the (m,n) pixel the parameter estimates are updated in the standard way. The pixel column counter is then incremented and the procedure is repeated for pixel $(m+1,n)$ and so on along the nth row until the edge (last column) of the data field is reached. The row counter is then incremented and the recursion continued at the first column of the $(n+1)$th row. The algorithm works though the 2D data field in this way in a raster scanning mode.

A key problem with 2D self-tuning is how to treat the edges of the data field. The procedure mentioned in the previous paragraph assumes that the end of one line joins naturally to the beginning of the line above it. In some cases this is exactly true. For example, the scanning radar example mentioned in Section 12.2 is a case in which the end of one scan links naturally to the beginning of the next. In other cases, however, such as image processing, this is not generally true. In such cases it is often appropriate to reset the covariance of the recursive estimator at the beginning of a new line, thus effectively restarting the estimator there.

12.6.1   Self-tuning prediction algorithm

Two-dimensional self-tuning prediction (like the 1D equivalent) can be performed in either explicit or implicit form. The explicit form of the 2D predictor is:

Algorithm: Explicit 2D Predictor

For each pixel (m,n):

Step (i)   Update the estimates $\{\hat{a}(i,j), \hat{c}(i,j)\}$ *of the ARMA model:*

$$y(m,n) = - \sum_{(i,j)\in\beta-(0,0)} \hat{a}(i,j)\, y(m-i,n-j) +$$

$$\sum_{(i,j)\in\alpha-(0,0)} \hat{c}(i,j)\, \epsilon(m-i,n-j) + \hat{e}(m,n) \quad (12.37)$$

Step (ii)   Use these estimates to synthesize $\{\hat{q}(i,j), \hat{g}(i,j)\}$ by solving the 2D MV prediction identity:

$$(12.38)$$

$$\hat{C}(w^{-1},z^{-1}) = \hat{A}(w^{-1},z^{-1})\, \hat{F}_1(w^{-1}) + w^{-k}\, \hat{G}(w^{-1},z^{-1})$$

$$= \hat{Q}(w^{-1},z^{-1}) + w^{-k}\, \hat{G}(w^{-1},z^{-1})$$

Step (iii)   Form the prediction of the k-step ahead pixel:

$$\hat{C}(w^{-1},z^{-1})\hat{y}(m+k,n) = \hat{G}(w^{-1},z^{-1})y(m,n) \quad (12.39)$$

Step (iv)   Increment the pixel column counter m to $m+1$, check for end of row; if end of current row, reset column counter and increment the row counter n to $n+1$. Loop back to step (i).   □

Consider now the equivalent implicit algorithm. With this approach the synthesis step is omitted and only two steps are required at each pixel: updating of the estimates $\hat{Q}(w^{-1},z^{-1})$, $\hat{G}(w^{-1},z^{-1})$ and prediction of the k-step ahead value. In detail, the implicit algorithm is:

Algorithm: Implicit 2D Predictor

For pixel (m,n):

(i) Update the estimate $\{\hat{q}(i,j), \hat{g}(i,j)\}$ for the model:

$$\hat{Q}(w^{-1},z^{-1})\hat{y}(m,n) = \hat{G}(w^{-1},z^{-1})\, \epsilon(m-k,n) \quad (12.40)$$

The last term involves prediction errors from the estimation process. If the coefficients $\{\hat{q}(i,j)\ \hat{g}(i,j)\}$ converge to the correct values, $\epsilon(m,n)$ corresponds to the prediction error $\bar{y}(m,n)$.

(ii) Form the k-step ahead prediction using

$$\hat{Q}(w^{-1},z^{-1})\hat{y}(m+k,n) = \hat{G}(w^{-1},z^{-1})\epsilon(m,n) \tag{12.41}$$

(iii) Increment the pixel column counter m to $m+1$, check for end of row; if end of current row; reset column counter and increment the row counter n to $n+1$. Loop back to step 1. $\qquad\square$

The following example illustrates the type of results obtained in self-tuning explicit prediction.

Example 12.3

Consider the process

$$A(w^{-1},z^{-1})y(m,n) = C(w^{-1},z^{-1})e(m,n) \tag{12.42}$$

$$A(w^{-1},z^{-1}) = 1-0.1w^{-1}-0.4wz^{-1} + 0.3z^{-1}+0.1w^{-1}z^{-1}$$

$$C(w^{-1},z^{-1}) = 1+0.1w^{-1}+0.3wz^{-1}-0.1z^{-1}-0.4w^{-1}z^{-1}$$

The data $y(m,n)$ is generated for an image area defined by $-9\leqslant n\leqslant60$ and $-9\leqslant m\leqslant50$. The predictions $\hat{y}(m+2,n)$, however, are calculated for the area $3\leqslant m\leqslant52$ and $1\leqslant n\leqslant50$.

Applying the identity (12.16) to the model (12.42) gives the polynomials

$$F_1(w^{-1}) = 1+0.2w^{-1}$$

$$G(w^{-1},z^{-1}) = 0.02+0.7w^3z^{-1}-0.36w^2z^{-1}-0.56wz^{-1}+0.02z^{-1}$$

The results of explicit self-tuning prediction applied to the data field generated by equation (12.42) are shown in Figures 12.15 and 12.16. Specifically, Figure 12.15 shows the parameter estimates of $A(w^{-1},z^{-1})$ and $C(w^{-1},z^{-1})$ and the corresponding $F_1(w^{-1})$ and $G(w^{-1}z^{-1})$ polynomials.

Note that the horizontal axis plots the total number of pixels currently processed. Thus the rows of the data field have been concatenated to form a total of 2500 pixels. In the diagram 50 pixels represent the length of one line of the data field.

Figure 12.16 compares the theoretical optimal cumulative loss function with that obtained using self-tuning. After approximately 100 pixels (i.e. two rows) the self-tuning cumulative loss is parallel with the optimal, indicating that the self-tuning algorithm has converged to the correct configuration. $\qquad\square$

Figure 12.15
Parameter and
predictor
estimates for the
2D predictor
$\hat{y}(m + k, n)$ of
Example 12.3.

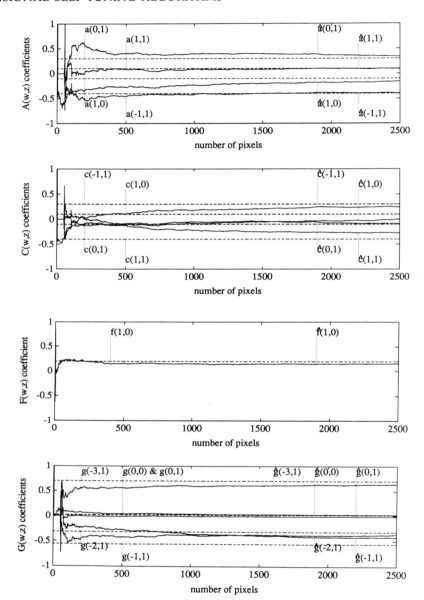

12.6.2   Self-tuning filtering/smoothing algorithm

Filtering is distinguished from smoothing only by setting $k=0$ in the smoothing algorithm. Otherwise, there is no practical difference between the two. In the self-tuning algorithm, the signal $\tilde{e}(m,n)$ in equation (12.31) will be replaced by the prediction error $\epsilon(m,n)$.

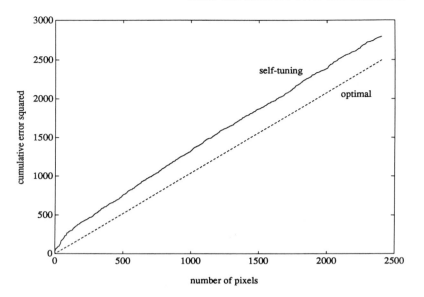

Figure 12.16
Cumulative loss
(optimal and
self-tuning) for
Example 12.3.

To ensure convergence, initial values other than zero must be set for $a(M,N)$ and $d(M,N)$. The algorithm is then:

Algorithm: Explicit 2D Fixed Lag Smoother

For pixel (m,n):

(i) Update the estimates $\{\hat{a}(i,j),\ \hat{d}(i,j)\}$ of the ARMA model:

$$y(m+k,n) = -\sum_{(i,j)\in\beta-(0,0)} \hat{a}(i,j)\, y(m+k-i,n-j) + \sum_{(i,j)\in\gamma_p} \hat{d}(i,j)\, \epsilon(m+k-i,n-j)$$
$$+ \hat{e}(m+k,n) \quad (12.43)$$

(ii) Use these estimates to synthesise $\{h(p,0)\}$ by solving the 2D smoothing identity:

$$\hat{a}(i,0) = \sum_{(p,j):p+j=i} \hat{h}(p,0)\, \hat{d}(j,0) \quad \{\hat{h}(0,0)=1\}$$

(iii) Calculate the smoothed signal value at the k-step back pixel:

$$\hat{s}(m,n) = y(m,n) - \frac{\hat{d}(M,N)}{\hat{a}(M,N)} \sum_{p=0}^{k} \hat{h}(p,0)\, \epsilon(m-k+p,n) \quad (12.44)$$

(iv) Increment the pixel column m to $m+1$ and check for the end of the row. If at end of current row, reset the column counter and increment the row counter n to $n+1$. Loop back to step 1.   □

Note that this algorithm is an explicit one; there is no implicit version for fixed lag smoothing.

12.7   APPLICATIONS

The 2D self-tuning algorithms described in the foregoing sections have a significant range of applications. The most direct application of 2D smoothing is to image restoration. The aim here is to take a noise-corrupted image and filter out the sensor noise. Such operations are used to prepare the raw image data for processing in a feature extraction algorithm or some related pattern recognition routine.

12.7.1   Image smoothing

A use of self-tuning 2D smoothing in image processing is for the restoration of noise-corrupted images prior to the application of a feature extraction technique. As an example of this kind of use, consider the photograph in Figure 12.17(a). The photograph shows a close-up of the human retina. Such images can be used for investigation of diseased retinas and also for research into the functions of the eye. The next photograph (Figure 12.17(b)) shows the result of adding white noise to the original photograph. Note that the main features of the image are heavily masked by the noise. Figure 12.17(c) shows the result of a self-tuning smoother applied to this noisy image. Note the recovery of the original photograph's main features. The self-tuner used in this case was a smoother with $k=0$, that is to say, a filter. The support used was $\beta=\alpha=R(2,2)$.

The image recovery in this example is rather good. This is not always the case. If a photograph (say, a portrait) with a wide range of different features is used, then the smoothing results are less impressive. Qualitatively, this is because such photographs differ markedly from an ARMA model of a 2D stochastic process. The example of Figure 12.17, however, can be plausibly represented by such a model. The secret of good results with self-tuning 2D smoothing is to select applications for which the assumed data generating mechanism is realistic.

Figure 12.17
The filtering of
noisy images
using self-tuning
2D smoothing:
(a) the original
noise-free
photograph of
the human
retina; (b) the
noise-corrupted
image; and (c)
the filtered
image obtained
by applying a
self-tuning zero
lag smoother to
the noisy image.

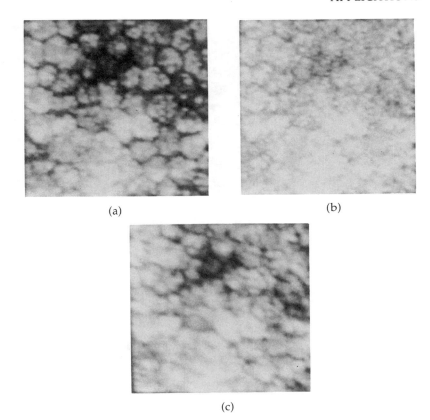

(a)   (b)

(c)

12.7.2 Artefact removal

The method of interference cancelling filters which was introduced in Chapter 11 can be extended for use in 2D signal processing in a very straightforward manner. As an example, consider the case of the 'vase on a table' used in Figure 12.2 as motivation for the idea of 2D filtering. Suppose that the original image (Figure 12.18(a)) is corrupted by a sinusoidal modulation of intensity, as shown in Figure 12.18(b). The 2D noise-cancelling filter assumes that the image of figure 12.18(b) can be modelled as

$$A(w^{-1}, z^{-1})y(m,n) = C(w^{-1}, z^{-1})e(m,n) + \mathcal{D}(m,n) \qquad (12.45)$$

where the interference process $\mathcal{D}(m,n)$ is modelled as a sinusoid of known frequency but unknown phase, i.e.

$$\mathcal{D}(m,n) = d_s \sin \omega m + d_c \cos \omega m \qquad (12.46)$$

Figure 12.18
Two-
dimensional
self-tuning
interference
cancellation: (a)
original image;
(b) image plus
sinusoidal
interference; and
(c) recovered
image.

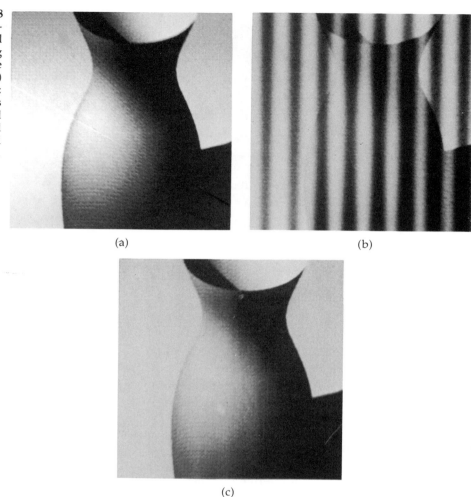

(a)

(b)

(c)

where ω is known. The 2D self-tuning noise cancelling filter then estimates the coefficients of the $A(w^{-1},z^{-1})$, $C(w^{-1},z^{-1})$ polynomials and the sinusoidal interference parameters d_s, d_c. The interference cancelling algorithm uses the estimates of d_s, d_c to synthesize the sinuosoidal disturbance and remove it from $y(m,n)$. The results of such a 2D interference cancelling filter is shown in Figure 12.18(c). The above example of artefact removal should be compared with the 1D 50 Hz interference cancelling filter discussed in Chapter 11.

The previous example serves to demonstrate the basic technique. A more realistic example of the potential use of 2D interference removal is shown in

Figure 12.19
Self-tuning
interference
cancellation of
corrupted
fingerprint: (a)
fingerprint
corrupted by
checkerboard
interference; and
(b) fingerprint
with
checkerboard
interference
adaptively
cancelled.

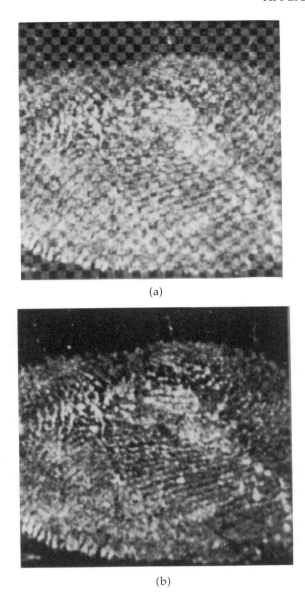

(a)

(b)

Figure 12.19. This shows a fingerprint which has been corrupted by a checker board pattern. Using the same methods as those outlined in the preceding paragraph, the checker board can be cancelled if the period of the checkering is known. The fingerprint recovered from Figure 12.19(a) in this way is shown in Figure 12.19(b).

12.8 SUMMARY

The key points made in this chapter are:

- One-dimensional self-tuning predictor, smoother and filter algorithms can be extended to 2D data
- An example of 2D data sources suitable for smoothing are images which require restoration to remove noise
- Many scanning sensors produce data which are effectively 2D in nature and advantage can be gained from self-tuning 2D prediction
- One-dimensional interference cancelling filters can also be extended to two dimensions and used for artifact removal in image restoration.

12.9 ASSIGNMENTS

Assignment 12.1

This is a theoretical assignment in which you are asked to extend 1D model order validation techniques to 2D ARMA models (e.g. equations (12.1b) and (12.2b)). Use the literature on model validation and model structure testing (see for example, 'System Identification' by Soderstrom and Stoica, Chapter 11) to develop a structure testing algorithm to determine the indices M,N for the 2D support $R(M,N)$.

Assignment 12.2

This is an algorithm development assignment. The raster scanning protocol assumed here is only one of the many which may be used in practice. Propose some alternative scanning protocols and associated supports. Then attempt to produce modified predictor and smoother algorithms for them. A straightforward starting point is to consider the so-called quarter plane support. You will need to read the existing theory on 2D models and supports. To this end, study the relevant sections of the book by Dudgeon and Merserau referred to in Section 12.10.

Assignment 12.3

This is both a theoretical and experimental assignment. Consider a 1D data source with periodic component of period T_p. Consider the datum at time step $T_p n+m$ to be a function of the shaded data shown in Figure A.12.1(a). Now stack data from successive periods as if they were rows in a 2D data

Figure A12.1

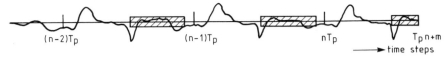

(a)

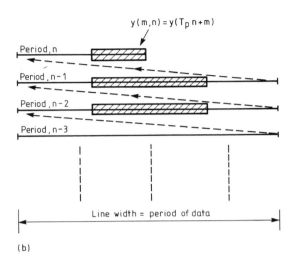

(b)

field, as shown in Figure A12.1(b). Consider how a 2D predictor could be applied to forecast k steps ahead in seasonal data. Note that conventional forecasting algorithms will separate out seasonal or periodic components from the stochastic parts of a data source. Your objective is to provide an algorithm which does not require this separation.

Assignment 12.4

This is a practical algorithm development assignment. The discussion in this chapter has not dealt with the edges of a data field. Propose practical ways of making a 2D smoother or predictor work at the edges of a 2D data field. Test your ideas using simulated data.

12.10 NOTES AND REFERENCES

The material for this chapter is based upon the following papers:

Wellstead, P.E., Wagner, G.R. and Caldas-Pinto, J.R.
    Two-dimensional adaptive prediction, smoothing and filter, *Proc. IEE*, Part F, **134** (3), 253, 1987.

Wellstead, P.E. and Caldas-Pinto J.R.
Self-tuning filters and predictors for two-dimensional systems, *Int. J. Control*, **42**(2), 457, 1985.

In addition, see the original research material in the PhD theses:

Caldas-Pinto J.R.
Two-Dimensional Self-Tuning Filtering, PhD Thesis, Control Systems Centre, UMIST, Manchester, 1983.
Wagner, G.R.
Self-Tuning Algorithms for Two-Dimensional Signal Processing, PhD Thesis, Control Systems Centre, UMIST, Manchester, 1985.

The view of 2D signal processing given here is specifically from the self-tuning systems aspect. A general mainstream treatment is given in:

Dudgeon, D.E. and Merserau, R.M.
Multidimensional Digital Signal Processing, Prentice-Hall, 1984.

See also the following text which includes a discussion on the stability of 2D polynomials:

Huang, T.S. (ed.)
Two-Dimensional Digital Signal Processing I, Springer, 1981.

The generalization of the prediction algorithm of section 12.4 that allows self-tuning prediction away from the current line has been reported in:

Heath, W.P. and Wellstead, P.E.
Self-tuning two dimensional predictors: a transfer function approach, *Proc. Sixth Multidimensional Signal Processing Workshop*, Monterey, California, USA, 1989.

The recursive estimators discussed here for 2D estimation are actually 1D estimators in which the parameters and data associated with the 2D process are concatenated in the θ and x vectors respectively. The estimator can be given a correct 2D formulation for the inclusion of forgetting factors. See:

Heath, W.P., Zarrop, M.B. and Wellstead, P.E.
Forgetting Factors for 2D Recursive Estimation, Control Systems Centre Report No. 724, UMIST, 1989.

For current papers and developments in multidimensional signal processing, the best source is the *IEEE Transactions on Acoustics, Speech and Signal Processing (ASSP)*.

13 Self-tuning Extremum Control

13.1 OUTLINE AND LEARNING OBJECTIVES

The idea of a self-tuning system which combines a recursive estimation algorithm with a synthesis algorithm can be applied in other areas besides feedback controller design. In this chapter we consider how this might be done in the case of static optimization of a performance function. This is an important practical problem since a number of industrial processes are such that their performance can be improved by adjusting plant parameters so as to maximize some performance criterion. The task of controlling a process to operate continually at the maximum point of a performance function is termed optimization or *extremum control*. Examples of systems which may involve extremum control include power generation systems and chemical and combustion processes.

Later in this chapter we will discuss in depth the extremum control of an automotive engine in order to illustrate the uses of self-tuning extremum control. The automotive application is a significant example of how extremum control can contribute to increased efficiency of an industrial process. As noted in Chapter 1, the use of self-tuning extremum control of the spark ignition system can lead to significant savings on fuel. Such economies are extremely important in industrial processes which use nonrenewable resources. Even when renewable resources are involved, such as in solar or wind power generation systems, the on-line optimization of system performance can be achieved by self-tuning extremum control.

In general, process optimization is a complex task which may involve significant computational effort and employ sophisticated algorithms which are capable of simultaneously optimizing a large set of parameters and complex performance criteria. Certain cases exist, however, in which the optimization problem can be reduced to a simple situation in which a *single* plant parameter $u(t)$ is manipulated so as to maximize a performance criterion as measured by the figure of merit $y(t)$ (see Figure 13.1). It is this simple but widely applicable case which is considered in this chapter. Section 13.2 outlines the nature of the extremum control problem. Section 13.3 describes some background ideas

Figure 13.1
Performance
function for a
single parameter
extremum
control problem
in which $y(t)$ is
the figure of
merit, and $u(t)$ is
the adjustable
parameter.

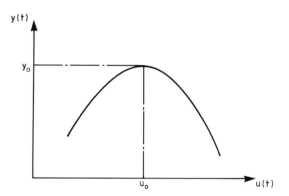

Figure 13.1
Performance function for a single parameter extremum control problem in which $y(t)$ is the figure of merit, and $u(t)$ is the adjustable parameter.

on performance optimization which prepare the way for the self-tuning extremum control material in Section 13.4.

In the remainder of the chapter, we treat implementation issues (Section 13.5) and a more detailed treatment of the application of self-tuning extremum control to automotive engine optimization.

The learning objective is to appreciate the problems of extremum control and to understand how self-tuning techniques can be applied in this important but neglected area of self-tuning technology.

13.2   PROBLEM OUTLINE

The aim of this chapter is to explore the use of self-tuning algorithms in a particular one-dimensional optimization problem. Accordingly, we begin by setting out the assumptions made concerning the performance function to be optimized.

Assumption (i)

The performance function is such that the relationship between the figure of merit $y(t)$ and the plant parameter $u(t)$ can be assumed to be quadratic. Thus:

$$y(t) = y_0 - a_0(u(t) - u_0)^2 \qquad (13.1)$$

where (see Figure 13.1) y_0 is the maximum attainable value of $y(t)$, u_0 is the value of $u(t)$ which maximizes $y(t)$ and a_0 is the sensitivity (or curvature parameter) of the quadratic curve.

The assumption of a quadratic relationship between $y(t)$ and $u(t)$ is important since later we will assume a quadratic model, based on equation (13.1), for estimation purposes. The quadratic assumption is in most cases acceptable for extremum controllers operating close to the optimum point.

Assumption (ii)

The variable $y(t)$ may be subject to zero mean noise $e(t)$ such that

$$y(t) = y_0 - a_0(u(t) - u_0)^2 + e(t) \tag{13.2}$$

In practice this noise may be due to measurement errors (sensor noise) or it may reflect random variations in value of the performance criterion. It is further assumed that this noise is white and uncorrelated with $u(t)$.

Assumption (iii)

The parameters y_0, u_0, a_0 which characterize the performance function are unknown but nominally constant.

Note that the figure of merit $y(t)$ and the adjustable parameter $u(t)$ are written as functions of the time step t. This is in order to emphasize that in extremum control the variable $u(t)$ functions as a control input which will be varied with time; likewise $y(t)$ functions as the process output. The role of an extremum controller is, as shown in Figure 13.2, to manipulate the input $u(t)$ as a function of the output measurements $y(t)$. In a general extremum control situation, the input $u(t)$ to the performance function $f(\cdot)$ actuates via a dynamical system $G_1(s)$. Likewise, the output $y(t)$ is sensed after it passes through a further dynamical system $G_2(s)$ as shown in Figure 13.3. For the purposes of

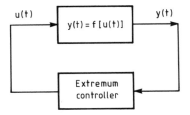

Figure 13.2 An extremum control system, where $f(\cdot)$ is a quadratic function.

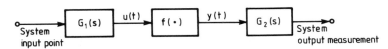

Figure 13.3 Extremum performance function with dynamical systems $G_1(s)$ and $G_2(s)$ operating on the input and output processes $u(t)$, $y(t)$, respectively.

this chapter, it is assumed that the relationship between $y(t)$ and $u(t)$ does not involve any dynamical components so that the transfer functions $G_1(s)$, $G_2(s)$ can either be ignored or do not exist. This is the case in most practical situations.

13.3   SOME OPTIMIZATION BACKGROUND

An off-line optimization approach to the problem in hand would be to use hill-climbing techniques to move from an initial position, by a series of steps, to the optimum point.

The simplest hill-climbing algorithms (see Figure 13.4) use no information about the shape of the performance curve. From an initial position u_1 a trial movement to $u_1 + h$ is made. The values of y_1 and y_2 are then compared. If y_2 is larger than y_1 then $u_2 = u_1 + h$ is adopted as the value of $u(t)$, otherwise $u_2 = u_1 - h$ is used. The process of making trial movements is continuous.

Figure 13.4 A simple hill-climbing algorithm in which steps of fixed length h are made until a decrease in y is sensed. The step direction is then reversed.

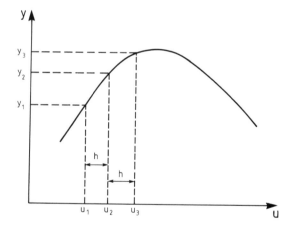

Eventually, and at a rate depending upon the size of h, the value of $u(t)$ lies in a neighbourhood of the optimal point u_0. The algorithm then operates so as to move backward and forward either side of the optimum. This form of approach assumes nothing about the performance curve parameters and is rather robust. Its drawbacks are:

(i) It can be very slow to converge if h (the step length) is small; this is a disadvantage in extremum control when the cost of nonoptimal behaviour is high.

(ii) If h is large, however, the algorithm will step substantial distances either side of the optimum when it is supposedly at the optimum.

(iii) The algorithm can make steps in the wrong direction unless the figure of merit is suitably filtered to remove the noise, modelled by $e(t)$ in equation (13.2).

The simple fixed step hill-climbing algorithm can be improved if the gradient dy/du is known. The gradient information allows the hill-climbing *rate* to be adjusted according to the size of the gradient. In this case the adjustment rule from u_1 to the new value u_2 of $u(t)$ is given by

$$u_2 = u_1 + \gamma \left[\frac{dy}{du} \right]_{u=u_1} \tag{13.3}$$

where γ is a gain parameter set by the user and which controls the convergence rate.

The advantage of the gradient algorithm is that the size of the adjustment in u at each step is dependent upon the size of the gradient. The adjustment step is large when the algorithm is far from the optimum and reduces accordingly as the optimum is achieved. The gradient hill-climber is generally reliable and is much faster than the fixed step length algorithm. Its disadvantages are:

(i) It requires a measurement of the gradient function and the gain γ must be selected by the user.

(ii) The gradient measurement can be very sensitive to additive noise, causing corresponding erratic behaviour.

The next stage of sophistication beyond the gradient method is to assume that the second derivative d^2y/du^2 is available. In this case the adjustment rule is:

$$u_2 = u_1 - \gamma \left[\frac{dy}{du} \bigg/ \frac{d^2y}{du^2} \right]_{u = u_1} \tag{13.4a}$$

This is known as a Gauss–Newton type algorithm and has particularly fast convergence properties in most cases. In particular, if the relationship between $u(t)$ and $y(t)$ is quadratic (as in equation (13.1)), then the noise-free algorithm may converge in one step, e.g.

Suppose

$$y(t) = y_0 - a_0(u_0 - u(t))^2$$

then

$$\left[\frac{dy}{du}\right]_{u_1(t)} = 2a_0(u_0 - u_1(t)) \qquad (13.4b)$$

and

$$\frac{d^2y}{du^2} = -2a_0$$

Hence, using (13.4a) with $\gamma = 1$, yields

$$u_2(t) = u_1(t) - \{u_1(t) - u_0\} = u_0 \qquad \text{as required}$$

The key disadvantage of the Gauss–Newton optimization algorithm is that it requires a knowledge of both the gradient and second derivative of the performance function. Even if these objects can be measured, the measurements may be heavily distorted by the effects of the additive noise $e(t)$. The aim of self-tuning extremum control is to obtain a performance comparable with the Gauss–Newton optimization algorithm even in the presence of noise.

13.4   SELF-TUNING EXTREMUM CONTROL

The derivatives required for Gauss–Newton hill-climbing are not generally known. The idea behind self-tuning optimization is, in effect, to use a recursive estimation procedure to identify these derivatives and at each adjustment step use the current best estimates to determine the new value of $u(t)$. This idea is illustrated in Figure 13.5; note that this structure is analogous to the self-tuning controller structure except that the controller synthesis rule is replaced by a rule which computes the extremum control signal $u(t)$. In terms of the hill-climbing algorithms of Section 13.3, this is an on-line procedure for computing the hill-climber step length. The key feature of self-tuning extremum control is the basic assumption of a model which describes the performance function. The form used here is the quadratic model defined by equation (13.1):

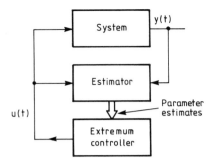

Figure 13.5 A block diagram of a self-tuning extremum controller. The estimator uses data $u(t)$, $y(t)$ to determine the parameters of the system's performance characteristic. The controller then uses the parameter estimates to determine the optimum operating point.

$$y(t) = y_0 - a_0(u(t) - u_0)^2 \qquad (13.1) \text{ bis}$$

This can be rewritten in expanded form as

$$y(t) = \theta_0 + \theta_1 u(t) + \theta_2 u^2(t) \qquad (13.5)$$

where $\theta_0 = y_0 - a_0 u_0^2$, $\theta_1 = 2a_0 u_0$, $\theta_2 = -a_0$.

The representation of equation (13.5) involves a constant coefficient θ_0. This coefficient can be eliminated by considering an incremental form of the quadratic model. This is obtained by writing equation (13.1) at times t and $t-1$ and subtracting the two resulting expressions. The resultant incremental quadratic model is given by

$$\Delta y(t) = \theta_1 \Delta u(t) + \theta_2 \Delta u^2(t) \qquad (13.6)$$

where

$$\Delta y(t) = y(t) - y(t-1)$$

$$\Delta u(t) = u(t) - u(t-1)$$

$$\Delta u^2(t) = u^2(t) - u^2(t-1)$$

Note that equation (13.6) involves incremental information concerning $u(t)$, $y(t)$. The incremental model has the advantage that only two parameters θ_1, θ_2 occur in the model. This reduces computation in the estimation phase of a self-tuning extremum controller. Moreover, only θ_1, θ_2 are required to compute the extremum controller output. Data differencing also has a beneficial effect upon the scaling of the estimator.

The basic self-tuning extremum control algorithm consists of using a recursive estimator to compute θ_1, θ_2 in the equation (13.6) and then using these estimates to compute an estimate of the location, u_0, of the optimum value. This estimated value of u_0 is used as the input $u(t+1)$ to the system at the next sample interval.

Using the incremental representation (13.6), the self-tuning extremum control recursion can be written as follows:

Algorithm: Explicit Self-tuning Extremum Controller

At time step t:

Step 1   Apply input $u(t)$ and measure $y(t)$, the system figure of merit.
Step 2   Use a recursive least squares estimator to update the estimates $\hat{\theta}_1$, $\hat{\theta}_2$ in the model

$$\Delta y(t) = \hat{\theta}_1 \Delta u(t) + \hat{\theta}_2 \Delta u^2(t) + \hat{e}(t) \tag{13.7}$$

Step 3   Calculate the new input $u(t+1)$ as

$$u(t+1) = -\frac{\hat{\theta}_1}{2\hat{\theta}_2} \tag{13.8}$$

Step 4   Increment the time counter $t \rightarrow t+1$ and loop back to Step 1.

□

The self-tuning recursion continues until $\hat{\theta}_1$ and $\hat{\theta}_2$ have converged to their theoretically correct values

$$\left.\begin{aligned} \theta_1 &= 2u_0 a_0 \\ \theta_2 &= -a_0 \end{aligned}\right\} \tag{13.9}$$

at which point the adjustment rule (equation (13.8)) will place the system at the optimum position.

13.5   COMPARISON WITH THE GAUSS–NEWTON ITERATION

The self-tuning extremum controller can be interpreted as a direct implementation of the Gauss–Newton iteration using *estimated* derivative information. To be specific, in equation (13.8) the adjustment is decided by dividing $\hat{\theta}_1$ by

$-2\hat{\theta}_2$. But $\theta_2 = -a_0$, so that (from equation (13.4b)) the quantity $-2\hat{\theta}_2$ is actually an estimate of d^2y/du^2. This point can be taken further by casting the adjustment rule (13.8) in a form which parallels equation (13.4a). Add and subtract $u(t)$ from equation (13.8) to give

$$u(t+1) = u(t) - \left[\frac{\hat{\theta}_1 + 2\hat{\theta}_2 u(t)}{2\hat{\theta}_2}\right] \tag{13.10}$$

Writing

$$\hat{\theta}_1 = 2\hat{a}_0\hat{u}_0$$

$$\hat{\theta}_2 = -\hat{a}_0$$

we have the modified form for the adjustment rule

$$u(t+1) = u(t) + \frac{2\hat{a}_0(\hat{u}_0 - u(t))}{2\hat{a}_0} \tag{13.11}$$

Comparing this equation with equations (13.4a) and (13.4b), the equivalence of the self-tuning adjustment rule (13.8) and the Gauss–Newton iteration can be seen.

13.6 MODIFIED SELF-TUNING EXTREMUM CONTROL

13.6.1 Single parameter algorithm

In certain cases the quadratic sensitivity parameter a_0 may be known either exactly or to within certain limits. If this is so, the two-parameter self-tuning optimizer may be simplified to a single parameter self-tuner. Equation (13.6) can be rewritten as

$$\Delta y(t) + a_0\Delta u^2(t) = 2u_0 a_0\Delta u(t) \tag{13.12}$$

Defining the auxiliary variable $\Delta w(t)$ as

$$\Delta w(t) = \Delta y(t) + a_0\Delta u^2(t) \tag{13.13}$$

the self-tuning algorithm becomes:

Algorithm: Single Parameter Explicit Self-tuning Extremum Controller

At time step t:

Step 1   Apply adjustment $u(t)$, measure $y(t)$ and compute the auxiliary variable $\Delta w(t)$ according to equation (13.13).

Step 2   Use a RLS estimator to update the estimate $\hat{\alpha}_1$ in the model

$$\Delta w(t) = \hat{\alpha}_1 \Delta u(t) + \hat{e}(t) \tag{13.14}$$

Step 3   Calculate the new adjustment $u(t+1)$ as

$$u(t+1) = \frac{\hat{\alpha}_1}{2a_0} \tag{13.15}$$

Step 4   Increment the time counter $t \rightarrow t+1$ and loop back to Step 1.

$\square$

The advantages of the simplified algorithm are a reduction in computational effort in the estimation process and increased speed of convergence. If, however, the value of a_0 is incorrect then the adjustment rule can be interpreted as a Gauss–Newton iteration with a variable step length, as follows. Suppose that the value of a_0 is incorrectly set at α, then

$$u(t+1) = \frac{\hat{\alpha}_1}{2\alpha} = u(t) - \gamma(t) \left[\frac{dy}{du} \right] \bigg/ \left[\frac{d^2y}{du^2} \right]$$

where

$$\gamma(t) = \frac{u(t) - (\alpha/a_0)u_0}{u(t) - u_0}$$

a variable gain which will take values close to unity if $\alpha \simeq a_0$. Clearly, unless $\alpha = a_0$, the adjustment will either be less than required so that the convergence process will tend to be like an overdamped step response, or more than required such that the estimate oscillates.

A further potential problem with fixing a_0 at an incorrect value α is that this effectively introduces an additional component of noise that is correlated with the data $\Delta u(t)$. Specifically, fixing a_0 at α will introduce the noise component $(a_0 - \alpha)\Delta u^2(t)$ in the estimation equation. Correlated noise of this kind is known to cause bias in least squares estimation (see Chapter 3).

Interestingly, however, the correlated noise is proportional to the increment $\Delta u^2(t) = u^2(t) - u^2(t-1)$, so that, if the algorithm converges to a constant $u(t)$, then the correlated component of noise will also vanish (see Section 13.9).

13.6.2 Implicit extremum control

An implicit self-tuning algorithm is, as noted in the previous Chapters 8 and 10, one in which the estimation phase is arranged to directly yield the controller parameters. Self-tuning extremum control can be cast into implicit form as follows. Note that equation (13.6) can be rearranged in the form:

$$\Delta u^2(t) = \theta_1 [2\Delta u(t)] + \theta_2 \Delta y(t) \tag{13.16}$$

where $\theta_1 = u_0$, $\theta_2 = -a_0^{-1}$.

The parameter θ_1 is now the optimum value of $u(t)$. Hence, in the corresponding self-tuning algorithm, the estimated model parameter $\hat{\theta}_1$ constitutes the control action $u(t+1)$.

13.6.3 Parameter constraints

In most extremum applications there will be some knowledge of the bounds on the parameters u_0, a_0. For example, the sensitivity parameter a_0 will have a minimum value of zero and some maximum value. Likewise bounds will be known for the optimum location. These bounds can be used to constrain the extremum controller action in one of two ways. Either the parameter estimates $\hat{\theta}_1$, $\hat{\theta}_2$ can be checked against known bounds and constrained if necessary, or the extremum control action $u(t)$ itself can be checked against bounds and constrained appropriately. Both approaches give similar results. Note that, because of the nondynamic nature of the basic quadratic model (13.1), bounding the extremum controller action will automatically bound $y(t)$. In this static case, therefore, the (usually problematic) stability problem is easily resolved.

13.7 SIMULATION EXAMPLES

In this section we illustrate the behaviour of the extremum controller in simulated conditions.

Example 13.1

The simulated performance function is defined by

Figure 13.6
Evolution of the
parameter
estimates for a
self-tuning
extremum
controller. The
underlying
system is
described by
$y(t) = 40 - 0.2$
$(u(t)-1)^2 + e(t)$,
where $e(t)$ is
unit variance
zero mean white
noise (Example
13.1).

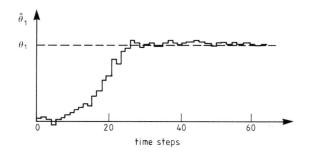

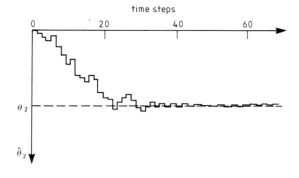

$$y(t) = 40 - 0.2(u(t) - 1)^2 + e(t) \tag{13.17}$$

where $e(t)$ is a unit variance, zero mean white noise process.

An incremental model (equation (13.7)) is used to describe the system. Figure 13.6 shows the evolution of the parameter estimates to their final values. The corresponding values of $u(t)$ and $y(t)$ are shown in Figure 13.7. The self-tuning controller achieves the optimum value of $y_0 = 40$ after about 25 time steps and holds it there with minor perturbations. These perturbations are caused by the addition of a test signal $u_{pert}(t)$ to the system as shown in Figure 13.8. The test perturbation used here is the sequence $\{-1,0, +1,0\}$ repeated periodically. Similar results are obtained with other configurations so that Figures 13.6 and 13.7 can be taken as typical. The need for a test perturbation and the selection of test signals are discussed below (Section 13.8).

□

Example 13.2

The next simulation shows the behaviour of the extremum controller when parameter estimates are unconstrained and enter inadmissible regions of

Figure 13.7
Extremum
control signal
$u(t)$ and process
output $y(t)$
corresponding to
simulation
Example 13.1.

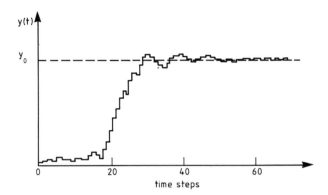

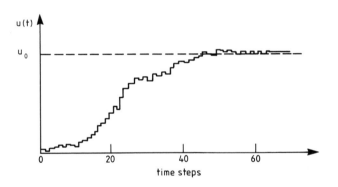

Figure 13.8
Self-tuning
extremum
controller,
showing the
addition of a
text perturbation
signal $u_{\mathrm{pert}}(t)$.

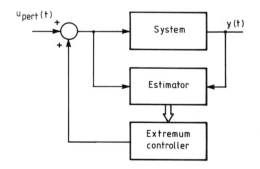

Figure 13.9
Illustrating the
behaviour of a
self-tuning
extremum
controller in
which the $\hat{\theta}_2$
estimate passes
through zero.
The
experimental
setup is that
described in
simulation
Example 13.1.

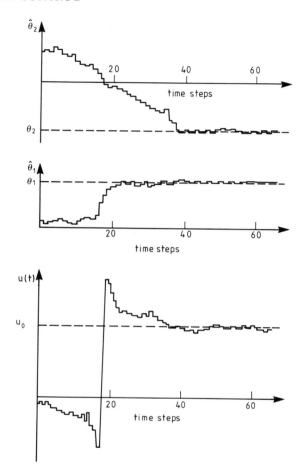

parameter space. For example, $\hat{\theta}_2$ may approach zero. If this is not prevented, then extremely large control values are generated when dividing by $\hat{\theta}_2$ (equation (13.8)). Figure 13.9 shows an example of this behaviour when the initial value of $\hat{\theta}_2$ is set with the incorrect sign. The estimate of $\hat{\theta}_2$ must therefore pass through zero in order to converge to its correct value. Note that as it does so a dramatic disturbance in $u(t)$ occurs when $\hat{\theta}_2$ is close to zero. This kind of behaviour is prevented by constraining the control signal or the parameter estimates.

□

13.8   IMPLEMENTATION ISSUES

In a practical extremum control situation, the self-tuning extremum algorithm of Section 13.4 will need some modifications. These modifications are required in order to meet practical requirements associated with existing control strategies and additional features which will improve the performance of the self-tuning algorithm itself. The main modification required in order to fit in with existing control strategies concerns the incorporation of *a priori* knowledge of the optimum location. In particular, consider the optimization of an industrial process. In almost all cases the factory manager responsible for the process will have a reasonable idea of the optimal setting for $u(t)$. It makes sense, therefore, to operate the self-tuner in a way which uses this knowledge. We call this the factory setting, $u_{fact}(t)$, and denote it as a function of time in order to emphasize that its value might change with time (or possibly some other parameter) as the process proceeds.

A further feature which is necessary in order to make the self-tuning extremum control work effectively is the addition of a test perturbation $u_{pert}(t)$. The test perturbation is necessary to ensure convergence of the extremum controller and has certain preferred forms. In particular, a three-level periodic test perturbation $\{\ldots -\gamma, 0 +\gamma, 0, -\gamma, \ldots\}$, as used in Example 13.1, has been found most satisfactory.

The self-tuning extremum controller incorporating factory settings and a test perturbation is illustrated in Figure 13.10. Note that in this form the extremum controller output is labelled $u_{adj}(t)$ in order to emphasize that it is an *adjustment* to be added to $u_{fact}(t)$ in order to locate $u(t)$ at the system optimum. The self-tuning algorithm now becomes:

Figure 13.10 Self-tuning extremum controller incorporating factory settings $u_{fact}(t)$ for the optimum location and test perturbation $u_{pert}(t)$.

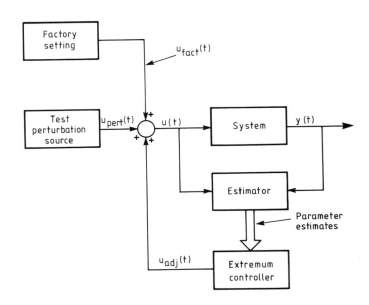

Algorithm: Explicit Self-tuning Extremum Controller with Factory Setting

At time step t:

Step 1   Apply adjustment $u(t) = u_{\text{fact}}(t) + u_{\text{eff}}(t)$ and measure $y(t)$. (Note: $u_{\text{eff}}(t) = u_{\text{adj}}(t) + u_{\text{pert}}(t)$.)

Step 2   Compute the RLS estimates $\hat{\theta}_1, \hat{\theta}_2$ using the model

$$\Delta y(t) = \hat{\theta}_1 \Delta u(t) + \hat{\theta}_2 \Delta u^2(t) + \hat{e}(t) \qquad (13.6)\text{bis}$$

where

$$\Delta u(t) = u_{\text{eff}}(t) - u_{\text{eff}}(t-1)$$

$$\Delta u^2(t) = u^2_{\text{eff}}(t) - u^2_{\text{eff}}(t-1)$$

Step 3   Calculate the new adjustment $u(t+1)$ as

$$u_{\text{adj}}(t+1) = -\frac{\hat{\theta}_1}{2\hat{\theta}_2} \qquad (13.18)$$

Step 4   Increment the time counter $t \to t+1$, and loop back to Step 1.

$\square$

The effect of the factory supervisor setting $u_{\text{fact}}(t)$ in this implementation is to 'back-off' a component of the signal $u(t)$, so that, as far as the self-tuner is concerned, the optimum is located at $u_0 - u_{\text{fact}}(t)$. Thus, after tuning, the algorithm should converge so that

$$u_{\text{adj}}(t) \to u_0 - u_{\text{fact}}(t)$$

and the effective extremum function shape is offset as shown in Figure 13.11
   A further practical point arises if the process being optimized has a dynamical component. The introduction of an adjustment $u(t)$ at step (i) will produce a transient component as well as the steady component of response $y(t)$. It is therefore usual to wait for a suitable period between the application of $u(t)$ and the measurement of $y(t)$ or take some other measure to ensure that the dynamics do not interfere with the extremum seeking process.

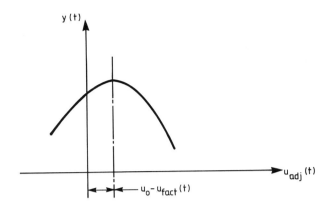

Figure 13.11
Effective process performance function when a factory setting $u_{\text{fact}}(t)$ is used for the approximate location of the optimum.

13.9 LOCAL CONVERGENCE

The ODE method of Chapter 6 can be used to examine the asymptotic properties of the extremum controller and hence check its local stability characteristics. We consider the two parameter case where the system is defined by equation (13.5) and the model used in the estimator is that given in equation (13.6). The output of the extremum controller is given by equation (13.8) in its basic form. This is an estimate of the extremum location u_0, to which is added a perturbation signal $u_{\text{pert}}(t)$. As noted previously the addition of a factory setting serves only to shift the effective location of the optimum. For this reason it will be ignored in this analysis. Thus the value of $u(t)$ is derived according to

$$u(t) = \hat{u}_0(t) + u_{\text{pert}}(t) \tag{13.19}$$

where

$$\hat{u}_0(t) = -\,\hat{\theta}_1/2\hat{\theta}_2 . \tag{13.20}$$

If the algorithm converges then

$$\hat{u}_0(t) \to u_0 = -\,\theta_1/2\theta_2$$

and

$$\Delta u(t) \to \Delta u_{\text{pert}}(t)$$

$$\Delta u^2(t) \to \Delta u_{\text{pert}}^2(t) - \frac{\theta_1}{\theta_2}\,\Delta u_{\text{pert}}(t)$$

$$\left.\right\} \tag{13.21}$$

Now, writing the system model (13.7) in vector form yields

$$\Delta y(t) = \Delta \mathbf{u}^{\mathrm{T}}(t)\, \hat{\boldsymbol{\theta}}$$

$$\Delta \mathbf{u}^{\mathrm{T}}(t) = [\Delta u(t),\, \Delta u^2(t)] \tag{13.22}$$

$$\hat{\boldsymbol{\theta}}^{\mathrm{T}} = [\hat{\theta}_1,\, \hat{\theta}_2]$$

The relevant ODE arrays associated with the recursion are

$$\mathbf{H}(\boldsymbol{\theta}) = \mathbf{R}^{-1}\, \frac{d\mathbf{f}(\boldsymbol{\theta})}{d\boldsymbol{\theta}} \tag{13.23}$$

where

$$\mathbf{f}(\boldsymbol{\theta}) = \bar{E}(\Delta \mathbf{u}(t)\, [\Delta y(t) - \Delta \mathbf{u}^{\mathrm{T}}(t)\boldsymbol{\theta}]) \tag{13.24}$$

$$\mathbf{R} = \bar{E}\{\Delta \mathbf{u}(t)\, \Delta \mathbf{u}^{\mathrm{T}}(t)\} \tag{13.25}$$

and

$$\bar{E}(\cdot) = \lim_{N \to \infty} \frac{1}{N} \sum_{t=1}^{N} E(\cdot)$$

is the generalized expectation operator.

If the estimates $\hat{\boldsymbol{\theta}}(t)$ associated with the optimizer converge to $\boldsymbol{\theta}^*$, then

$$\mathbf{f}(\boldsymbol{\theta}^*) = 0$$

and $\mathbf{H}(\boldsymbol{\theta}^*)$ has all its eigenvalues in the closed left half plane.

Using equations (13.21) it can be shown that

$$\mathbf{f}(\boldsymbol{\theta}) = \begin{bmatrix} f_1(\boldsymbol{\theta}) \\ f_2(\boldsymbol{\theta}) \end{bmatrix}$$

with

$$f_1(\boldsymbol{\theta}) = (a_0 - \theta_2)\, \rho_{12} - 2a_0 \left[u_0 + \frac{\theta_1}{2\theta_2} \right] \rho_{11} \tag{13.26}$$

$$f_2(\boldsymbol{\theta}) = (a_0 - \theta_2)\, \rho_{22} - 2a_0 \left[u_0 + \frac{\theta_1}{2\theta_2} \right] \rho_{12} - \frac{\theta_1}{\theta_2} f_1(\boldsymbol{\theta}) \tag{13.27}$$

where

$$\rho_{ij} = E\left[\Delta u^i_{\text{pert}}(t)\, \Delta u^j_{\text{pert}}(t)\right] \qquad (13.28)$$

It follows that $\theta_1 = -2a_0u_0$, $\theta_2 = a_0$ is the unique locally stable convergence point as required provided that

$$\rho_{11}\rho_{22} > \rho_{12}^2 \qquad (13.29)$$

This last condition is a persistent excitation requirement on the test perturbation without which the desired convergence properties are not guaranteed. The $\mathbf{H}(\theta^*)$ matrix has both its eigenvalues at minus one. This occurs independently of the perturbation signal, provided that it is persistently exciting.

Figures 13.12 show the ODE phase portraits for Example 13.1 for (a) binary $u_{\text{pert}} = \{\ldots -1, +1, -1, +1, \ldots\}$ and (b) three-level $u_{\text{pert}} = \{\ldots -1, 0, +1, 0, -1, \ldots\}$. Note that (a) gives rise to a parameter identifiability problem. The trajectories converge only to the line $\theta_2 = -2\theta_1$, but the correct optimum is still achieved. Case (b) gives parameter consistency. Finally, we note that this local convergence analysis can be extended to justify using quadratic models to approximate a wide class of concave functions (see Section 13.12 Assignments).

13.10 AUTOMOTIVE APPLICATION

In Chapter 1, the self-tuning optimization of automotive petrol engine performance was presented as an example of the application of self-tuning ideas to a typical engineering system. The aim in that initial discussion was to motivate the reader by highlighting the uses and advantages of self-tuning. In this section we return to this application and describe how self-tuning extremum control can be used to provide self-adaptive optimization of automotive engine performance.

The need for performance optimization arises because in a spark ignition engine the variations in engine torque are a nonlinear function of variables such as the spark angle, the air/fuel ratio, the engine speed and load. For a constant speed and load, the engine performance function (as measured by engine torque) will vary with spark angle and air/fuel ratio. Typically, the engine performance function will resemble the dome-shaped characteristic shown in Figure 13.13, where the top of the dome corresponds to the maximum value of the performance function. In most currently produced petrol engines the air/fuel ratio cannot be independently controlled (although this situation

Figure 13.12 (a)
ODE phase
portrait for
Example 13.1
(binary
perturbation).
(b) ODE phase
portrait for
Example 13.1
(three-level
perturbation).

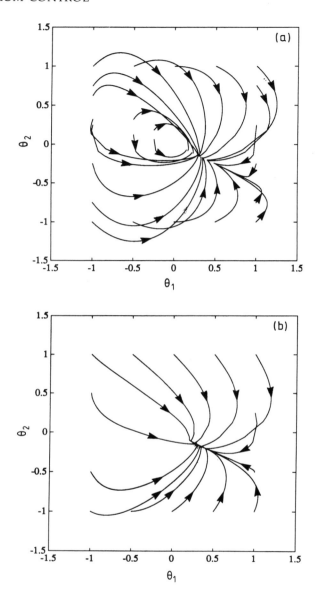

Figure 13.12 (a) ODE phase portrait for Example 13.1 (binary perturbation). (b) ODE phase portrait for Example 13.1 (three-level perturbation).

is expected to change as fuel injection systems develop). The spark ignition angle, however, can be readily controlled using the electronic ignition systems installed in modern engine control units (ECUs). Assume, for the moment, that the air/fuel ratio is constant at the value corresponding to the cross-section XX indicated in Figure 13.13 through the performance function. The problem is now reduced to one-dimensional extremum control, where the control variable $u(t)$ is the spark ignition angle (measured in degrees before

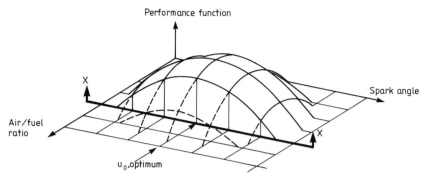

Figure 13.13 The performance function of a spark ignition engine at constant speed and load. For a constant air/fuel ratio the performance function becomes the one-dimensional object obtained from the cross-section XX.

top dead centre (°btdc)) and the performance function $y(t)$ is some measure (assumed quadratic) of engine torque:

$$y(t) = y_0 - a_0(u(t) - u_0)^2 \qquad\qquad (13.1)\text{bis}$$

In a real engine the load and speed will vary as the vehicle driver varies the throttle and gear setting. This in turn will cause the performance function and hence the parameters y_0, u_0, a_0 to vary with time. Figure 13.14 shows how the performance function (here measured directly in terms of engine brake torque) might vary with changes in throttle setting. The task of the self-tuning extremum controller is to track the time variations and continually operate the engine at a predetermined optimum value of the performance function. In a practical ECU the self-tuning extremum controller will be required to calculate the optimum spark angle position for a set of load and speed conditions. As indicated in Chapter 1, this set of spark angle settings is stored in the form of map values indexed against speed and manifold pressure (which gives a measure of engine load). Figure 13.15 illustrates a typical engine spark angle map. In a conventional system this map will be a predetermined object based on a series of engine trials at the factory. The aim of a self-tuning extremum controller is to provide an *adjustment* map. This map contains the estimated optimal locations obtained by the self-tuning extremum controller under the assumption that the engine is already running at the factory setting $u_{\text{fact}}(t)$ for the current load and speed. The self-tuning extremum controller for this situation is shown in block diagram form in Figure 13.16. Note that this uses the factory setting idea introduced in Section 13.8 but extended so that the factory settings are associated with a map of settings one for each of a matrix of load and speed settings. The adjustment map will have the same form and will be a measure of how far away from the optimum the engine factory

Figure 13.14
Illustrating the
variations
in engine
performance
function (given
here as brake
torque) with
throttle angle
setting.

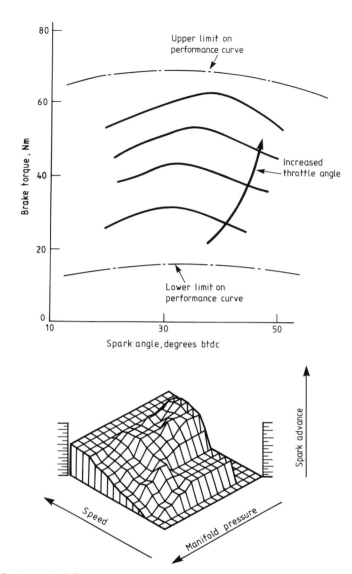

Figure 13.15   A typical factory produced map of spark angle setting for different values of load (as determined by inlet manifold pressure) and speed. Note that the speed and load have been quantized into a matrix of discrete values with a spark setting for each value within the matrix.

settings are for the particular engine. The discussion in Chapter 1 gives figures which indicate that nominally identical engines can have rather different optimized spark angle maps.

The typical behaviour of a self-tuning extremum controller applied to a spark angle optimization process is shown in Figures 13.17 and 13.18. The

Figure 13.16
Self-tuning
extremum
controller for
adaptive spark
angle control in
a spark-ignition
engine.

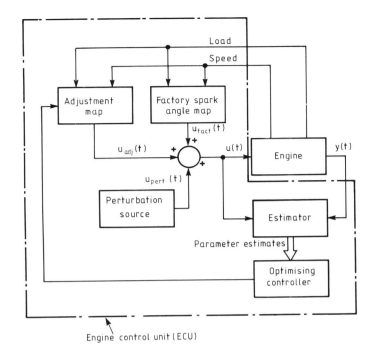

Engine control unit (ECU)

engine used in this trial was a medium sized petrol engine of the type found in family saloon cars. Figure 13.17 shows the evolution of the spark angle applied to the engine when the engine self-tuner was initialized with $u(0) = 20°$ before top dead centre (°btdc). The anticipated location of the optimum was at 32°btdc. Note that the plots incorporated a three-level test perturbation of the form $\{-2°, 0, 2°, 0, \ldots\}$ and that tuning is achieved in 40 cycles of the self-tuner in each case. In this application, one cycle corresponds to one firing of a cylinder. Thus, for a four-stroke four-cylinder engine rotating at 1200 rev/min, the tuning-in time is approximately 1 s. This is typical of the two-parameter optimizer in its standard form (Section 13.4).

Figure 13.18 illustrates a similar self-tuning test, in which no parameter constraints were placed on θ_2. As a result, when θ_2 approaches zero (optimization step 160) the spark angle becomes erratic before the transient forces a rapid retuning. The use of parameter constraints (which were used in the experiment illustrated in Figure 13.17) removes this form of behaviour (see Section 13.6.3). The tuning-in performance can also be improved beyond that shown in Figures 13.17 and 13.18. A useful technique in this respect is the use of variable amplitude perturbation signals. This procedure involves sensing when a change in the optimum location has occurred and temporarily increasing the size of the perturbation signal $u_{\text{pert}}(t)$ for a short period while

Figure 13.17
Illustrating the adaptive behaviour of a self-tuning extremum controller for spark angle optimization from the initial angle setting of 20°btdc. The actual optimum was thought to be 32°btdc.

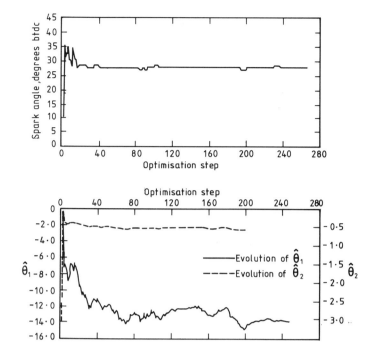

the extremum controller retunes. Figure 13.19 shows the improvement in retuning time which can be obtained in this way. In particular, for the example shown, the retuning time is reduced from 95 steps to about 60 steps.

13.11   SUMMARY

This chapter has been concerned with self-tuning extremum control. The following main points have been covered:

- The background, motivation and advantages of extremum control
- The description of a basic self-tuning extremum control algorithm and its relationship to traditional hill-climbing methods of optimization
- The modifications and implementation issues which are associated with practical self-tuning extremum control
- A basic treatment of the convergence properties of self-tuning extremum control using ODE methods
- A discussion of the use of self-tuning extremum control applied to spark ignition engine optimization

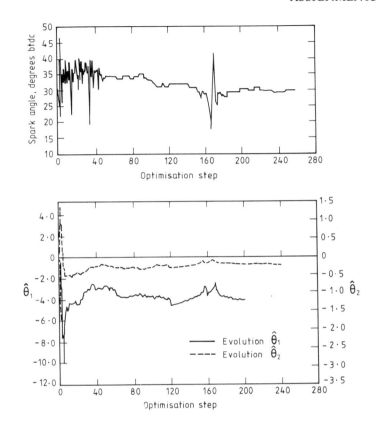

Figure 13.18 Illustrating the adaptive behaviour of a self-tuning extremum controller for spark angle optimization without using parameter constraints. Note the erratic behaviour when θ_2 approaches zero (optimization step 160).

13.12   ASSIGNMENTS

Assignment 13.1

Design a certainty equivalent self-tuning extremum controller to find a stationary point u_0 of the function $f(u)$. A datum is given by

$$y(t) = f(u(t)) + e(t)$$

in the usual notation and one of the following quadratic models is used:

(a) $y(t) = \theta_0 + \theta_1 u(t) + \theta_2 u^2(t)$

(b) $\Delta y(t) = \theta_1 \Delta u(t) + \theta_2 \Delta u^2(t)$

(c) As in (a), but θ_2 is fixed

(d) As in (b), but θ_2 is fixed

Figure 13.19
Illustrating the
use of variable
perturbation
signals to
improve the
convergence rate
of a self-tuning
extremum
controller for
spark angle
estimation:
(a) fixed
perturbation
amplitude;
(b) variable
perturbation
amplitude.

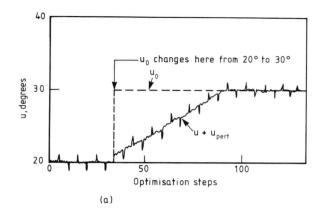

(a)

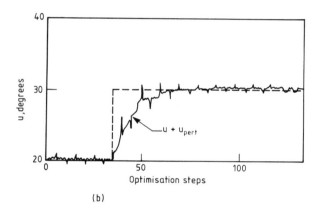

(b)

Use the ODE approach to analyze the local convergence properties in each case, commenting on the following:

(i) What happens if

$$f(u_0+u) = f(u_0-u)$$

i.e. $f(\cdot)$ is a symmetric function in some interval enclosing the stationary point?

(ii) What happens if the dither signal becomes very small?

(iii) Does fixing the θ_2 parameter always help?

Assignment 13.2

This is an algorithm development assignment. Given a process output $y(t)$ which is a quadratic function of two inputs $u_1(t)$, $u_2(t)$, develop the corresponding explicit self-tuning extremum control algorithm. Investigate different forms of algorithm including implicit forms and fixed parameter forms.

Assignment 13.3

The material in this chapter assumes that there are no dynamics associated with the system which is to be subjected to extremum control. Investigate how you would apply a self-tuning extremum control to a system in which linear dynamics appear:

(a) in front of the quadratic function (Wiener model);

(b) at the output of the quadratic function (Hammerstein model).

Assignment 13.4

Local convergence theory indicates that the self-tuning extremum control based on a quadratic model may converge to the optimum even when the function $f(\cdot)$ is nonquadratic (see Assignment 13.1). Test this with various nonquadratic functions. For example, try

(a) the gaussian function

$$y(t) = (2\pi\sigma_u^2)^{-\frac{1}{2}} \exp\left(-\frac{(u(t) - \mu_u)^2}{2\sigma_u^2}\right)$$

(b) the Rayleigh distribution function

$$y(t) = \frac{u(t)}{b^2} \exp\left(\frac{u^2(t)}{2b^2}\right); \qquad u(t) > 0$$

(c) the Cauchy distribution function

$$y(t) = \frac{a}{\pi}\left(\frac{1}{u^2(t) + a^2}\right)$$

In each case suggest a transformation of variables which will allow a quadratic model to apply correctly.

Assignment 13.5

The convergence analysis of Section 13.9 tells us that the perturbation signal $u_{pert}(t)$ is necessary for convergence. Convince yourself of this and devise extremum control algorithms where u_{pert}

(a) is a function of the estimation prediction error;

(b) minimizes the covariance matrix $P(t)$ in some sense.

13.13   NOTES AND REFERENCES

A general reference on optimization is:

Walsh, G.R.
Methods of Optimization, Wiley, 1975.

An introduction to early work on extremum control is given in:

Blackman, P.F.
Extremum-seeking regulators, in *An Exposition of Adaptive Control*, ed. J. Westcott, Pergamon Press, 1962.

See also:

Eveleigh, V.W.
Adaptive Control and Optimisation Techniques, McGraw Hill, 1967.

A classic applications paper is:

Draper, C.S. and Li, Y.
Principles of Optimising Control Systems, ASME Publications, 1954.

The idea of self-tuning extremum controllers is quite new. A good survey indicating the possibilities is:

Sternby, J.
Extremum control system – an area for adaptive control, *Proc. Joint Automatic Control Conference, San Francisco, CA, USA, August 1980.*

The material presented here is taken from:

Wellstead, P.E. and Scotson, P.G.
Self-tuning extremum control, *Proc. IEE, Part D,* **137**(3), 165–75, 1990.

Scotson, P.G. and Wellstead, P.E.
    Self-Tuning Optimisation of Spark Ignition Automotive Engines, *Proc. American Control Conf., Pittsburg, PA, USA, 1989.*

The convergence properties of self-tuning extremum controllers are considered in:

Bozin, A. and Zarrop, M.B.
    Convergence and robustness of an adaptive extremum controller, *Proc IEEE Colloquium on Adaptive Signal Estimation and Control, Edinburgh, September, 1989.*

The following Control System Centre thesis describes the automctive application area in depth:

Scotson, P.G.
    PhD Thesis, Control Systems Centre, UMIST, 1986.

The motivating force behind the extremum control work described here was its application to automotive engine control. An appraisal of this area of adaptive control is given in:

Wellstead, P.E.
    Application of adaptive techniques to internal combustion engine control, *The Benefits of Electronic Control Systems for Internal Combustion Engines,* Mechanical Engineering Publications, pp. 11–22, 1989.

14 Frequency Domain Self-tuning

14.1 OUTLINE AND LEARNING OBJECTIVES

The self-tuning system considered in this chapter is a form of signal processing algorithm. It is designed to deal with the problems associated with random vibration testing of components, machines and structures which are naturally subject to vibrations, shocks and random impacts throughout their working life. Testing of this kind is clearly important for almost all man-made objects. At a superficial level, consumer products need testing in laboratory conditions in order to ensure that they can withstand the rigours of everyday use. At a more fundamental level, the buildings we live in, the vehicles we drive and the aircraft we fly in must be tested. Testing ensures that they and their components can withstand the random vibrations which they will receive throughout their life cycle.

The size and nature of the objects which must be tested vary considerably and vibration test equipment must be able to self-tune to match the characteristics of each new test specimen. This self-tuning capability must be matched by an ability to adapt to changes in dynamics during tests to destruction. In Chapter 1 a commercial self-tuning instrument which performs random vibration testing was introduced. The aim of this chapter is to explain the theory and techniques upon which such instruments are based. At a basic level the self-tuning algorithms for vibration testing differ from those treated elsewhere in this book in the form of model used. Previously, we have considered only system models which have a parametric form. The corresponding self-tuning algorithms have estimated the coefficients of these parametric models (i.e. the coefficients of the A, B, C, D, $\mathscr{D}$ polynomials). In vibration testing, however, the accepted way of representing desired system characteristics is in terms of a *nonparametric model*. The model takes the form of a desired shape of the power spectrum associated with a particular vibration pattern. For this reason the type of self-tuner associated with vibration testing is often referred to as a nonparametric self-tuner or a frequency domain self-tuner. This latter term arises because the power spectrum is a frequency domain characterization of a random signal. Since the power spectrum object forms the basis of frequency domain self-tuners, Section 14.2 describes the background to power spectra and their importance in vibration testing. The basic idea of vibration testing is also outlined in Section 14.2. This preliminary description is extended in Section 14.3 in which the overall algorithm is detailed. An important part of self-tuning vibration testing concerns the special methods used to generate random signals with desired power spectra. This is described in detail in Section 14.3.1. A self-tuning frequency domain self-tuner is described in Section 14.4, together with the corresponding convergence theory. This frequency domain self-tuner is unusual in that it has a complete global stability theory associated with it. Simulated results which support the convergence analysis are discussed in Section 14.5. The chapter closes (Section 14.6) with the application of frequency domain self-tuning to the testing of civil engineering structures.

The learning objective is to understand the basis of nonparametric, frequency domain self-tuning and to understand its relevance and areas of application.

14.2 BACKGROUND

The form of self-tuner to be discussed in this chapter is rather different in form from those treated in the rest of the book. For this reason it will be useful to review some background on random signal modelling and motivate the need for such self-tuners (see also Chapter 2 and Appendix). (Note that we use the symbol τ below to denote continuous time in seconds and f to denote frequency in Hertz. As only continuous time signals are considered here the subscript c is not needed to differentiate them from discrete time signals.)

Frequency domain self-tuning is based on the idea that the vibrations experienced by an object in normal use can be characterized by a zero mean stationary random process $y(\tau)$ with power spectrum $S_{yy}(f)$. Further, a stationary random process can often be represented by an ARMA model, i.e. a white noise process fed through a linear transfer function. In the vibration testing application the transfer function represents the dynamics of the vibration generating mechanism, while the intensity is governed by the magnitude of the driving noise $e(\tau)$ (Figure 14.1).

If $F(s)$ is the transfer function associated with the vibration source and the intensity of the vibration is reflected in the white noise variance σ_c^2, then the corresponding power spectrum $S_{yy}(f)$ is given by

$$S_{yy}(f) = |F(jf)|^2 \sigma_c^2 \tag{14.1}$$

Figure 14.1 A vibration signal $y(\tau)$ with a required power spectrum is obtained by filtering white noise through a linear dynamical system $F(s)$.

In terms of the vibration signal itself the power spectrum is defined as the Fourier transform of the autocorrelation function $R_{yy}(\tau')$, thus

$$S_{yy}(f) = \int_{-\infty}^{+\infty} R_{yy}(\tau') \exp(-2\pi jf\tau') \, d\tau' \tag{14.2}$$

and

$$R_{yy}(\tau') = E\{y(\tau+\tau')y(\tau)\} \tag{14.3}$$

In a practical application the actual vibrations $y(\tau)$ experienced by an object can be measured experimentally, and the corresponding power spectrum shape, $S_{yy}(f)$, estimated using the defining relationships (equations (14.2) and (14.3)) or some equivalent estimator. The estimated power spectrum obtained from practical measurements then becomes a model for the vibrations to be applied in laboratory conditions. Specifically, random signal testing requires that a system be devised such that test items can be excited by a controlled noise source in such a way that the vibrations sensed during vibration tests have a power spectrum $S_{yy}(f)$ which matches that actually determined in practical measurements.

Typical vibration test equipment is shown in schematic form in Figure 14.2. It takes the form of a movable platform or shaker table coupled by a shaft to a linear actuator. The aim is to provide an input $u(\tau)$ to the actuator, such that the displacement $y(\tau)$ of the platform can take any one of a number of predefined forms. The platform is equipped with fixing lugs so that various test specimens can be rigidly fixed to the platform and hence subjected to the same vibration sequence $\{y(\tau)\}$.

Because the test specimens can vary significantly in their size and dynamics, the input signal $u(\tau)$ which gives a desired power spectrum with one test item will give a different output power spectrum for another test item. In general, it is necessary to retune the spectrum of $u(\tau)$ for each new test item. Similarly, when the shape of the desired output spectrum is redefined (in order to simulate some different environmental vibration source) the $u(\tau)$ object must

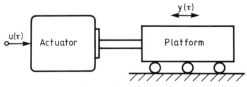

Figure 14.2 Schematic representation of a vibration test system in which the actuator input $u(\tau)$ must be selected to give required vibrations $y(\tau)$ on the shaker platform.

again be retuned to a new power spectrum. Before self-tuning algorithms of the type outlined in this chapter became available, the manual retuning of vibration sources was a time consuming and imprecise activity.

The basic idea for a frequency domain self-tuning system for vibration control is illustrated in Figure 14.3. In this figure, the reference power spectrum $S_{rr}(f)$ represents a model of the desired shape which the spectrum of the test specimen should have. A comparison is made with the spectrum $S_{yy}(f)$ of the test specimen. An adjustment mechanism is used to modify the input $u(\tau)$ to the vibration test system.

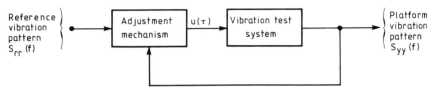

Figure 14.3   Illustrating the basic idea of an adjustment mechanism to modify platform vibration patterns, $S_{yy}(f)$ to match reference vibration pattern $S_{rr}(f)$.

The self-tuning problem, therefore, is one of designing a recursive algorithm to adjust the power spectrum $S_{uu}(f)$ of $u(\tau)$ to force the power spectrum $S_{yy}(f)$ of the output $y(\tau)$ to converge to the desired form of the reference spectrum $S_{rr}(f)$. Note that, because of the specification of a desired *power* spectral shape, there is no restriction on the phase relationship betwen $u(\tau)$ and $y(\tau)$. It follows that an algorithm which operates only by adaption of power spectral properties will suffice. This neatly avoids the problems associated with self-tuning of phase relations in a parametric scheme (e.g. nonminimum phase behaviour, unknown time delays, etc.). Further, the chosen form for $S_{rr}(f)$ rarely has a simple underlying parametric form. In fact, $S_{rr}(f)$ will usually be obtained by estimating the actual power spectrum associated with the particular vibration pattern to be simulated. Coupled with the difficulties often associated with model selection in parametric self-tuning, this strengthens the argument that a natural approach here is via a nonparametric frequency domain self-tuning algorithm.

Figure 14.2 shows a single-degree-of-freedom vibration table. Although applications exist which require two or three degrees of freedom, almost all the significant theoretical problems are contained in the single-degree-of-freedom case discussed here.

14.3   THE VIBRATION CONTROL ALGORITHM

The block diagram of Figure 14.3 can be put in the form of a normal feedback control system in which an output object $S_{yy}(f)$ is compared with a reference object $S_{rr}(f)$. A controller then applies an input to the system which forces the output power spectrum towards correspondence with the reference spectrum. However, the vibration control algorithm differs from the normal feedback controller in that the reference and output are frequency domain objects while the signal applied to the system is a time signal. The consequence of this is that the vibration control loop must contain a mechanism for transforming between the time signals $y(\tau)$, $u(\tau)$ and the frequency domain objects $S_{yy}(f)$, $S_{uu}(f)$. The mechanisms for this transformation are (Figure 14.4):

(a) A test signal synthesizer which constructs a control signal $u(\tau)$ corresponding to a desired control signal power spectrum $S_{uu}(f)$.

(b) A power spectrum estimator which reconstructs the power spectrum $S_{yy}(f)$ corresponding to the measured output $y(\tau)$.

The remaining key element (Figure 14.4) is the adjustment mechanism which compares the reference spectrum $S_{rr}(f)$ with the estimated output spectrum $\hat{S}_{yy}(f)$ in order to generate the input power spectrum $S_{uu}(f)$. The adjustment mechanism involves a recursive updating procedure which is discussed separately in Section 14.4. The remainder of this section is devoted to a discussion of the test signal synthesizer and the power spectrum estimator.

14.3.1 Test signal generation

Traditional vibration test equipment uses a white noise source, filtered and amplified to give a required power spectrum. In a digital self-tuning system, the test signal can be generated using discrete Fourier transform techniques to synthesis a periodic pseudo-noise signal. The algorithm reported here generates pseudo-noise with an approximate gaussian amplitude.

The pseudo-noise generator works as follows:

Figure 14.4
Block diagram of a frequency domain self-tuning system to modify the platform vibrations $y(\tau)$ to match a desired reference power spectrum $S_{rr}(f)$.

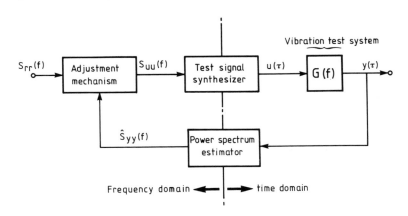

Algorithm: Pseudo-noise Generator

Step (i)   Select the desired power spectrum $S_{uu}(f)$ for pseudo-noise signal $u(\tau)$.

Step (ii)   Since the signal $u(\tau)$ is to be generated on a digital computer, replace the continuous $S_{uu}(f)$ by the discrete equivalent $S_{uu}(k)$ ($k = 0, 1, \ldots, N/2$), where $S_{uu}(k)$ is an equispaced set of spectral ordinates picked such that

$$S_{uu}(k) \cong S_{uu}(f_k), k = 0, 1, \ldots, \frac{N}{2}$$

and

$$f_k = k/T$$

(14.4)

where T is measured in seconds and gives the period of the final pseudo-noise signal. The choice of T and the even integer N determine the spectral resolution.

Step (iii)   Generate the Fourier series $U(k)$ ($k = 0, 1, \ldots, N-1$), for a sequence of real numbers $u(l)$ ($l = 0, 1, \ldots, N-1$) such that the power spectrum of $U(k)$ is $S_{uu}(k)$ and the phase of $U(k)$ is randomized as follows. Let

$$U(k) = a(k) + jb(k)$$

Define

$$a(k) = [S_{uu}(k)]^{\frac{1}{2}} r_k m_k$$

$$b(k) = [S_{uu}(k)]^{\frac{1}{2}} r_k (1 - m_k)$$

$$k = 1, 2, \ldots, \frac{N}{2} - 1$$

(14.5)

where r_k takes values ± 1 as a random function of k and m_k takes values $0,1$ as a random function of k and is statistically independent of r_k.

Equation (14.5) ensures that the modulus squared of $U(k)$ is equal to $S_{uu}(k)$ for all given values of k. The binary random sequences r_k, m_k have the purpose of scattering the phase of $U(k)$ so that it is a random function of k. The remaining values of $U(k)$ need to be specially set to ensure that $U(k)$ has a real inverse Fourier series. This is done by selecting the remaining $a(k)$ and $b(k)$ according to

$$a(0) = [S_{uu}(0)]^{\frac{1}{2}}, \qquad b(0) = 0$$

$$a\left[\frac{N}{2}\right] = S_{uu}\left[\frac{N}{2}\right]^{\frac{1}{2}}, b\left[\frac{N}{2}\right] = 0$$

$$\left.\begin{array}{l} a(N-k) = a(k) \\ b(N-k) = -b(k) \end{array}\right\} k = 1, 2, \ldots, \frac{N}{2} - 1$$

(14.6)

Step (iv)   The pseudo-noise sequence $u(l)$ is then obtained by inverse discrete Fourier transformation

$$u(l) = \sum_{k=0}^{N-1} U(k) \exp(j2\pi kl/N), \qquad l = 0,1,\ldots,N-1 \tag{14.7}$$

Step (v)   The final pseudo-random signal $u(\tau)$ is obtained by outputting the sequence $u(l)$ through a zero order hold from a digital/analogue converter with sample interval $\tau_s = T/N$. Thus

$$u(\tau) = u(\tau_s l) \qquad \text{for } \tau_s l \leq t < \tau_s(l+1) \tag{14.8}$$

The main steps in this noise generation algorithm are indicated in schematic form by Figure 14.5.

$\square$

Note that the signal $u(\tau)$ will have period T s and power spectrum $S_{uu}(f)$ at the specified frequency ordinates. Furthermore, because of the randomized phase of $U(f)$ (and invoking the central limit theorem) the summation in (14.7) will give $u(\tau)$ an almost gaussian amplitude distribution. The term 'almost' needs some clarification; this we give by stating that for N reasonably large (>100, say) the amplitude distribution of $u(\tau)$ is very close to gaussian.

The test signal synthesizer then produces a periodic signal of duration T, the power spectrum of which can be changed every T seconds according to a new updated spectrum for $u(\tau)$. It is therefore the period T s which is the

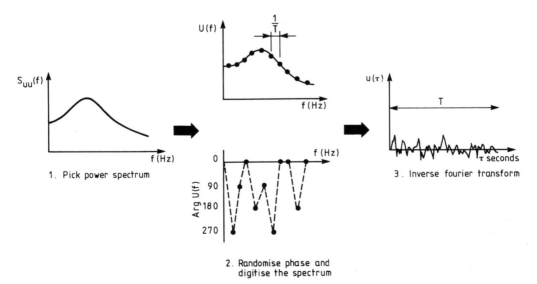

Figure 14.5   The main steps in the pseudo-noise generation algorithm.

basic iteration time in frequency domain self-tuning and is equivalent (in convergence studies) to the sampling interval in parametric self-tuning algorithms. To this end, the power spectral quantities at the ith iteration are denoted $S_{uu}(f)_i$ and $S_{yy}(f)_i$ and correspond to time data $u(\tau)$ and $y(\tau)$ in the period $T(i-1) \leqslant \tau < Ti$ s.

14.3.2 Power spectrum estimator

The power spectrum of $y(\tau)$ at the ith iteration is obtained by sampling the output of the system at a rate $f_s = 1/\tau_s$ Hz over the period T of the signal $u(\tau)$ and in synchronism with the sample-and-hold device which outputs $u(\tau)$. In this way a sequence of data points $y(l)$ $(l = 0, \ldots, N-1)$ is obtained.

The Fourier series of this sequence is then obtained as

$$Y(f_k) = \frac{1}{N} \sum_{l=0}^{N-1} y(l) \exp\left[-j\frac{2\pi kl}{N}\right] \tag{14.9}$$

$$f_k = \frac{kf_s}{N} = \frac{k}{T}, \qquad k = 0, \ldots, N/2$$

The power spectrum $P_{yy}(f_k)_i$ of $y(\tau)$ over the ith cycle is then calculated as

$$P_{yy}(f_k)_i = |Y(f_k)|^2, \qquad k = 0,1, \ldots, N/2 \tag{14.10}$$

Here the notation P_{yy} is used to denote an *intermediate* calculation of the power spectrum, which is sometimes referred to as the periodogram. In the presence of noise, further processing of the $P_{yy}(f_k)_i$ sequence is required to generate an acceptable estimate of the power spectrum $S_{yy}(f_k)$ (see later).

Equation (14.10) defines the periodogram of $y(\tau)$ and, in the noise free case,

$$P_{yy}(f_k)_i = |G(f_k)|^2 S_{uu}(f_k)_i \tag{14.11a}$$

However, if the vibration rig is subject to extraneous noise (Figure 14.6) then (14.11a) is replaced by

$$P_{yy}(f_k)_i = |G(f_k)|^2 S_{uu}(f_k)_i + P_{nn}(f_k)_i \tag{14.11b}$$

Figure 14.6 The vibration test system with extraneous noise corrupting the output $y(\tau)$.

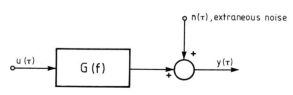

where the term $P_{nn}(f_k)_i$ is the periodogram of the ith segment of extraneous noise. The noise periodogram introduces a χ^2 distributed noise component into the scheme. The mean value of the noise periodogram can be removed by separate estimation. The variability of the noise contribution can be reduced in a number of ways. A popular method, however, is exponential smoothing via the recursion

$$S_{yy}(f_k)_i = (1 - \rho)\, S_{yy}(f_k)_{i-1} + \rho P_{yy}(f_k)_i \tag{14.12}$$

where the parameter ρ is a forgetting factor analogous to the forgetting factor λ employed in recursive estimation (see Chapter 4) and lies between 0 and 1. If the self-tuning loop converges such that $y(\tau)$ is a stationary process, then the variance reduction achieved by using ρ is clear from the equation

$$\text{var}\,\{S_{yy}(f_k)\} = \left(\frac{\rho}{2-\rho}\right) \text{var}\,\{P_{yy}(f_k)\} \tag{14.13}$$

Thus, the random errors of estimation are decreased roughly in proportion to ρ. However, the value of ρ has a critical influence on the convergence of the self-tuning loop. More will be said of smoothing the power spectral estimator in connection with the self-tuning recursion.

14.4   SELF-TUNING ADJUSTMENT MECHANISM

The function of the adjustment mechanism block in Figure 14.4 is to compute a new input power spectrum for $u(\tau)$, based on the estimated power spectrum of $y(\tau)$ obtained from the previous iteration of the signal synthesizer and power spectrum estimator. Thus, at time $\tau = iT$ the adjustment mechanism uses the estimated power spectrum $S_{yy}(f_k)_i$ and the reference spectrum $S_{rr}(f_k)$ to compute $S_{uu}(f_k)_{i+1}$, the power spectrum for $u(\tau)$ over the time interval $iT \leqslant \tau < (i+1)T$.

An *ad hoc* recursion which can be used as an adjustment mechanism is given by

$$S_{uu}(f_k)_{i+1} = \frac{S_{rr}(f_k)\, S_{uu}(f_k)_i}{S_{yy}(f_k)_i} \tag{14.14}$$

The rationale behind this choice is that, if the forgetting factor ρ is unity and there is no extraneous noise in the system, then the recursion will cause the spectrum of $y(\tau)$ to assume the desired shape $S_{rr}(f_k)$ in one iteration. In general, however, convergence of nonparametric self-tuners using (14.14) is not straightforward to analyze because of the nonlinear (quotient) form of the recursion. Nonetheless, the recursion is a popular one and several practical self-tuning vibration controllers use it. Technically, however, it can be improved upon in several ways. Most significantly, adjustment mechanisms can be proposed which are globally stable. In the next section we consider how the frequency domain self-tuner can be placed in a suitable framework for analysis so that new algorithms with provable convergence properties are obtained.

14.4.1 Orthogonality property

Consider a frequency domain self-tuning algorithm using the signal generator outlined in Section 14.3.1 and the spectral estimator of Section 14.3.2. The discrete Fourier transform (14.9) is an orthogonal transform of the sequence $\{y(l),\ 0 \leqslant l \leqslant N-1\}$. It follows, therefore, that at the frequencies f_k $\{k = 0, \ldots, (N/2)-1\}$ the quantities $S_{yy}(f_j)_i$ and $S_{yy}(f_k)_i$ are orthogonal for all $j \neq k$.

Remarks

(1) This has the important practical consequence that it is possible to specify independently the reference power spectrum values $S_{rr}(f_k)$ for each f_k.

(2) This nonparametric orthogonality property is analogous to the least-squares orthogonality property which is an important component of some parametric self-tuning analyses (see Chapter 6).

14.4.2 Basic self-tuning recursion

The important theoretical consequence of the orthogonality property is that the time evolution of nonparametric self-tuners can be analyzed in terms of the behaviour of one spectral ordinate. With this justification we simplify the notation by relabelling the quantities in (14.11b), (14.12) and (14.14) as follows:

$$r = S_{rr}(f_k), \qquad y_i = S_{yy}(f_k)_i, \qquad u_i = S_{uu}(f_k)_i$$

$$g = |G(f_k)|^2, \qquad p_i = P_{yy}(f_k)_i, \qquad n_i = P_{nn}(f_k)_i$$

The *ad hoc* recursion becomes

$$y_i = (1-\rho)y_{i-1} + \rho(gu_i + n_i)$$

$$u_{i+1} = r(u_i/y_i)$$

$$\left.\begin{array}{c} \\ \\ \\ \end{array}\right\} \qquad (14.15a)$$

where $\{n_i\}$ is a serially uncorrelated χ^2 distributed noise sequence which is uncorrelated with u_j and y_j for $i \leq j$. The difference equations (14.15a) can be linearized about the fixed point $y_i = r$ to show that, in the noise-free case, y_i will converge locally for $0 < \rho < 1$. However, little can be said about the global convergence properties of the recursion and this remains a significant drawback despite its practical uses.

14.4.3   Logarithmic self-tuning recursion

The problem with the basic recursion leads us to seek related algorithms with more tractable convergence properties. One way to do this is to follow the classical procedure and work in terms of logarithmic frequency response quantities. Note that such quantities are not constrained to be positive. In addition, the spectral smoothing procedure (14.12) is replaced by

$$S_{yy}(f_k)_i = [S_{yy}(f_k)_{i-1}]^{1-\rho} [P_{yy}(f_k)_i]^\rho \qquad (14.15b)$$

Again, redefining variables,

$$r = \log(S_{rr}(f_k)), \qquad y_i = \log(S_{yy}(f_k)_i), \qquad p_i = \log(P_{yy}(f_k)_i)$$

$$g = \log|G(f_k)|^2, \qquad u_i = \log(S_{uu}(f_k)_i)$$

the difference equations which define the recursion become

$$u_{i+1} = u_i + r - y_i$$

$$y_i = (1-\rho)y_{i-1} + \rho(u_i + g + e_i) \qquad (14.15c)$$

where $\{e_i\}$ is an uncorrelated noise sequence with zero mean.

Remark

With $e_i \equiv 0$ and $\rho = 1$, (14.15c) is the same as the *ad hoc* recursion (14.14) and is simply the Newton descent step for minimizing

$$J = (y - r)^2 \qquad (14.16)$$

with respect to u subject to the constraint

$$y = u + g \tag{14.17}$$

The inclusion of the forgetting factor ρ is a relaxation procedure, such that the evolution of y_i is governed by the equation

$$[1 - 2(1-\rho)z^{-1} + (1-\rho)z^{-2}]\, y_i = \rho r + \rho(1-z^{-1})e_i \tag{14.18}$$

The characteristic roots are $(1-\rho) \pm j(\rho(1-\rho))^{\frac{1}{2}}$ with magnitude $(1-\rho)^{\frac{1}{2}}$ (i.e. stable oscillatory behaviour for $0 < \rho < 1$). Thus, we have a globally stable algorithm with (on average) $y_i \rightarrow r$ as $i \rightarrow \infty$. After the transient has decayed the variability of y_i is given by

$$E[y_i - r]^2 = \frac{2\rho\sigma_e^2}{4 - 3\rho} \tag{14.19}$$

where $\sigma_e^2 = \text{var}\ \{e_i\}$.

In addition to providing a stable recursion, the logarithmic algorithm also has a simple control engineering interpretation as a digital feedback control loop with integral action only (Figure 14.7). Note that in this formulation the tools of classical deterministic control theory can be used to analyze situations in which r varies (by treating it as a servo input) and g varies (by treating it as a time-varying disturbance or offset). Likewise, proportional control action can be added on an *ad hoc* basis to improve the transient convergence behaviour of the system.

14.4.4   Optimized self-tuning recursion

The variance expression (14.19) suggests that the use of a variable forgetting factor ρ which decreases with time would be advantageous. With ρ initially large the recursion would converge rapidly to the required region. As ρ

Figure 14.7 The logarithmic self-tuning recursion interpreted as a feedback loop with integral action.

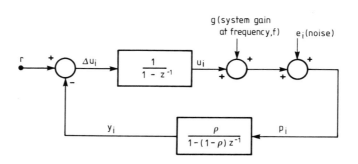

decreases the variability of y_i also decreases. In the following analysis we investigate how a variable forgetting factor may be obtained with optimal behaviour.

Given the measurement equation

$$p_i = x_i + u_i + e_i \tag{14.20}$$

where x_i is an unknown but constant system state satisfying

$$x_{i+1} = x_i = g \tag{14.21}$$

the estimate of y_i is obtained from the smoothing equation

$$y_i = (1-\rho_i)y_{i-1} + \rho_i p_i \tag{14.22}$$

where $\rho_i \in [0,1]$.

Combining (14.20) and (14.22) gives

$$y_i = \hat{y}_i + \rho_i e_i \tag{14.23}$$

where

$$\hat{y}_i = (1-\rho_i)y_{i-1} + \rho_i(x_i + u_i) \tag{14.24}$$

In the absence of noise, the solution is to select u_i such that

$$\hat{y}_i = r \tag{14.25}$$

By certainty equivalence, replace x_i by the best estimate $\hat{x}_i$, giving the control law

$$r = (1-\rho_i)y_{i-1} + \rho_i(\hat{x}_i + u_i) \tag{14.26}$$

or, by rearranging,

$$u_i = \frac{1}{\rho_i}(r-(1-\rho_i)y_{i-1}) - \hat{x}_i \tag{14.27}$$

The best linear filter for x_i takes the form

$$\hat{x}_{i+1} = \hat{x}_i + \alpha_i(y_i - (1-\rho_i)y_{i-1} - \rho_i(\hat{x}_i + u_i)) \tag{14.28}$$

Denote $\tilde{x}_i = \hat{x}_i - x_i$ and $V_i = E\{\tilde{x}_i^2\}$. Then it can be shown that

$$\tilde{x}_{i+1} = \tilde{x}_i(1 - \alpha_i\rho_i) + \alpha_i\rho_i e_i \tag{14.29}$$

$$V_{i+1} = V_i(1 - \alpha_i\rho_i)^2 + (\alpha_i\rho_i)^2\,\sigma_e^2 \tag{14.30}$$

By direct minimization, α_i is determined by

$$\frac{1}{\alpha_i\rho_i} = 1 + \frac{1}{\alpha_{i-1}\rho_{i-1}} \tag{14.31}$$

and

$$V_i = \frac{\sigma_e^2\alpha_i\rho_i}{1 - \alpha_i\rho_i} \tag{14.32}$$

so that $\alpha_i\rho_i \to 0$ and $V_i \to 0$ as $i \to \infty$. Also, this implies

$$\lim_{i \to \infty} E\left[\frac{y_i - r}{\rho_i}\right]^2 = \sigma_e^2 \tag{14.33}$$

Thus, the ratio $(y_i - r)/\rho_i$ remains well behaved as $\rho_i \to 0$, provided α_i is chosen according to (14.31). This is important because the control law can be written in the form

$$u_{i+1} = u_i + \left[\frac{1}{\rho_i} - 1\right](y_{i-1} - r) - \left[\frac{1}{\rho_{i+1}} - 1 + \alpha_i\right](y_i - r) \tag{14.34}$$

Inverting the logarithmic transform gives

$$(S_{uu})_{i+1} = (S_{uu})_i(S_{rr})\,((1/\rho_{i+1}) - (1/\rho_i) + \alpha_i)W$$

where

$$W = (S_{yy})_{i-1}\,((1/\rho_i) - 1)/(S_{yy})_i\,((1/\rho_{i+1}) - 1 + \alpha_i) \tag{14.35}$$

and the dependence upon frequency has been dropped for convenience.

14.5 SIMULATION STUDIES

In this section we consider computer simulations of the self-tuning recursions proposed in Section 14.4. We begin with a comparison of the *ad hoc* recursion (14.15a) with the equivalent logarithmic recursion (14.15c) for various values of forgetting factor. For comparison purposes the plots are made in terms of log variables. In each case the system parameters r, g were set at unity and, except where indicated, all variables are expressed in logarithmic terms. Where noise is added it is for qualitative purposes only and in each case a uniformly distributed, white sequence of variance 2 is used. The initial values of y and u were chosen to be zero.

14.5.1 *Ad hoc* recursion

Figure 14.8 shows the evolution of y_i for ρ varying from 0.5 to 0.02. Note the characteristic shape of the responses, including a 'nonminimum phase' response for the smallest value of ρ. Despite repeated experiments it was not possible to find combinations of parameters which make the recursion unstable; this observation coincides with practical experimental experience. The unfortunate aspect of the recursion is the extremely long convergence time when ρ is small. For example, one would typically wish to have ρ of the order of 0.02 in order to obtain reasonable smoothing in a noisy environment,

Figure 14.8 The evolution of y_i for the *ad hoc* recursion for various values of ρ, and with no extraneous noise.

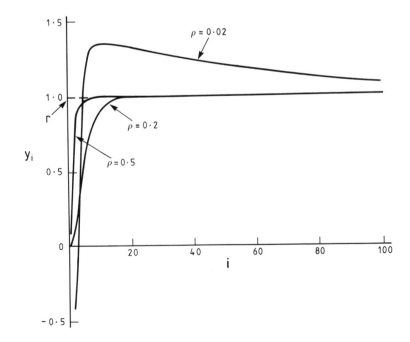

Figure 14.9 The evolution of antilog (y_i) for the *ad hoc* recursion, where $\rho = 0.02$ and with no extraneous noise.

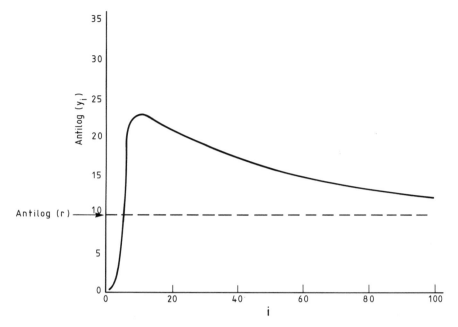

Figure 14.10 The evolution of y_i for the logarithmic recursion for various values of ρ and with no noise present.

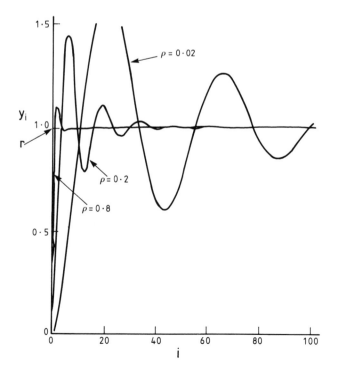

yet with such a value the *ad hoc* recursion would still have 20% error after 100 self-tuning cycles (Figure 14.9).

14.5.2   Log recursion

The main point of the log recursion (14.15c) is that it gives response characteristics which can be analysed by simple linear dynamical theory. Figure 14.10 shows typical behaviour of the simple log recursion in the noise-free case and for various values of ρ. As predicted by (14.18), the transient behaviour of the recursion becomes progressively more oscillatory as ρ decreases. Yet, as indicated by the variance expression (14.19), a low value of

Figure 14.11
Evolution of y_i for logarithmic self-tuning recursion with additive noise $(\sigma_{nn} = 2)$ and (a) $\rho = 0.2$, (b) $\rho = 0.02$.

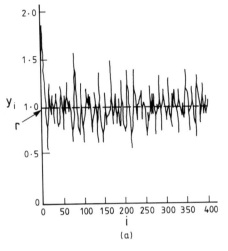

(a)

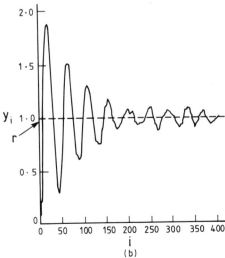

(b)

ρ is desirable for smoothing purposes. This point is illustrated in Figures 14.11(a) and 14.11(b) which show the log recursion for different values of ρ, but with additive noise. Note the trade-off between noise suppression and transient behaviour. The need for this trade-off can be mitigated (as hinted earlier) by adding proportional action to the equation for computing u_{i+1} (equation (14.15c)). With proportional action added, the update for u_{i+1} becomes

$$u_{i+1} = u_i + (r - y_i) + k_{\mathrm{p}}(y_{i-1} - y_i) \qquad (14.36)$$

where k_{p} is, by analogy with digital feedback control terminology, the proportional gain. The improving influence of k_{p} upon the transient behaviour of the noise-free log recursion is shown in Figure 14.12. Moreover, the addition of proportional action apparently does not influence the noise-suppressing properties which are controlled by ρ. For example, see Figure 14.13 which shows the log recursion for various values of k_{p}, but in the presence of disturbing noise.

To summarize, the log recursion (14.15c) has well-defined (linear) dynamical properties which can be improved by the addition of a term analogous to proportional action in conventional feedback systems. Moreover, the proportional term can be used to influence the transient behaviour of the self-tuning recursion in a manner which is apparently independent of the noise-smoothing properties of the loop.

14.5.3   Optimized self-tuning recursion

The performance of the optimized self-tuning recursion is now examined with a view to illustrating its noise-suppressing properties and the interplay between α_i and ρ_i in the setting up of the algorithms. First consider some straightforward choices of ρ_i. If $\rho_i = (1/i)^{\frac{1}{2}}$ then this gives a relatively short memory to the smoothing algorithm with a correspondingly slow convergence of the log power spectrum y_i. From (14.31), if $\rho_i = (1/i)^{\frac{1}{2}}$ then $\alpha_i = \rho_i$; this fact is verified in the plots of Figure 14.14. An impression of the long term evolution of y_i for this choice of ρ_i is given in Figure 14.15(a) and this can be compared with the convergence of y_i for the other choices $\rho_i = 1/i$ and $\rho_i = (1/i)^2$ in Figures 14.15(b) and 14.15(c) respectively. Clearly, the faster ρ_i decreases the more quickly y_i converges. However, if ρ_i decreases more quickly than $1/i$, then α_i is obliged to diverge as i increases in order to satisfy (14.31). Intuitively, what this means is that, in order for the state estimator recursion to get a sufficiently large update, then α_i must grow in order to counter the diminishing influence of ρ_i in (14.28). Such a line of thought leads one to surmise that ρ_i could be allowed to decrease as rapidly as desired to achieve

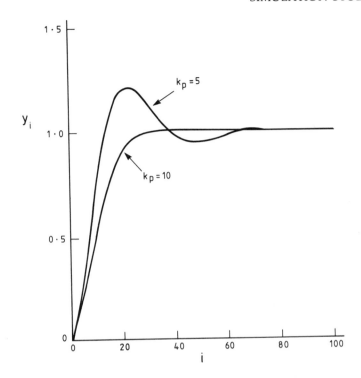

fast convergence of y_i to the reference value, while keeping α_i finite, so long as no state information is required. For example, Figure 14.16(a) shows the evolution of $\hat{x}_i$ for $\rho_i = (1/i)^2$ and $\alpha_i = i$ (from equation (14.29)). As required, $\hat{x}_i$ converges towards the true value of $g = 1$. If α_i is clamped at unity, the state estimate $\hat{x}_i$ fails to converge to the correct value (Figure 14.16(b)). However, in both cases the convergence properties of y_i were indistinguishable from one another.

The interplay between α_i and ρ_i is thus clear through their individual roles in respectively influencing $\hat{x}_i$ and y_i. The 'best' choice of ρ_i sequences, however, is less clear. A superficial 'best' choice would be to make ρ_i decrease as fast as possible; however, this would certainly mean clamping α_i in some way to prevent numerical overflow. Thus, for practical purposes the fastest that ρ_i can decrease is as $1/i$. Now, if one inserts this choice of ρ_i (and the corresponding $\alpha_i = 1$) in (14.34), then the equation can be rewritten as

$$u_{i+1} = u_i + 2(r-y_i) + (i-1)(y_{i-1} - y_i) \tag{14.37}$$

which is the same form as the logarithmic recursion with an integral gain of 2 and $k_p = i-1$. Thus we are left with an interpretation of the optimized

Figure 14.13
Evolution of y_i
for logarithmic
recursion with
added
proportional
action $k_p = 10$
and 20 and
additive noise
$(\sigma_{nn} = 2)$.

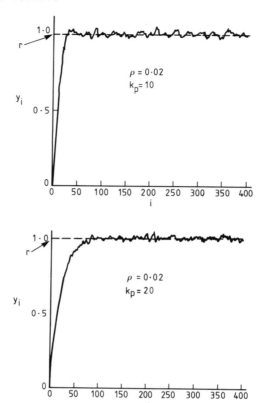

recursion as a time-varying feedback law in which the proportional gain is increasing linearly.

A further link exists between the optimized recursion and power spectral smoothing algorithms. The most widely used spectral estimator uses a running-average smoothing algorithm in order to achieve a consistent estimate of power or cross-spectral density. In the nomenclature of this book, a running-average smoother is given by (cf. equation (14.22)).

$$y_i = \frac{1}{i} \sum_{j=0}^{i} p_j$$

$$= \left[1 - \frac{1}{i}\right] y_{i-1} + \frac{1}{i} p_i \tag{14.38}$$

which corresponds to the choice $\rho_i = 1/i$ in the optimized self-tuning recursion.

Thus, a running-average smoother as used in spectral estimation corresponds to the fastest converging choice of ρ_i which is not associated with a diverging

Figure 14.14
Optimized
self-tuning
recursion,
showing
evolution of y_i
and ρ_i, when
$\alpha_i = (1/i)^{\frac{1}{2}}$, and
noise is present
$(\sigma_{nn} = 2)$.

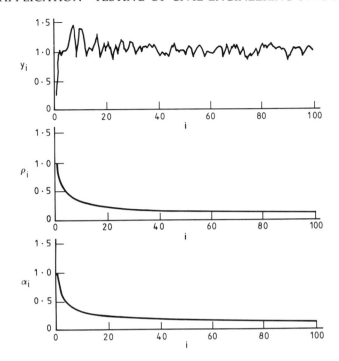

α_i sequence. Moreover, it is (as one would expect) associated with the best linear estimate of the system gain x_i.

14.6 APPLICATION—TESTING OF CIVIL ENGINEERING STRUCTURES

Frequency domain self-tuning is widely used in random vibration testing. We have already mentioned a general purpose commercial product which implements these algorithms (see Section 1.4.4). A significant specialist application of self-tuning random vibration testing is the testing of civil engineering structures with random vibrations. This topic has been pioneered in Europe by the Laboratorio Nacional de Engenharia Civil (LNEC), Portugal. The ideas and results quoted here were kindly supplied by this organization. Civil engineering structures are subjected to a variety of environmental vibrations but in view of Portugal's history and geography, vibrations associated with marine structures and earthquake prone regions are particularly important.

The influence of the sea upon a marine or coastal structure is determined using scale models of the structure under test. The model is placed in a large hydraulic tank and subjected to a wave motion which simulates the wave patterns which the full scale structure will experience. Figure 14.17 shows a

Figure 14.15
Evolution of y_i for the optimized self-tuning recursion, with noise present ($\sigma_{nn} = 2$) and (a) $\rho_i = \alpha_i = (1/i)^{\frac{1}{2}}$; (b) $\rho_i = (1/i)$, $\alpha_i = 1$; and (c) $\rho_i = (1/i)^2$, $\alpha_i = 1$.

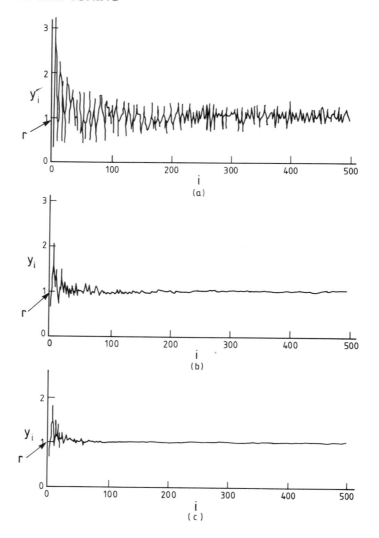

typical wave testing machine in which waves are generated by moving a paddle backward and forward in a water channel. A transducer is used to measure the water pressure variations, $y(\tau)$, just in front of the model. A frequency domain self-tuning controller can then be used to manipulate the power spectrum of a control signal $u(\tau)$ in order to force the power spectrum of $y(\tau)$ into correspondence with a reference power spectrum which is representative of the wave shape experienced by the full-sized structure.

The effect of earthquakes upon civil engineering structures is tested by mounting a model of the structure (or indeed the structure itself) upon a shaker table. The capacity and design of shaker tables can vary greatly. The

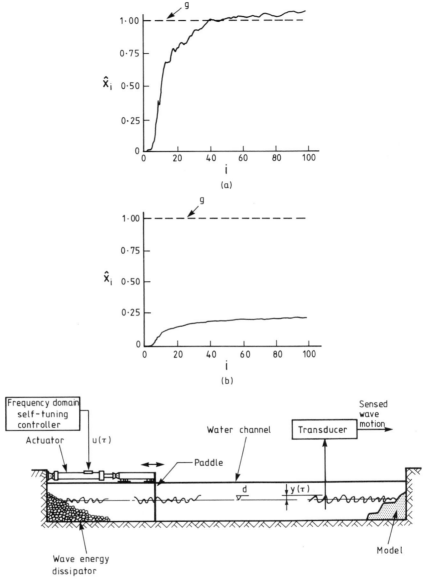

Figure 14.16
Evolution of the
state estimate $\hat{x}_i$
for optimized
recursion with
$\rho_i = (1/i)^2$ and
(a) $\alpha_i = i$ and (b)
$\alpha_i = 1$.

Figure 14.17 Illustrating a system used for testing models of marine and coastal
structures. The paddle is moved in such a way that it produces wave motion with a
desired form. The sensed wave motion $y(\tau)$ is fed back to the self-tuning control and
used to modify the actuation signal $u(\tau)$. (Figure courtesy of LNEC.)

smallest tables are one-degree-of-freedom devices for testing small components
(like the camera in Figure 1.27). Such small tables are driven by electromagnetic
actuators. At the other extreme are the large multi-axis shaker tables which

Figure 14.18
Schematic view of a LNEC shaker table showing a 5.5 m × 4.5 m platform (P), torque tubes (TT), connecting rods (CR), actuators (AC) and spherical swivels (S). (Figure courtesy of LNEC.)

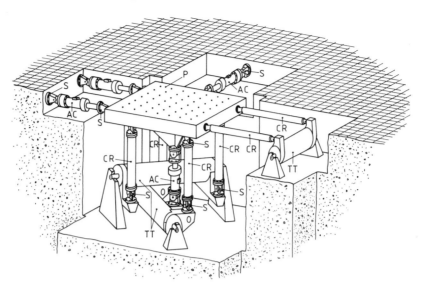

are used on large civil engineering specimens. Figure 14.18 shows a recent LNEC design. This is a three-degrees-of-freedom table which is driven by hydraulic activators. The main platform is 5.5 m × 4.5 m and the actuators have force capacities in the range 300 to 1500 kN, depending upon the configuration of the shaker table. The use of these tables for earthquake testing can involve applying random vibrations in a combination of up to three of the directions x, y, z. In general, therefore, the earthquake vibration testing problem requires a multivariable extension of the algorithms discussed here.

Under certain conditions, the random vibrations associated with an earthquake can be approximated by a velocity power spectrum that is flat within a certain range. The actuators associated with a shaker table are position controlled, hence the reference spectrum corresponding to a flat velocity spectrum resembles the $1/(f)^2$ shape shown in Figure 14.19(a), where the excitation range is 1 to 8 Hz. The remaining sequence of plots in Figure 14.19 show the measured output spectrum $S_{yy}(f)$ for a test item attached to a single-degree-of-freedom shaker table at LNEC. Note that, after one cycle of the self-tuning algorithm, the output spectrum still contains a major anti-resonance at approximately 5 Hz (Figure 14.19(b)). This is associated with the vibrational modes of the test item. However, after two cycles (Figure 14.19(c)), the anti-resonance is substantially removed. After three cycles (Figure 14.19(d)) the output spectrum is closely tuned to the desired reference shape. The adjustment mechanism used in this example was based on the *ad hoc* recursion (14.14). Despite the lack of global convergence results for this recursion, its rapid convergence rates make it practically useful.

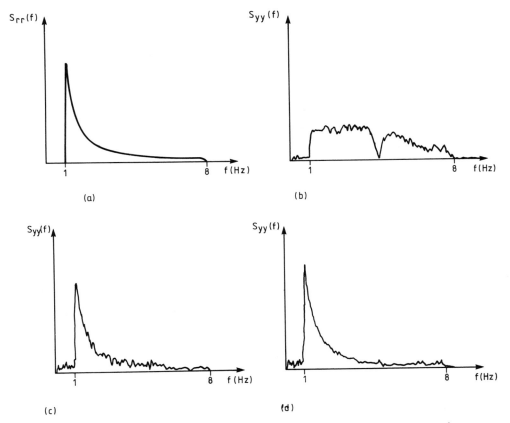

Figure 14.19 Self-tuning results for tuning an output response spectrum to a reference spectrum corresponding to a 'flat velocity' earthquake spectrum (results courtesy of LNEC.) (a) Reference spectrum, (b) output spectrum $S_{yy}(f)$ (after one cycle), (c) (after two cycles) and (d) after three cycles.

14.7 SUMMARY

The chapter has described a special form of self-tuning system for adaptive vibration testing. The main points concerning the algorithm are:

- It provides a method for tuning the power spectrum of a signal to a desired shape
- The techniques employed are nonparametric, so that issues of correct parametrization of the underlying system are avoided
- Embodied within the technique is a very useful but little known pseudonoise generation algorithm
- The method has been described in a one-degree-of-freedom form, but can be extended to the multidimensional case

14.8 ASSIGNMENTS

Assignment 14.1

This is a short algorithmic/software assignment. Prepare a self-tuning algorithm which automatically adjusts the amplitude of a measured sine wave vibration applied to a system such that the amplitude of the output sine wave amplitude corresponds to some reference amplitude.

Assignment 14.2

This is a theoretical research assignment. Take the one-degree-of-freedom vibration testing algorithm described above and derive the extension to two degrees of freedom.

Assignment 14.3

This is a theoretical research assignment. Take the *ad hoc* recursion (14.14) and attempt to prove its convergence properties. Start with local convergence (which should be easy) and then try to establish global properties.

Assignment 14.4

This is an applied research assignment. The algorithms described in this chapter are, as stated, widely used in vibration testing of mechanical components. The same idea, however, can be used to test the acoustic properties of materials and enclosures. The research assignment is a practical one and requires you to use a hi-fi audio amplifier and loudspeaker combined with a test signal synthesizer of the type described in Section 14.3.1 in order to generate an audible noise spectrum with the required properties. The spectrum of the noise generated should be self-tuned such that the noise sensed by a microphone some distance away and in close proximity to a test item has the shape of the reference spectrum. Basically, the assignment is to redevelop the vibration control algorithms for audible noise testing.

14.9 NOTES AND REFERENCES

The material in this chapter is based on:

Wellstead, P.E. and Zarrop, M.B.
  Self-tuning regulators; non-parametric algorithms, *Int. J. Control*, **37**(4), 787–807, 1983.

The generation of pseudo-noise with a desired power spectrum is a key feature of the nonparametric self-tuning algorithm. It is also a technique which can be generally applied in system identification and related applications where a test signal is required which has exactly specified power spectral (or autocorrelation) properties. Unfortunately, however, books on system identification and signal processing rarely mention the technique. This is surprising since pseudo-noise generated in this way has many advantages over more commonly used pseudo-noise sources. Specifically, (i) pseudo-noise sequences of arbitrary length can be generated, (ii) the power spectrum is *exactly* specified by the user, (iii) the amplitude probability distribution can also be specified within limits. To probe further on this topic the following papers are recommended:

Borgman, L.E.
   Proc. ASCE (Waterways and Harbours Div.), **WW4**, 557, 1969.

This paper is probably the first to use this form of pseudo-noise to generate test signals with prescribed spectra. In this case specific wave shapes were required for testing in navigable waterways.
   Also:

Schroeder, M.R.
   Synthesis of low peak factor signals and binary sequences with low autocorrelation. *IEEE Trans.*, **16**, 84, 1970.

This is an authoritative article on pseudo-noise generation. It includes a method for constraining the test signal probability destribution function. This is important since, as noted in Chapter 4, the tails of the Gaussian density mean that high amplitude signal components are occasionally generated which may damage physical equipment.

Wellstead, P.E.
   Pseudo noise test signals and the fast Fourier transform, *Electron. Lett*, **11**, 202, 1975.

This paper, written without knowledge of the previous two, introduced this form of pseudo-noise to the control and system identification community as an alternative to pseudo-random binary noise.
   The techniques of power spectrum estimation as described here are detailed in a number of texts. For example, see:

Bendat, J.A. and Piersol, A.G.
   Engineering Applications of Correlation and Spectral Analysis, Wiley, 1980.

This is an excellent engineering text with many examples and applications drawn from the rich experiences of these distinguished authors.
   The description of the civil engineering application of frequency domain self-tuning was made possible by courtesy of the Director of LNEC (Laboratorio Nacional de Engenharia Civil) Portugal. The work of LNEC is further described in, for example:

Jervis Pereira, J.M. and Carvalhal, F.J.
   Adaptive control techniques for the simulation of seismic actions, *Proc. Design of Concrete Structures Seminar – The Use of Model Analyses*, Watford, UK, 1984.

Appendix: Probability and Random Processes

The basic assumption underlying this book is that no physical process can be completely known and that every such system operates in a noisy environment and is subject to disturbance and change. The basic tool that is employed when 'uncertainty' plays a significant role is Probability Theory.

This appendix briefly summarizes the basic material on probability, random processes and estimation that is required for an understanding of the main text. An attempt has been made to place particular results within a coherent framework but no claim is made to completeness or rigour and the reader is encouraged to refer to more comprehensive treatments of stochastics when he or she feels the need (see Section A.11).

A.1 PROBABILITY

An *event* A is any outcome of interest from an experiment. The *probability* $P(A)$ (or prob(A)) of this event can be thought of as the limit, as the experiment is repeated indefinitely, of the ratio of the number of times A occurs to the number of trials. This definition implies that

$$0 \leqslant P(A) \leqslant 1 \tag{A.1}$$

If A_1, A_2, ..., are events, no two of which can occur together (mutually *exclusive*), then the event that A_1 or A_2 or ... occurs is given by

$$P(A_1 \text{ or } A_2 \text{ or} \ldots) = P(A_1) + P(A_2) + \ldots \tag{A.2}$$

and it assumed that

$$P(A_1) + P(A_2) + \ldots = 1 \tag{A.3}$$

if the list of events covers all possible outcomes of the experiment.

If two outcomes A,B do not preclude each other, then

$$P(A \text{ or } B) = P(A) + P(B) - P(A \text{ and } B) \tag{A.4}$$

in an obvious notation. Clearly the last term vanishes for exclusive events to yield a special case of (A.2).

A key concept in estimation theory is that of *conditional probability*. The probability of A occurring, given that B has occurred, is denoted by $P(A|B)$ and is defined by

$$P(A \text{ and } B) = P(A|B)P(B) \tag{A.5}$$

If A, B are *independent* events, so that the occurrence of B has no effect on the likelihood of A occurring, then we can replace $P(A|B)$ by the unconditional probability $P(A)$ so that (A.5) becomes

$$P(A \text{ and } B) = P(A)P(B) \tag{A.6}$$

The event 'A and B' is the same as 'B and A' and (A.5) leads to the useful relationship

$$P(B|A) = P(A|B)P(B)/P(A) \tag{A.7}$$

A.2   RANDOM VARIABLES

The events generally of interest to engineers are the occurrences of specific numerical values of physical variables (temperature, voltage, etc). A variable whose value is determined by the (uncertain) outcome of an experiment is called a *random variable* (abbreviated to rv). The probabilistic behaviour of a rv X is completely specified by its probability *distribution* function (pdf) $F_X(x)$ or its probability *density* function $f_X(x)$ defined by

$$F_X(x) = P(X \leqslant x) = \int_{-\infty}^{x} f_X(u)\mathrm{d}u \tag{A.8}$$

so that $f_X(x)$ is the gradient of $F_X(x)$. It follows that

$$\int_{-\infty}^{\infty} f_X(x)\mathrm{d}x = 1 \tag{A.9}$$

Note that $F_X(x)$ is a probability but that $f_X(x)$, although positive, can take large values. It also follows that

$$P(x \leqslant X \leqslant x + \delta x) \simeq f_X(x)\delta x \qquad (A.10)$$

for small δx, provided the density function is bounded. This is true of a continuously distributed rv for which the probability is zero of any specific value occurring. If X takes a set of *discrete* values of x_1, x_2, ..., with probabilities $p(x_1)$, $p(x_2)$, ..., then the density function can be cast in the form

$$f_X(x) = \sum_i p(x_i)\, \delta(x - x_i) \qquad (A.11)$$

where $\delta(\cdot)$ is the Dirac delta function. In general, a density function may be partly continuous and partly of the 'blip' structure (A.11).

For two rvs X,Y the probabilistic structure is specified by the *joint pdf* $F_{XY}(x,y)$ or *joint density* $f_{XY}(x,y)$ defined by

$$F_{XY}(x,y) = P(X \leqslant x \text{ and } Y \leqslant y) = \int_{-\infty}^{y} \int_{-\infty}^{x} f_{XY}(u,v)\mathrm{d}u\mathrm{d}v \qquad (A.12)$$

It follows that, for X alone,

$$F_X(x) = F_{XY}(x,\infty); \qquad f_X(x) = \int_{-\infty}^{\infty} f_{XY}(x,y)\mathrm{d}y \qquad (A.13)$$

and similarly for Y. The extension to more than two rvs is straightforward.

If X,Y are independent rvs, then (A.5) and (A.12) imply that

$$F_{XY}(x,y) = F_X(x)F_Y(y); \qquad f_{XY}(x,y) = f_X(x)\, f_Y(y) \qquad (A.14)$$

A.3 EXPECTATIONS AND MOMENTS

The *expectation* (average, mean) or first *moment* of a rv X is defined as the weighted average

$$E_X[X] = m_X = \int_{-\infty}^{\infty} x f_X(x)\mathrm{d}x \qquad (A.15)$$

A measure of the dispersion or size of fluctuation of X around its mean value is given by the second central moment or *variance* σ_{XX} (or σ_X^2 or var X) defined by

$$\sigma_{XX} = E_X[X - E_X[X]]^2 = E_X[X^2] - (E_X[X])^2 \tag{A.16}$$

The *standard error* (or standard deviation) σ_X is defined as the positive square root of the variance and has the same dimensions as X. The moments give useful but only partial information about the pdf in general.

A useful (but often conservative) relationship is given by the *Chebyshev inequality*

$$P(|X| \geqslant \epsilon) \leqslant \frac{1}{\epsilon^2} E_X[X^2] \tag{A.17}$$

where ϵ is any positive constant.

The extension to moments involving more than one rv is straightforward, by simply calculating the weighted average using the appropriate joint density. This leads to

$$E_{XY}[g(X) + h(Y)] = E_X[g(X)] + E_Y[h(Y)] \tag{A.18}$$

and, if X,Y are independent,

$$E_{XY}[g(X)h(Y)] = E_X[g(X)]E_Y[h(Y)] \tag{A.19}$$

where g,h are arbitrary functions (subject to the existence of the moments). A useful second order moment that gives some indication of the degree to which X and Y are related is the *covariance*

$$\sigma_{XY} = E_{XY}[(X - m_x)(Y - m_Y)] = E_{XY}[XY] - m_X m_Y \tag{A.20}$$

The rvs are *uncorrelated* if σ_{XY} is zero and it is easy to see from (A.19) that this is implied by independence. Uncorrelated rvs, however, are not necessarily independent.

A useful measure of correlation is given by the dimensionless scalar *correlation coefficient*

$$\rho_{XY} = \sigma_{XY}/\sigma_X \sigma_Y \tag{A.21}$$

which always lies between -1 and $+1$, taking one of the extreme values if X and Y are linearly related.

Generalizing (A.20) to the case of n rvs $\mathbf{X} = (X_1, X_2, \ldots, X_n)^{\mathrm{T}}$ leads to the *covariance matrix* (cov $\mathbf{X}$) whose (i,j) element is the covariance between X_i and X_j $(i \neq j)$ and whose (i,i) diagonal element is var X_i.

A.4 UNIFORM AND GAUSSIAN DISTRIBUTIONS

(a) $X \sim U(a,b)$

The *uniform* distribution is a useful way of expressing ignorance. If a rv X is known to lie in some finite interval $[a,b]$ but no other information is available, the density function can be taken as constant over the interval (see Figure A.1). Scientific calculators usually have the facility for generating a sequence of uniformly distributed rvs.

(b) $X \sim N(m,\sigma^2)$

The *gaussian* or *normal* probability density function is shown in Figure A.2 and has the form

$$f_X(x) = (2\pi\sigma^2)^{-\frac{1}{2}} \exp[-(x-m)^2/2\sigma^2] \qquad (A.22)$$

Figure A.1 The uniform distribution.

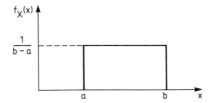

Figure A.2 The gaussian distribution.

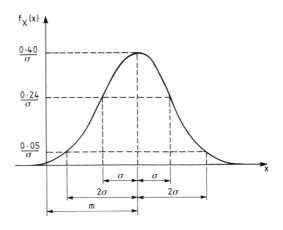

where the two parameters m, σ are the mean m_X and the standard error σ_X respectively. There is a 95% probability that a value of X lies within two standard errors of the mean, i.e.

$$P(|X-m| \leqslant 2\sigma) \simeq 0.95 \tag{A.23}$$

Other important properties are:

(i) Any linear combination of gaussian rvs is gaussian.

(ii) The sum of an increasing number of independent rvs with arbitrary pdfs tends to be gaussian (*central limit theorem*).

In nature, many random variables are approximately gaussian distributed, no doubt reflecting the fact that 'noise' is often the superposition of random elements from many physical sources.

For n rvs $\mathbf{X} = (X_1, X_2, \ldots, X_n)^T$ the multivariate gaussian density is

$$f_\mathbf{x}(\mathbf{x}) = (2\pi)^{-n/2} |\Sigma|^{-\frac{1}{2}} \exp[-\tfrac{1}{2}(\mathbf{x}-\mathbf{m})^T \Sigma^{-1}(\mathbf{x}-\mathbf{m})] \tag{A.24}$$

where $\mathbf{m}$ is the mean of $\mathbf{X}$ and Σ is its covariance matrix. In this case, if the rvs are uncorrelated (i.e. Σ is diagonal), then $X_1, \ldots, X_n$ are independent.

A.5   CONDITIONAL EXPECTATION AND ESTIMATION

The concept of conditional probability introduced in Section A.1 is now translated into the language of random variables by introducing the conditional density (cf. equation (A.5))

$$f_{X|Y}(x|y) = f_{XY}(x,y)/f_Y(y) \tag{A.25}$$

so that $f_{X|Y}(x|y)\mathrm{d}x$ denotes the probability that X lies in the interval $(x, x + \mathrm{d}x]$ given the datum $Y = y$.

The conditional density can be used to form various moments, in particular the *conditional expectation*

$$E_{X|Y}[X|Y=y] = \int_{-\infty}^{\infty} x f_{X|Y}(x|y)\mathrm{d}y = m_{X|y} \tag{A.26}$$

(cf. equation (A.15)).

For jointly distributed gaussian rvs X, Y

$$f_{X|Y}(x|y) = (2\pi\sigma^2_{X|y})^{-\frac{1}{2}} \exp[(x-m_{X|y})^2/2\sigma^2_{X|y}] \tag{A.27a}$$

so that the conditional distribution remains gaussian with conditional mean

$$m_{X|y} = m_X + \sigma_{XY}(y-m_Y)/\sigma^2_Y \tag{A.27b}$$

and conditional variance

$$\sigma^2_{X|y} = \sigma^2_X - \sigma^2_{XY}/\sigma^2_Y \tag{A.27c}$$

Writing

$$g(Y) = E_{X|Y}[X|Y] \tag{A.28}$$

emphasizes that the conditional expectation will usually depend on the given datum (as in A.27b) and is therefore itself a rv. Using (A.25) it follows that

$$E_Y E_{X|Y}[\cdot] = E_{XY}[\cdot] \tag{A.29}$$

so that

$$E_{XY}[g(Y) - X] = 0 \tag{A.30}$$

Any function $g(Y)$ of the datum rv satisfying (A.30) is an *unbiased estimator* of X in that it coincides with X on average. Otherwise the left-hand side of (A.30) is the estimator *bias*.

Casting (A.28) in the form

$$E_{X|Y}[g(Y) - X] = 0 \tag{A.31}$$

it follows that

$$E_{XY}[(g(Y) - X)h(Y)] = 0 \tag{A.32}$$

for arbitrary h. This expresses an important property of the conditional expectation that carries over to the multivariate case: the estimation error is *orthogonal* to (uncorrelated with) the data. This is shown geometrically in Figure A.3.

Using (A.32) leads to the equality

$$E_{XY}[g(Y) - X + h(Y)]^2 = E_{XY}[g(Y) - X]^2 + E_Y[h(Y)]^2 \tag{A.33}$$

Figure A.3 The geometry of MMSE estimation.

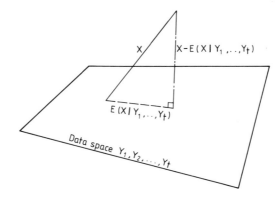

Clearly the right-hand side is minimized by setting h to zero, so that we conclude that the conditional expectation is the best estimator of X if we define 'best' as the choice of $\hat{X}(Y)$ that minimizes the cost function

$$J(\hat{X}) = E_{XY}[\hat{X}(Y) - X]^2 \qquad (A.34)$$

the mean square error. The conditional expectation is therefore the *minimum mean square error* (MMSE) estimator of X and, being an unbiased estimator, achieves the minimum variance $J(g)$. In general, the conditional expectation may be nonlinear in the data. If we restrict $\hat{X}(Y)$ to be linear Y, it turns out that $J(\hat{X})$ is minimized by (A.27b). Hence, if we want the best (MMSE) *linear estimator* of X (whatever its pdf), we need only write down $m_{X|Y}$ (or its relevant multivariate extension) under a gaussian assumption.

A.6   RANDOM PROCESSES

A *stochastic process* (sp) $\{X(t)\}$ may be thought of as a complete collection (*ensemble*) of possible time trajectories of a signal $X(t)$. For a particular experiment, one of these trajectories (a *realization* of the sp) occurs. For a particular time $t = t_1$, $X(t_1)$ is a rv (with its own pdf) because its value will differ according to the realization (see Figure A.4). Note, therefore, that statistical operations (such as taking expectations) are performed across realizations (for any value of time) as with a single rv. Ensemble averaging is not equivalent to time averaging over a particular trajectory except under special conditions. Note also that time t is merely a label for each rv and is not itself random. We are therefore in a multiple rv situation but a special notation is used to emphasize the importance of the t label. The 'two rv' case is shown in Table A.1 and can be easily extended to higher dimensions.

Figure A.4
Realizations of
the ensemble
$\{X(t)\}$.

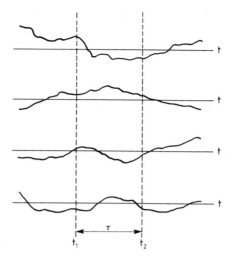

Table A.1   Basic statistical operations for sps and jointly distributed rvs, showing notational differences

(a) Random variables	(X_1, X_2)
(b) Stochastic process	$\{X(t),\ t\ =\ 1,2\}$

Distribution function

(a) $P(X_1 \leq x_1,\ X_2 \leq x_2) = F_{X_1 X_2}(x_1, x_2)$

(b) $P(X(t_1) \leq x_1,\ X(t_2) \leq x_2) = F_X(x_1, t_1;\ x_2, t_2)$

Marginal density

(a)
$$f_{X_1}(x_1) = \int_{-\infty}^{\infty} f_{X_1 X_2}(x_1, x_2)\, dx_2$$

(b)
$$f_X(x_1, t_1) = \int_{-\infty}^{\infty} f_X(x_1, t_1;\ x_2, t_2)\, dx_2$$

Expectation

(a)
$$E_{X_1 X_2}[g(X_1, X_2)] = \int_{-\infty}^{\infty} \int_{-\infty}^{\infty} g(x_1, x_2)\, f_{X_1 X_2}(x_1, x_2)\, dx_1 dx_2$$

(b)
$$E_X[g(X(t_1), X(t_2))] = \int_{-\infty}^{\infty} \int_{-\infty}^{\infty} g(x_1, x_2)\, f_X(x_1, t_1;\ x_2, t_2)\, dx_1 dx_2$$

Given the importance of first and second order moments, we define the *mean function*

$$m_X(t) = E_X[X(t)] \tag{A.35}$$

(cf. equation (A.15)) and the *autocovariance function* (acf)

$$r_{XX}(t_1,t_2) = E_X\{[X(t_1) - m_X(t_1)]\,[X(t_2) - m_X(t_2)]\} \tag{A.36}$$
$$= r_{XX}(t_2,t_1)$$

(cf. equation (A.20)) and note that

$$\text{var } X(t) = r_{XX}(t,t) \tag{A.37}$$

The quantity $E_X[X(t_1)X(t_2)]$ is referred to as the *autocorrelation function* and coincides with the acf if the sp mean is zero.

The *cross-covariance function* (ccf) between two sps $\{X(t)\}$ and $\{Y(t)\}$ is defined as

$$r_{XY}(t_1,t_2) = E_{XY}\{[X(t_1) - m_X(t_1)]\,[Y(t_2) - m_Y(t_2)]\} \tag{A.38}$$
$$\neq r_{XY}(t_2,t_1)$$

The sp $\{X(t)\}$ is said to be (weakly) *stationary* (ws) if

(i) the mean function is constant (m_X); and

(ii) the acf can be expressed as $r_{XX}(t_1-t_2)$.

These properties imply that the process has settled down in a statistical sense and therefore stationarity is a basic form of *stochastic stability*. Note that, for any wssp,

$$\text{var } X(t) = r_{XX}(0) \tag{A.39a}$$
$$r_{XX}(\tau) = r_{XX}(-\tau) \tag{A.39b}$$
$$|r_{XX}(\tau)| \leq r_{XX}(0) \tag{A.39c}$$

Wssps for which expectations can be relaced by time averages over a single realization are said to be *ergodic*. If the discrete-time sp $\{X(t)\}$, $t = 1,2, \ldots\}$ is ergodic, then

$$m_X = \lim_{N \to \infty} \frac{1}{N} \sum_{t=1}^{N} X(t) \qquad\qquad \text{(A.40a)}$$

$$r_{XX}(\tau) = \lim_{N \to \infty} \frac{1}{N} \sum_{t=1}^{N} X(t)X(t+\tau) - m_X^2 \qquad (\tau = 0,1,\ldots) \qquad \text{(A.40b)}$$

with obvious changes of definition for continuous-time sps.

A.7   POWER SPECTRA AND WHITE NOISE

Engineers are usually familiar with noise analysis via power spectra, i.e. a *frequency domain* approach. The *spectral density function* for a wssp $\{X(t)\}$ is defined as the Fourier Transform of its acf as follows.

Continuous time:

$$S_{XX}(\omega) = \int_{-\infty}^{\infty} R_{XX}(\tau) \exp(-j\omega\tau) \, d\tau \qquad\qquad \text{(A.41a)}$$

$$R_{XX}(\tau) = \frac{1}{2\pi} \int_{-\infty}^{\infty} S_{XX}(\omega) \exp(j\omega\tau) \, d\omega \qquad\qquad \text{(A.41b)}$$

Discrete-time:

$$S_{XX}(\omega) = \sum_{k=-\infty}^{\infty} R_{XX}(k) \exp(-j\omega k) \qquad |\omega| \leq \pi \qquad \text{(A.41c)}$$

$$R_{XX}(k) = \frac{1}{2\pi} \int_{-\pi}^{\pi} S_{XX}(\omega) \exp(j\omega k) \qquad\qquad \text{(A.41d)}$$

Note that, in discrete time, frequencies for which $|\omega| > \pi/\tau_s$ are aliassed with those within the range. The sampling interval τ_s is normalized to unity in equation (A.41c).

The input/output relation for a *discrete-time linear system* may be expressed in the form

$$y(t) = H(z^{-1})x(t) \qquad\qquad \text{(A.42)}$$

where z is the unit forward shift operator. If $\{x(t)\}$ is a wssp and H is a stable transfer function (in the sense that zero input asymptotically leads to zero output for arbitrary initial conditions), then $\{y(t)\}$ is a wssp and

$$S_{YY}(\omega) = |H(\exp(j\omega))|^2 S_{XX}(\omega) \qquad |\omega| \leq \pi \qquad \text{(A.43a)}$$

Similarly, for a *continuous-time linear system*

$$S_{YY}(\omega) = |H(j\omega)|^2 S_{XX}(\omega) \qquad \text{(A.43b)}$$

The output spectral density function takes a particularly simple form if

$$S_{XX}(\omega) = \text{constant} = \alpha \qquad \text{(A.44)}$$

A wssp $\{X(t)\}$ with this property is called *white noise* and implies that the process is *uncorrelated* in the sense that, in discrete-time,

$$R_{XX}(0) = \text{var } X(t) = \alpha \qquad \text{(A.45a)}$$

$$R_{XX}(k) = 0 \qquad (k \neq 0) \qquad \text{(A.45b)}$$

and in continuous time

$$R_{XX}(\tau) = \alpha\delta(\tau) \qquad \text{(A.46)}$$

using the relations (A.41). Equation (A.46) indicates that the continuous-time white noise has infinite variance and therefore does not exist. All continuous processes are correlated over small enough time intervals. White noise is a useful idealization, however, if the noise correlation time is very small compared to the significant time constants of interest in the system.

Because it is uncorrelated, discrete-time white noise is useful as a basic building block in both simulating and analysing stochastic processes. For *simulation* purposes there are many computer programs that claim to produce a discrete-time zero mean, white noise sequence $\{X(t)\}$ of unit variance. Given the finite data sequence $X(1), X(2), \ldots, X(N)$ we can (and should) test this claim. First, form the estimates

$$\hat{m}_x = \frac{1}{N} \sum_{t=1}^{N} X(t) \qquad \text{(A.47)}$$

$$R_{XX}(\tau) = \frac{1}{N} \sum_{t=1}^{N-\tau} X(t)X(t+\tau) \qquad (\tau = 0, 1, \ldots, M) \qquad \text{(A.48)}$$

where M should not exceed $N/4$. If

$$N\hat{m}_X^2 < 4R_{XX}(0) \qquad \text{(A.49)}$$

then accept the claim that $\{X(t)\}$ is zero mean. Secondly, if

$$N^{\frac{1}{2}}|R_{XX}(\tau)| < 2R_{XX}(0) \qquad (A.50)$$

for at least 95% of the lags $\tau = 1, 2, \ldots, M$, then accept the claim that $\{X(t)\}$ is white noise.

4.8   ARMA PROCESSES

An important class of sps is generated by passing zero mean white noise $\{e(t)\}$ through a rational polynomial transfer function. In discrete time this *autoregressive moving average* — ARMA (n,m) — model takes the form

$$y(t) + a_1 y(t-1) + \ldots + a_n y(t-n) = e(t) + c_1 e(t-1) + \ldots + c_n e(t-m) \quad (A.51a)$$

or equivalently

$$A(z^{-1})y(t) = C(z^{-1})e(t) \qquad (A.51b)$$

If A has its zeroes inside the unit circle, then the transfer function C/A is stable and $\{y(t)\}$ is a wssp with spectral density function

$$S_{YY}(\omega) = \sigma_e^2 \left| \frac{C(\exp(j\omega))}{A(\exp(j\omega))} \right|^2 \qquad (A.52)$$

where σ_e^2 denotes the white noise variance.

The covariance structure of the sp is given by (A.41d) and can be cast in the form of a complex integral by substituting $z = \exp(j\omega)$ in (A.52). The theory of residues then gives

$$R_{YY}(k) = \sigma_e^2 \operatorname{res}\left[z^{k-1} \frac{C(z)}{A(z)} \frac{C(z^{-1})}{A(z^{-1})} \right] \qquad (k = 0,1,2, \ldots) \qquad (A.53)$$

wher $\operatorname{res}[\cdot]$ denotes the sum of the residues of $[\cdot]$ at its poles within the stability region (the unit circle). Note that $k = 0$ gives the variance of $y(t)$.

Conversely, tracing the path (A.52) back to (A.51) leads to the *Spectral Factorization Theorem*:

If $\{y(t)\}$ is a wssp with rational spectral density $S_{YY}(\omega)$, then there exists a rational transfer function $H = C/A$ with poles inside the stability region and zeroes inside or on the boundary of the stability region such that (A.51) and (A.52) hold.

In particular, if $y = y_1 + y_2$ where $\{y_i(t)\}$ is a wssp of the form (A.51b) for $i = 1,2$:

$$A_i(z^{-1})y_i(t) = C_i(z^{-1})\, e_i(t)$$

and the white noise sequences $\{e_1(t)\}$, $\{e_2(t)\}$ are not correlated with each other, then

$$S_{YY}(\omega) = S_{Y_1 Y_1}(\omega) + S_{Y_2 Y_2}(\omega)$$

$$= \sigma_1^2 \left| \frac{C_1\,(\exp(j\omega))}{A_1\,(\exp(j\omega))} \right|^2 + \sigma_2^2 \left| \frac{C_2\,(\exp(j\omega))}{A_2\,(\exp(j\omega))} \right|^2 \qquad (A.54)$$

Hence the wssp $\{y(t)\}$ has a rational spectral density and the theorem states that the sp has an ARMA representation of the form (A.51b). The relevant polynomials A,C and white noise variance σ_e^2 are derived from (A.54) where the left-hand side is replaced by (A.52).

The above brief discussion is valid for continuous-time wssps with the obvious comment that the stability region refers to the open left half of the complex plane (Re $s < 0$).

If the poles of the transfer function lie outside the stability region, then the sp is *nonstationary*. An important example is the *random walk* (or Brownian motion) described by

$$y(t) = y(t-1) + e(t) \qquad (A.55)$$

so that $\{y(t)\}$ integrates up the white noise process. From (A.55) the mean function $m_Y(t)$ is zero and therefore constant, but

$$\operatorname{var} y(t) = \operatorname{var} y(t-1) + \operatorname{var} e(t) \qquad (A.56)$$

so that the variance of $y(t)$ increases without limit. The model (A.55) is a useful way of expressing random drift.

A.9   INNOVATIONS AND LINEAR ESTIMATION

The geometric view of estimation illustrated in Figure A.3 leads to an elegant and useful approach. Let $y(1)$, $y(2)$, ..., $y(t)$ be t data points from which a signal $x(l)$ is to be estimated. Denoting the MMSE linear estimate of $y(k)$ using $y(1), \ldots, y(k-1)$ by $\hat{y}(k|k-1)$ we note that the sequence $\{\tilde{y}(k)\}$ defined by

$$\tilde{y}(k) = y(k) - \hat{y}(k|k-1) \tag{A.57}$$

is uncorrelated. These are the *innovations* (or new information). The signal $\tilde{y}(k)$ represents what is new in $y(k)$ and is not present in previous data. Parallelling the Gram–Schmidt orthogonalization procedure, we can write the required estimate of x as

$$\hat{x}(l|t) = \sum_{i=1}^{t} \bar{c}_i y(i) = \sum_{i=1}^{t} c_i \tilde{y}(i)$$

where c_i is the projection of x onto $\tilde{y}(i)$. As the innovations sequence is white, we have

$$\hat{x}(l|t) = \sum_{i=1}^{t} [Ex(l)\tilde{y}(i)] [E\tilde{y}(i)^2]^{-1} \tilde{y}(i) \tag{A.58}$$

or, in recursive form,

$$\hat{x}(l|t) = \hat{x}(l|t-1) + [Ex(l)\tilde{y}(t)] [E\tilde{y}(t)^2]^{-1} \tilde{y}(t) \tag{A.59}$$

When suitable structure is introduced, these relations can be shown to lead to the Kalman filter ($l = t$) and the smoother of Section 11.2 ($l < t$).

A.10 STOCHASTIC CONVERGENCE

The estimates generated by a recursive estimation algorithm form a sequence of rvs and it is usually of interest to know how the sequence behaves and in particular whether it converges in some sense. In general, it is too restrictive to demand that every realization of the sequence converges, but there are a number of weaker forms of convergence, two of which are relatively easy to use.

The sequence $\{x(t)\}$ tends to the rv x in the *mean-square* sense (ms) if

$$E|x(t) - x|^2 \to 0 \qquad \text{as } t \to \infty \tag{A.60}$$

The sequence $\{x(t)\}$ tends to the rv x *in probability* if, for any positive, ϵ,

$$P[|x(t) - x| > \epsilon] \to 0 \qquad \text{as } t \to \infty \tag{A.61}$$

written

$$\text{plim } x(t) = x \qquad \text{as } t \to \infty \tag{A.62}$$

Using the Chebyshev inequality (A.17) it follows that ms convergence implies convergence in probability. The converse is not true.

A.11   NOTES AND REFERENCES

A number of texts have influenced the material in this appendix.

Larson, H.J.
Introduction to Probability Theory and Statistical Inference, 3rd ed, Wiley, 1982.

This is a leisurely introductory book on basic probability without stochastic processes and avoids the mathematical depths.

Papoulis, A.
Probability, Random Variables and Stochastic Process, McGraw-Hill, 1965.

This work has been popular for 25 years — quirky, but stimulating.

Kailath, T.
Lectures on Linear Least-Squares Estimation, CISM Courses and Lectures No. 140, Springer, 1976.

An excellent short text on signal estimation (including filtering, prediction and smoothing) emphasizing the innovations approach.

Shiryayev, A.N.
Probability, Graduate Texts in Mathematics No. 95, Springer, 1984.

A superb text for those who require a rigorous treatment of probability and random processes.
Checking the correlation properties of random sequence in the context of ARMA model building is discussed in the compact text:

Anderson, O.D.
Time Series Analysis and Forecasting: the Box–Jenkins Approach, Butterworths, 1975.

Index

Index compiled by Geoffrey C. Jones

SELF-TUNING FRIEND
Adaptive Systems Design and Simulation
Software for use with Matlab™ and Simulink™

The Self-Tuning Friend is the software designed to accompany 'Self-Tuning Systems: Control and Signal Processing by Wellstead and Zarrop. The Self-Tuning Friend is a toolbox for use with Matlab and Simulink for the development and simulation of self-adaptive control and signal processing systems.

- The Self-Tuning Friend functions just like a normal Matlab toolbox and provides more than 45 special commands for the recursive estimation, self-tuning control and self-tuning signal processing methods described in the book. In addition to the commands, the Self-Tuning Friend also provides over 30 demonstrations which illustrate the commands and implement the examples given in the book

- The Simulink version provides a special library of self-tuning blocks which can be mixed with regular Simulink blocks in order to validate a proposed self-adaptive system. For example, the picture to the right shows a Simulink simulation of a self-tuning pole-assignment controller. Note the special blocks for recursive estimation, controller synthesis and tuneable control.

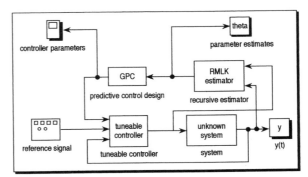

- Educational and volume users are offered special prices.

- A classroom/laboratory teaching kit is available.

For further details, telefax Janice Prunty, on (+44)-061-200-4647 or write to:
Janice Prunty, Control Systems Centre, UMIST
P.O. Box 88
Manchester M60 1QD
United Kingdom

TM   Matlab and Simulink are trademarks of The Mathworks, Inc.